OXFORD
UNIVERSITY PRESS

ASPIRE
SUCCEED
PROGRESS

Cambridge IGCSE® & O Level
Essential
Physics

Third Edition

W0044127

Jim Breithaupt

Darren Forbes

Editor: Lawrie Ryan

Patrick Akande

Mohamed Ghoneim

Ghanapathi Ramanathan

OXFORD
UNIVERSITY PRESS

OXFORD
UNIVERSITY PRESS

Great Clarendon Street, Oxford, OX2 6DP, United Kingdom

Oxford University Press is a department of the University of Oxford. It furthers the University's objective of excellence in research, scholarship, and education by publishing worldwide. Oxford is a registered trade mark of Oxford University Press in the UK and in certain other countries

© Jim Breithaupt and Oxford University Press 2021

The moral rights of the authors have been asserted

First published in 2021

British Library Cataloguing in Publication Data
Data available

978-1-382-00621-7 (standard)

10 9 8 7

978-1-382-00620-0 (enhanced)

10 9 8 7 6 5 4

The manufacturing process conforms to the environmental regulations of the country of origin.

Printed in Great Britain by Bell and Bain Ltd, Glasgow.

Acknowledgements

I would like to thank my family for their support in the preparation of the book, particularly my wife Marie, for secretarial support and encouragement. I am also grateful to the publishing team at Oxford University Press, in particular Oliver Thornton who initiated the production of the book and Eleanor Walter and Stephen Holford who coordinated and edited the production of this edition. *Jim Breithaupt*

The publisher and authors would like to thank the following for permission to use photographs and other copyright material:

Cover: Mircea Costina/Shutterstock

Photos: p3: Hank Morgan/Science Photo Library; p4t: Mega Pixel/Shutterstock; p4b: Anci Valiart/Shutterstock; p6: Cody Images/Science Photo Library; p7: Kaj R. Svensson/Science Photo Library; p10: Reuters Photographer/Adobe Stock Images; p12 & p18: Martyn Chillmaid; p20: Colin Cuthbert/Science Photo Library; p26: Brian Lawrence/Alamy Stock Photo; p28: flo Co. Ltd./Alamy Stock Photo; p31: DoD photo by Corporal Branden P. O'Brien, U.S. Marine Corps; p32: Konrad Bąk/Alamy Stock Photo; p42: Crispin Thruston/Action Images/Adobe Stock Images; p44: Trinity Mirror/Mirrorpix/Alamy Stock Photo; p46: Barrett & MacKay/Age Fotostock; p50: TRL Ltd./Science Photo Library; p51: Dr Jeremy Burgess/Science Photo Library; p52: Richard R. Hansen/Science Photo Library; p53: VisualCommunications/E+/Getty Images; p54t: Public Health England/Science Photo Library; p54b: Robert Brook/Science Photo Library; p56: Alex Bartel/Science Photo Library; ;p57: Mikkel Juul Jensen/Science Photo Library p58t: Martin Bond/Science Photo Library; p58b: David Hay Jones/Science Photo Library; p59: Y.Derome/Publiphoto Diffusion/Science Photo Library; p60: G. Brad Lewis/Science Photo Library; p61: Steve Allen/Science Photo Library; p64: Mark Turnball/Science Photo Library; p65: Bertl123/Shutterstock; p66: NASA/Science Photo Library; p67: Maximilian Stock Ltd/Science Photo Library; p71: Image Source/DigitalVision/Getty Images; p70: DR. Morley Read/Science Photo Library; p72b: Louise Murray/Science Photo Library; p72t: GIRODJL/Shutterstock; p73: Colin Cuthbert/Science Photo Library; p74: John Henshall/Alamy Stock Photo; p76: Sheila Terry/Science Photo Library; p78: Andrew Lambert Photography/Science Photo Library; p79: Clive Freeman/Biosym Technologies/Science Photo Library; p83: Sam Ogden/Science Photo Library; p88: David Taylor/Science Photo Library; p89: Mark Burnett/Science Photo Library; p96: Mark Boulton/Alamy Stock Photo; p98t: Jim Zipp/Science Photo Library; p98b: Laura Stone/Shuttestock; p101: Steve Bloom Images/Alamy Stock Photo; p102: Studio-Annika/iStockphoto; p104: Cordelia Molloy/Science Photo Library; p106: Matt Meadows/Science Photo Library; p110: bacalao64/iStockphoto; p114: Jpl-Caltech/Stsci/Vassar/Nasa/Science Photo Library; p118t: Andrew Lambert Photography/Science Photo Library; p118b: David R. Frazier Photolibrary, Inc./Alamy Stock Photo; p120: kitzzeh/Shutterstock; p122: Adam Hart-Davis/Science Photo Library; p124: Jim Breithaupt; p127: Deep Light Productions/Science Photo Library; p128l: Ton Kinsbergen/Science Photo Library; p128r: Paschon/Shutterstock; p131: Mauro Fermariello/Science Photo Library; p134: David Parker/Science Photo Library; p136: Mediscan/Alamy Stock Photo; p137: ER Productions/Corbis Premium RF/Alamy Stock Photo; p138: Pgia/iStockphoto; p142: Darwin Dale/Science Photo Library; p146: Keith Kent/Science Photo Library; p150: jovannig/iStockphoto; p154: Adrienne Hart-Davis/Science Photo Library; p158: Alex Bartel/Science Photo Library; p164: Peter Menzel/Science Photo Library; p165: Sputnik/Science Photo Library; p172t: Oleksiy Mark/Shutterstock; p172b: Patrick Dumas/Eurelios/Science Photo Library; p174: Martyn F. Chillmaid/Science Photo Library; p180: Jim Dowdalls/Science Photo Library; p198: Andrew Lambert Photography/Science Photo Library; p196b: David J. Green/Alamy Stock Photo; p196t: Leslie Garland Pictures/LGPL/Alamy Stock Photo; p197: Emmeline Watkins/Science Photo Library; p218: Sputnik/Science Photo Library; p223: Emilio Segre Visual Archives/American Institute Of Physics/Science Photo Library; p227: Martyn F. Chillmaid/Science Photo Library; p238: Creativemarc/iStockphoto; p239: Getty Images; p240: NASA; p242: Getty Images.

All artwork by: Aptara, GreenGate Publishing Services & OUP

This Student Book refers to the Cambridge IGCSE® Physics (0625) and Cambridge O Level Physics (5054) Syllabuses published by Cambridge Assessment International Education.

This work has been developed independently from and is not endorsed by or otherwise connected with Cambridge Assessment International Education.

Contents

Introduction v
Syllabus matching grid vi

Unit 1 Motion **2**

1.1 Making measurements 2
1.2 Distance–time graphs 6
1.3 More about speed 8
1.4 Acceleration 10
1.5 More about acceleration 12
1.6 Free fall 14
Summary and Practice questions 16

Unit 2 Forces and their effects **18**

2.1 Mass and weight 18
2.2 Density 20
2.3 Force and shape 22
2.4 Force and motion 24
2.5 More about force and motion 26
2.6 Momentum 28
2.7 Explosions 30
2.8 Impact forces 32
Summary and Practice questions 34

Unit 3 Forces in equilibrium **36**

3.1 Moments 36
3.2 Moments in balance 38
3.3 The Principle of Moments 40
3.4 Centre of gravity 42
3.5 Stability 44
3.6 More about vectors 46
Summary and Practice questions 48

Unit 4 Energy **50**

4.1 Energy transfers 50
4.2 Conservation of energy 52
4.3 Fuel for electricity 54
4.4 Nuclear energy 56
4.5 Energy from wind and water 58
4.6 Energy from the Sun and the Earth 60
4.7 Energy and the environment 62
4.8 Energy and work 64
4.9 Power 66
Summary and Practice questions 68

Unit 5 Pressure **70**

5.1 Under pressure 70
5.2 Pressure at work 72
5.3 Pressure in a liquid at rest 74
5.4 Solids, liquids and gases 76
5.5 More about solids, liquids and gases 78
5.6 Gas pressure and temperature 80
5.7 Evaporation 82
5.8 Gas pressure and volume 84
Summary and Practice questions 86

Unit 6 Thermal physics **88**

6.1 Thermal expansion 88
6.2 Thermometers 90
6.3 Specific heat capacity 92
6.4 Change of state 94
6.5 Thermal conduction 96
6.6 Convection 98
6.7 Infrared radiation 100
6.8 More about infrared radiation 102
6.9 Thermal energy at work 104
Summary and Practice questions 106

Unit 7 Waves **108**

7.1 Wave motion 108
7.2 Transverse and longitudinal waves 110
7.3 Wave properties – reflection and refraction 112
7.4 Wave properties – diffraction 114
Summary and Practice questions 116

Unit 8 Light **118**

8.1 Reflection of light 118
8.2 Refraction of light 122
8.3 Refractive index 124
8.4 Total internal reflection 126
8.5 The converging lens 128
8.6 Applications of the converging lens 130
8.7 Correction of sight defects 132
8.8 Electromagnetic waves 134
8.9 Applications of electromagnetic waves 136
8.10 More about electromagnetic waves 138
Summary and Practice questions 140

Unit 9 Sound **142**

9.1 Sound waves 142
9.2 Properties of sound 144
9.3 The speed of sound 146
9.4 The uses of ultrasound 150
Summary and Practice questions 152

Contents

Unit 10 Magnetism   **154**

10.1 Magnets   154
10.2 Magnetic fields   156
10.3 More about magnetic materials   158
Summary and Practice questions   160

Unit 11 Electric charge   **162**

11.1 Static electricity   162
11.2 Electric fields   164
11.3 Conductors and insulators   166
11.4 Charge and current   168
Summary and Practice questions   170

Unit 12 Electrical energy   **172**

12.1 Batteries and cells   172
12.2 Potential difference   174
12.3 Resistance   176
12.4 More about resistance   178
12.5 Electrical power   180
Summary and Practice questions   182

Unit 13 Electric circuits   **184**

13.1 Circuit components   184
13.2 Series circuits   186
13.3 Parallel circuits   188
13.4 More about series and parallel circuits   190
13.5 Sensor circuits   192
13.6 Switching circuits   194
13.7 Electrical safety   196
13.8 More about electrical safety   198
Summary and Practice questions   200

Unit 14 Electromagnetism   **202**

14.1 Magnetic field patterns   202
14.2 The motor effect   204
14.3 The electric motor   206
14.4 Electromagnetic induction   208
14.5 The alternating current generator   210
14.6 Transformers   212
14.7 High-voltage transmission of electricity   214
Summary and Practice questions   216

Unit 15 Radioactivity   **218**

15.1 Observing nuclear radiation   218
15.2 Alpha, beta and gamma radiation   220
15.3 The discovery of the nucleus   222
15.4 More about the nucleus   224
15.5 Half-life   226
15.6 Radioactivity at work   228
Summary and Practice questions   230

Unit 16 Space   **232**

16.1 The Earth in space   232
16.2 The Solar System   234
16.3 Planetary orbits   236
16.4 The life history of a star   238
16.5 The expanding universe   240
16.6 The beginning and future of the Universe   242
Summary and Practice questions   244

Glossary   246
Index   252

Introduction

This book is designed specifically for Cambridge IGCSE® Physics 0625. Experienced teachers have been involved in all aspects of the book, including detailed planning to ensure that the content gives the best match possible to the syllabus.

Using this book will ensure that you are well prepared for studies beyond the IGCSE level in pure sciences, in applied sciences or in science-dependent vocational courses. The features of the book outlined below are designed to make learning as interesting and effective as possible:

LEARNING OUTCOMES

- These are at the start of each spread and will tell you what you should be able to do at the end of the spread.

- **S** Some outcomes will be needed only if you are taking a supplement paper and these are clearly labelled, as is any content in the spread that goes beyond the syllabus.

EXAM TIP

Experienced teachers give you suggestions on how to avoid common errors or give useful advice on how to tackle questions.

KEY POINTS

These summarise the most important things to learn from the spread.

For **answers**:

www.oxfordsecondary.com/essential-igcse-science

PRACTICAL

These show the opportunities for practical work. The results are included to help you if you do not actually tackle the experiment or are studying at home.

SUMMARY QUESTIONS

These questions are at the end of each spread and allow you to test your understanding of the work covered in the spread.

At the end of each unit there is a double page of examination-style questions written by the author.

At the end of the book you will also find an extensive glossary of key terms.

Assessment structure

Paper 1: Multiple Choice (Core)

Paper 2: Multiple Choice (Supplement)

Paper 3: Theory (Core)

Paper 4: Theory (Supplement)

Paper 5: Practical Test

Paper 6: Alternative to Practical

The Enhanced Online Book supports this student book by offering high-quality digital resources that help to build scientific and examination skills in preparation for the assessment. If you purchase access to the digital course, you will find a wealth of additional resources to help you with your studies and revision:

- Tailored exam skills support
- A worksheet and interactive quiz for every unit
- On Your Marks activities to help you achieve your best
- Full practice papers with mark schemes

Syllabus matching grid

Topic number in this book	Topic title	IGCSE Syllabus section
1.1	Making measurements	1.1
1.2	Distance–time graphs	1.2
1.3	More about speed	1.2
1.4	Acceleration	1.2
1.5	More about acceleration	1.2
1.6	Free fall	1.2
2.1	Mass and weight	1.3
2.2	Density	1.4
2.3	Force and shape	1.5
2.4	Force and motion	1.5
2.5	More about force and motion	1.1/1.5
2.6	Momentum	1.6
2.7	Explosions	1.6
2.8	Impact forces	1.6
3.1	Moments	1.5
3.2	Moments in balance	1.5
3.3	The Principle of Moments	1.5
3.4	Centre of gravity	1.5
3.5	Stability	1.5
3.6	More about vectors	1.1
4.1	Energy transfers	1.7
4.2	Conservation of energy	1.7
4.3	Fuel for electricity	1.7
4.4	Nuclear energy	1.7
4.5	Energy from wind and water	1.7
4.6	Energy from the Sun and the Earth	1.7
4.7	Energy and the environment	1.7
4.8	Energy and work	1.7
4.9	Power	1.7

Topic number in this book	Topic title	IGCSE Syllabus section
5.1	Under pressure	1.8
5.2	Pressure at work	1.8
5.3	Pressure in a liquid at rest	1.8
5.4	Solids, liquids and gases	2.1
5.5	More about solids, liquids and gases	2.1
5.6	Gas pressure and temperature	2.1/2.2
5.7	Evaporation	2.2
5.8	Gas pressure and volume	2.1
6.1	Thermal expansion	2.2
6.2	Thermometers	2.1
6.3	Specific heat capacity	2.2
6.4	Change of state	2.2
6.5	Thermal conduction	2.3
6.6	Convection	2.3
6.7	Infra-red radiation	2.3
6.8	More about infra-red radiation	2.3
6.9	Thermal energy at work	2.3
7.1	Wave motion	3.1
7.2	Transverse and longitudinal waves	3.1
7.3	Wave properties – reflection and refraction	3.1
7.4	Wave properties – diffraction	3.1
8.1	Reflection of light	3.2
8.2	Refraction of light	3.2
8.3	Refractive index	3.2
8.4	Total internal reflection	3.2
8.5	Lenses	3.2
8.6	Applications of the converging lens	3.2
8.7	Correction of sight defects	3.2
8.8	Electromagnetic waves	3.2/3.3

Topic number in this book	Topic title	IGCSE Syllabus section
8.9	Applications of electromagnetic waves	3.2/3.3
8.10	More on communications	3.3
10.1	Magnets	4.1
10.2	Magnetic fields	4.1
10.3	More about magnetic materials	4.1
11.1	Static electricity	4.2
11.2	Electric fields	4.2
11.3	Conductors and insulators	4.2
11.4	Charge and current	4.2
12.1	Batteries and cells	4.2
12.2	Potential difference	4.2
12.3	Resistance	4.2/4.3
12.4	More about resistance	4.2/4.3
12.5	Electrical power	4.2
13.1	Circuit components	4.3
13.2	Series circuits	4.3
13.3	Parallel circuits	4.3
13.4	More about series and parallel circuits	4.3
13.5	Sensor circuits	4.3
13.6	Switching circuits	4.3/4.5
13.7	Electrical safety	4.4

Topic number in this book	Topic title	IGCSE Syllabus section
13.8	More about electrical safety	4.2/4.4
14.1	Magnetic field patterns	4.5
14.2	The motor effect	4.5
14.3	The electric motor	4.5
14.4	Electromagnetic induction	4.5
14.5	The alternating current generator	4.5
14.6	Transformers	4.5
14.7	High-voltage transmission of electricity	4.5
15.1	Observing nuclear radiation	5.1/5.2
15.2	Alpha, beta and gamma radiation	5.1
15.3	The discovery of the nucleus	5.1
15.4	More about the nucleus	5.1/5.2
15.5	Half-life	5.1/5.2
15.6	Radioactivity at work	5.1/5.2
16.1	The Earth in space	6.1
16.2	The Solar System	6.1
16.3	The life history of a star	6.2
16.4	Planetary orbits	6.1
16.5	The expanding universe	6.2
16.6	The beginning and future of the Universe	6.2

1.1 Making measurements

- Use a metre rule to measure length accurately
- Use a clock to measure a time interval
- Use a measuring cylinder to measure liquid volume accurately
- Use and describe the use of a stopwatch to measure the period of oscillation of a pendulum

SI units

quantity	unit	unit symbol
length	metre	m
time	second	s
mass	kilogram	kg
electric current	ampere	A
temperature	kelvin	K

MORE ABOUT ACCURACY

A measurement result is said to be **accurate** if it is close to the **true value** if known. If this is not known, the measurements could be repeated several times (repeatable) to see if the repeat results are in close agreement. If there are large differences, the reliability of the instruments and how they are used should be checked.

Measurements at work

Many people need to make accurate measurements in their jobs. For example:

- a builder needs to measure lengths and distances accurately; stairs that are too short would be dangerous to climb and a door that doesn't fit exactly into the doorway would jam
- an air traffic controller needs to know the distance to an aircraft and how far it is above the ground as well as its estimated time of arrival
- a doctor or nurse giving an injection to a patient needs to fill the syringe with exactly the correct volume of the liquid to be injected.

A measurement has to be made accurately using a reliable instrument calibrated in agreed units. A reliable instrument must give the same reading every time the same quantity is measured. It must also pass the test of giving the correct reading when it is checked using a known quantity. In addition, it must be used correctly to obtain an accurate measurement. As we shall see below, even a simple metre rule or a stopwatch can be used inaccurately.

The units for scientific measurements are internationally agreed and are referred to as **SI units** (Système International). All scientific units you will meet in this course are derived from the five base units that are listed in the table on the left. Each unit is defined in terms of an agreed standard. For example, the unit of **mass** is the **kilogram** and is defined by a standard mass of 1 kilogram in the International Bureau of Weights and Measures in Paris.

Measuring length

We can use a **metre** rule graduated in millimetres to measure lengths to within a millimetre. For example, suppose we measure the **length** of a bar. The bar is placed alongside the rule with one end at the zero mark of the rule. Reading the position of the other end of the bar on the millimetre scale gives the length of the bar. However, common errors in such a measurement include:

- inaccurate setting of the end of the bar at the zero mark of the rule
- movement of the bar after setting one end at the zero mark
- 'line of sight' errors, as shown in Figure 1.1.1, when observing each end of the bar against the scale. The observer must be positioned so that the line of sight is at right angles to the scale.

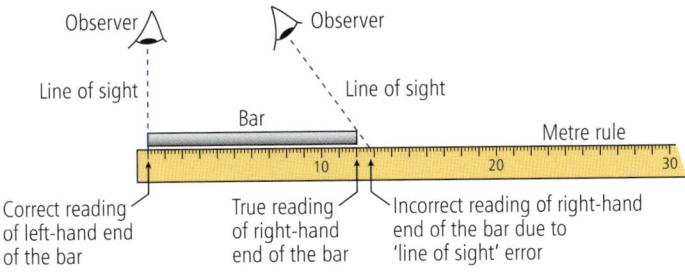

Figure 1.1.1 A line of sight error

Extension

Using a micrometer

A micrometer is used to measure lengths of up to 30 mm to within 0.01 mm, as shown by the digital micrometer in use in the photo. A digital micrometer gives a direct read-out equal to the width of the micrometer gap.

An engineer using a digital micrometer

Measuring volume

The volume of a liquid can be measured using a measuring cylinder. The scale is usually graduated in cubic centimetres (cm^3). The liquid is poured into the empty measuring cylinder. Then the level of the liquid in the measuring cylinder can be read on the scale, as shown in Figure 1.1.2. The meniscus of the liquid surface is where it is in contact with the glass surface. The liquid level is at the bottom of the curved meniscus and the line of sight from the observer must be along this level.

Note

1 1 cubic metre (m^3) = 1 000 000 cm^3

2 The volume of an irregular solid can also be measured using a measuring cylinder. See Topic 2.2.

KEY POINTS

- The average thickness of two or more flat objects, e.g. coins or sheets of paper, can be found by measuring their total thickness. Dividing this measurement by the number of objects gives their average thickness.

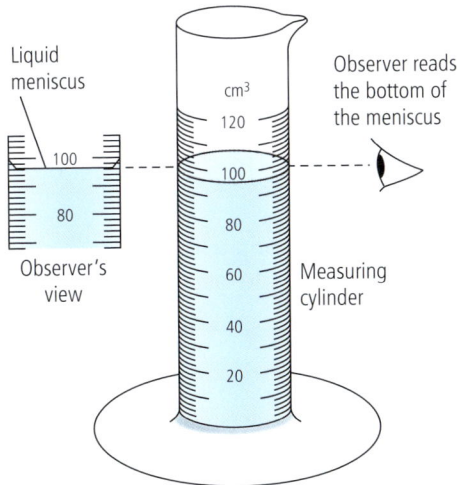

Figure 1.1.2 Using a measuring cylinder

CONTINUED

Measuring time intervals

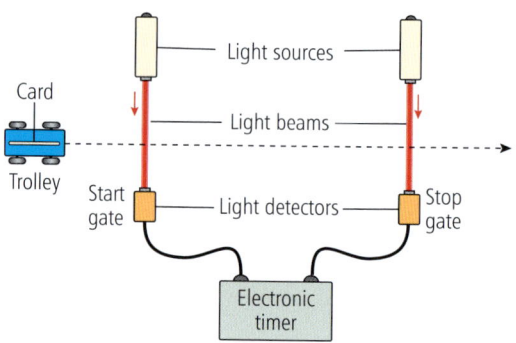

Figure 1.1.3 Using gates

To measure a time interval, we need to use an appropriate timing device. Such devices may be mechanical or electronic and they include ordinary clocks and watches, stopclocks, stopwatches and electronic light-gate timers. Examples of their use are listed in this table.

timing device	example of use
ordinary clock or watch	pay-to-park car parking
stopclock or stopwatch	timing oscillations, such as a swinging pendulum
electronic light-gate timer	100 m race

Timing devices need to be checked periodically to ensure they do not run too slow or too fast. Assuming a timing device is accurate, common errors in its use include:

- line of sight errors in reading the position of a pointer against the dial (if the line of sight from the observer to the pointer is not at right angles to the dial)

- reaction time errors when using a stopwatch or stopclock; such errors can be reduced with practice to about 0.1 s. Even so, timings of the same event by different people would be likely to differ by at least 0.1 s

- misalignment of the light beams when using a light-gate timer (if the light beams at the 'start' gate and the 'stop' gate are not parallel to each other).

Figure 1.1.4 Analogue and digital timers. The digital timer (top) gives a read-out of the timing. The timer below it is an example of an analogue measuring device as it has a pointer that must be read against the scale

Measurement of the period of a pendulum

A pendulum oscillates repeatedly with a constant time period. This is the time interval between successive passes in the same direction through the centre of the oscillations (i.e. its position when at rest). To measure its period:

- place a reference marker directly under the pendulum's rest position and observe the pendulum as it repeatedly moves past the marker

- ensure the line of sight to the marker is at right angles to the plane in which the pendulum oscillates

- time how long the pendulum takes to undergo a measured number of at least 10 oscillations from when it passes the marker in a certain direction

- repeat the procedure at least two more times and calculate the average time for the measured number of oscillations.

To calculate the time period, divide the average time for the measured number of oscillations by the number of oscillations.

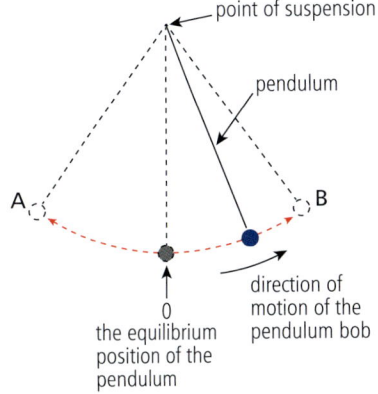

Figure 1.1.5 The time period of the oscillations is the time taken for the moving pendulum to travel from 0 to B, then down past 0 and up to A, and finally back to 0.

SUMMARY QUESTIONS

1 Copy and complete the following sentences using words from the list:

measuring cylinder   metre rule   millimetre scale

    a A _____ with a _____ is used to measure lengths to within a millimetre.

    b A _____ is used to measure volume.

    c A _____ should be used to measure the diameter of a coin.

2 Explain how you would use a stopwatch to measure someone's pulse rate in beats per minute.

3 A student used a digital stopwatch three times to time 10 oscillations of a pendulum. The timings were 13.52 s, 13.64 s and 13.58 s.

    a Calculate **i** the average time for 10 oscillations, **ii** the time period of the oscillations.

    b Estimate the accuracy of the timings, giving a reason for your estimate.

KEY POINTS

- A metre rule graduated in millimetres may be used to measure lengths that are too long to measure with a micrometer.

- A measuring cylinder is used to measure liquid volume.

- A stopwatch or stopclock is used to time oscillations.

1.2 Distance–time graphs

- Interpret a distance–time graph to tell if an object is stationary
- Interpret a distance–time graph to tell if an object is moving at constant speed
- Calculate the speed of a body

Capturing the land speed record and becoming the first land vehicle to break the sound barrier

On a car journey

Some roads have marker posts every kilometre. If you are a passenger in a car on such a road, you can use these posts to check the speed of the car. You need to time the car as it passes each post. This table shows some measurements made on a car journey.

Data for distance–time graph

distance / metres (m)	0	1000	2000	3000	4000	5000	6000
time / seconds (s)	0	40	80	120	160	200	240

Figure 1.2.1 shows the readings on a graph of distance plotted against time.

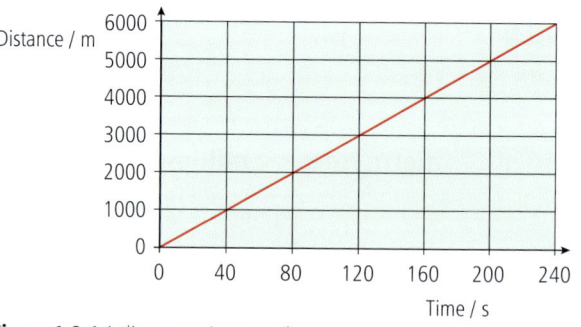

Figure 1.2.1 A distance–time graph

The graph shows that:

- the car took 40 s to go from each marker post to the next and so its speed was **constant**.
- the car travelled a distance of 25 m every second (= 1000 m ÷ 40 s) and so its speed was 25 m/s.

If the car had travelled faster, it would have gone further than 1000 m every 40 s and so the graph would have been steeper. In other words, the gradient of the graph would have been greater.

The gradient on a distance–time graph represents speed.

- If the graph is flat, the gradient is zero and therefore the speed is zero.
- If the graph is straight and not flat, the gradient of the graph is constant and not zero. Therefore the speed is constant.
- If the graph is curved, the gradient of the graph changes and therefore the speed changes.

The steeper a distance–time graph is, the greater the speed it represents.

Speed

For an object moving at constant speed, we can calculate its speed using the equation:

$$\text{speed} = \frac{\text{total distance travelled}}{\text{total time taken}}$$

The scientific unit of speed is metres per second, usually written as metre/second or m/s.

Note For an object moving at changing speed, the equation above gives its **average speed**.

Long-distance journeys

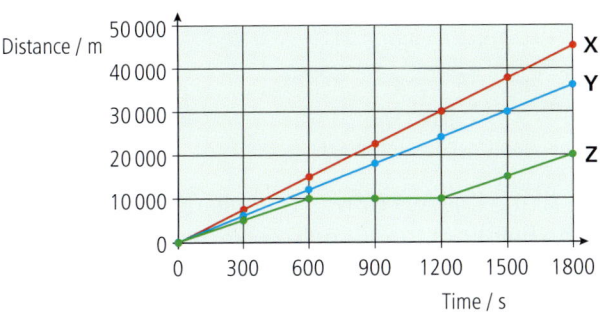

Figure 1.2.2 Comparing distance–time graphs

Long-distance vehicles are fitted with recorders that are used to check that drivers do not drive for too long. The information from a recorder may be used to plot a distance–time graph. Figure 1.2.2 shows distance–time graphs for three lorries: X, Y and Z on the same route.

- X went fastest because it travelled furthest in the same time.
- Y travelled more slowly than X. From the graph, you can check that it travelled 30 000 m in 1500 s. So its speed was 20 m/s (= 30 000 m ÷ 1500 s).
- Z stopped for some of the time. This is shown by the flat section of the graph. Its speed was zero during this time.

EXAM NOTE

The word equation for speed may be written as $v = \dfrac{s}{t}$ where v = speed, s = distance and t = time.

Rearranging this equation gives

$$s = v\,t \qquad \text{or} \qquad t = \frac{s}{v}$$

WORKED EXAMPLE

A car takes 300 s to travel a distance of 6600 m.

Calculate the average speed of the car.

Solution

$$v = \frac{s}{t} = \frac{6600\text{ m}}{300\text{ s}} = 22\text{ m/s}$$

KEY POINTS

- The steeper a distance–time graph is, the greater the speed it represents.
- $\text{speed (metre/second; m/s)} = \dfrac{\text{distance travelled (metre; m)}}{\text{time taken (second; s)}}$

SUMMARY QUESTIONS

1 Figure 1.2.3 shows the distance–time graphs for two model vehicles A and B plotted on the same axes. The two vehicles travelled the same distance.

   a Describe the motion of each vehicle in terms of its speed.

   b Calculate the average speed of each vehicle.

2 A vehicle on a road travels 1800 m in 60 s. Calculate:

   a the speed of the vehicle in m/s

   b how far it would travel at this speed in 300 s.

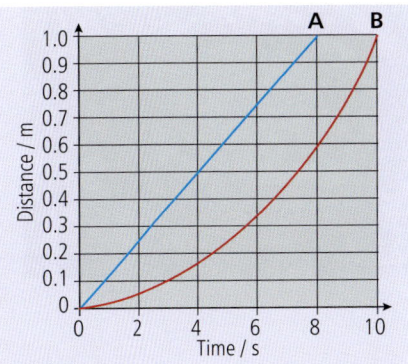

Figure 1.2.3

1.3 More about speed

LEARNING OUTCOMES

- Interpret a distance–time graph to tell if an object's speed increases or decreases
- Measure the speed of an object
- Distinguish between speed and velocity

Using distance–time graphs

For an object moving at constant speed, we saw in Topic 1.2 that:

- the distance–time graph is a straight line
- the speed of the object is represented by the gradient of the line.

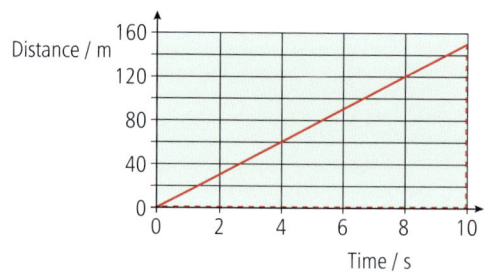

Figure 1.3.1 A distance–time graph for constant speed

To find the gradient of the line, we draw a triangle under the line, as in Figure 1.3.1. The height of the triangle represents the distance travelled and the base represents the time taken.

So the gradient of the line which represents the object's speed $=$ $\dfrac{\text{the height of the triangle}}{\text{the base of the triangle}}$

For a moving object with changing speed, the distance–time graph is not a straight line. Figure 1.3.2 shows two examples. In both cases, the gradient of the line changes. This tells us that the speed of the object changes. For example, the gradient of the line in graph **a** increases because it becomes steeper as the time increases. Therefore the speed of the object that gave this line must be increasing. In other words, the object is accelerating.

Note When an object slows down, we say it **decelerates**.

EXAM TIP

If the speed–time graph of an object has a flat section, the gradient of this section is zero. The object's speed in this section is zero as its position doesn't change.

A corkscrew ride

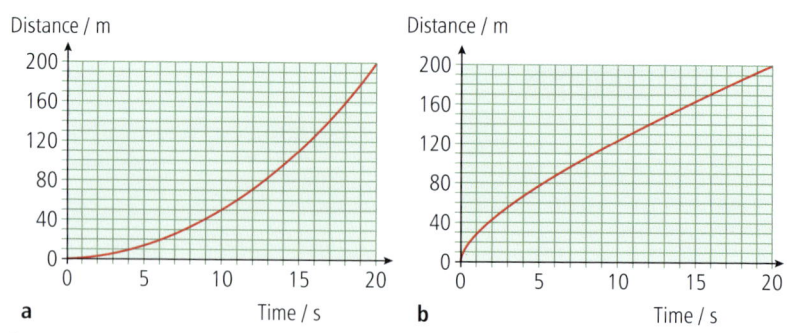

Figure 1.3.2 Distance–time graphs for changing speed

Measuring speed

1 Release a free-wheeling model vehicle or trolley at the top of a sloped runway. Adjust the runway so the trolley runs down as slowly as possible after being released at the top.

2 Use a stopwatch and a metre rule to make the measurements needed to plot a distance–time graph.

From your graph:

a Determine whether or not the speed of the trolley increased, decreased or remained constant as it moved down the runway.

b Calculate the average speed of the model vehicle over the first metre after it was released.

Speed and velocity

When you visit a fairground, the hardest rides are the ones that throw you around. Your speed and your direction of motion keep changing. We use the word **velocity** for speed in a given direction. An exciting ride would be one that changes your velocity often and unexpectedly!

Velocity is speed in a given direction.

- An object moving at constant speed along a straight line has a constant velocity.
- An object moving steadily around in a circle has a constant speed. Its direction of motion changes as it goes around so its velocity is not constant.
- Two moving objects can have the same speed but different velocities. For example, a car travelling north at 30 m/s on a straight road has the same speed as a car travelling south at 30 m/s. But their velocities are not the same because they are moving in opposite directions.

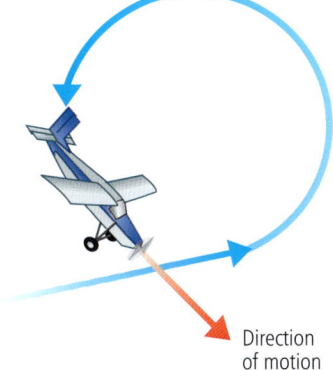

Direction of motion

Figure 1.3.3 Speed and velocity

1 Find the speed of the object graphed in Figure 1.3.1.

2 Describe how the speed of the object changes in Figure 1.3.2b.

3 Copy and complete the following sentences below using words from the list:

acceleration   speed   velocity

a An object moving steadily around in a circle has a constant _____.

b If the velocity of an object increases by the same amount every second, its _____ is constant.

c When an object moves in a straight line with zero acceleration, its _____ is constant.

1.4 Acceleration

LEARNING OUTCOMES

- Interpret speed and time data to tell if an object is accelerating
- Interpret speed and time data to tell if the acceleration of an accelerating object is constant
- **S** Use speed and time data to calculate the acceleration of an object undergoing constant acceleration

On a test track

EXAM NOTE

The word equation for acceleration may be written as $a = \dfrac{\Delta v}{\Delta t}$ where a = acceleration, Δv = change of velocity and Δt = time taken for the change of velocity.

On a test track

A car maker claims their new car 'accelerates more quickly than any other new car'. A rival car maker is not pleased by this claim and issues a challenge. Each car in turn is tested on a straight track with a speed recorder fitted. The results are shown in the table below. They show that the cars reach different speeds 6 seconds after the start which means that one accelerates more than the other.

time from standing start / seconds (s)	0	2	4	6	8	10
speed of car X / metres/second (m/s)	0	5	10	15	20	25
speed of car Y / metres/second (m/s)	0	6	12	18	18	18

Which car accelerates more? The results are plotted on the speed–time graph in Figure 1.4.1. You can see that in the first 6 seconds, the speed of Y increases more than the speed of X. So Y accelerates more than X in the first 6 seconds.

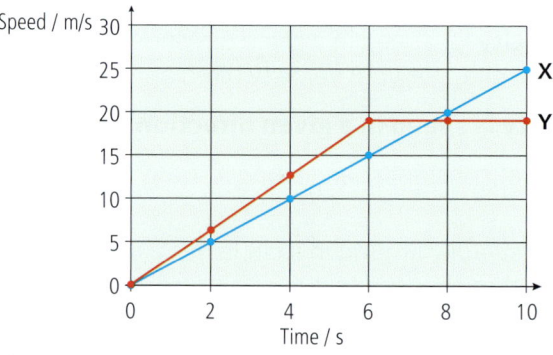

Figure 1.4.1 Speed–time graphs

Any object with an increasing speed is accelerating. Any object with a decreasing speed is decelerating. Figure 1.4.2 shows how the speed of car Y changed in the final 12 seconds of its journey on the test track. The graph shows that:

- the car moved at a constant speed of 18 m/s for the first 2 s
- the car's speed decreased from 18 m/s to 0 m/s in the next 10 s
- the car's speed was zero after 12 s.

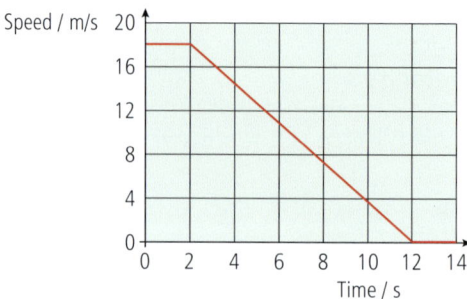

Figure 1.4.2 Decreasing speed

Supplement

The acceleration of an object is its change of velocity per second. The unit of **acceleration** is the metre per second squared, abbreviated to m/s².

We can work out the acceleration of an object using the equation:

$$\text{acceleration} = \frac{\text{change in velocity}}{\text{time taken for the change}}$$

In Figure 1.4.1, notice that:

- the velocity of X increases by 2.5 m/s every second. So the acceleration of X is constant and is equal to 2.5 m/s²
- the velocity of Y increases steadily for the first 6 s then remains constant. So Y's acceleration is constant for the first 6 s and then Y has zero acceleration from 6 s onward.

For an object moving with constant acceleration along a straight line, we can write the above word equation in symbols. Suppose the object's speed increases from an initial speed u to speed v in time t, as shown in Figure 1.4.3.

For motion along a straight line, acceleration, $a = \dfrac{\text{change of speed}}{\text{time taken}}$

$$= \frac{v - u}{t}$$

Multiplying both sides of this equation by t gives: $at = v - u$

Rearranging this equation gives $\boldsymbol{v = u + at}$

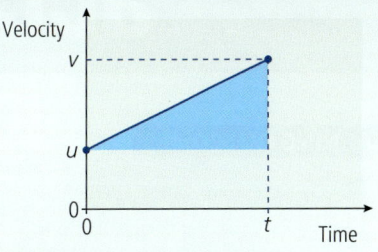

Figure 1.4.3 Uniform acceleration

A motorcycle accelerates from rest for 15 s at a constant acceleration of 2.2 m/s². Calculate its velocity at the end of this time.

Solution

Initial velocity $u = 0$, acceleration $a = 2.2$ m/s².

At time $t = 15$ s, velocity $v = u + at$
$= 0 + 2.2 \times 15 = 33$ m/s.

Note An object moving at constant acceleration is sometimes said to have a **uniform** acceleration.

KEY POINTS

- Acceleration is change of velocity per second.
- The unit of acceleration is the metre/second² (m/s²).

SUMMARY QUESTIONS

1 The table below shows how the speed of a train changed between two stations.

time from start / seconds (s)	0	40	80	120	160	200	240
speed of train / metres/second (m/s)	0	5	10	15	15	15	0

   a Use the data in the table to plot a graph of speed on the y-axis against time on the x-axis.

   b How long did the train spend **i** moving at constant speed, **ii** accelerating, **iii** decelerating?

   c Calculate the initial acceleration of the train.

2 The velocity of a car increased from 8 m/s to 28 m/s in 8 s without changing direction. Calculate:

   a its change of velocity and **b** its acceleration.

3 A jet plane accelerated along a straight runway from rest to a speed of 140 m/s in 50 s when it took off. Calculate: **a** its change of velocity and **b** its acceleration.

1.5 More about acceleration

Investigating acceleration

We can use a motion sensor linked to a computer to record how the speed of an object travelling along a straight line changes. Figure 1.5.1 shows how we can do this using a trolley as the moving object. The computer can also be used to display the measurements as a speed–time graph.

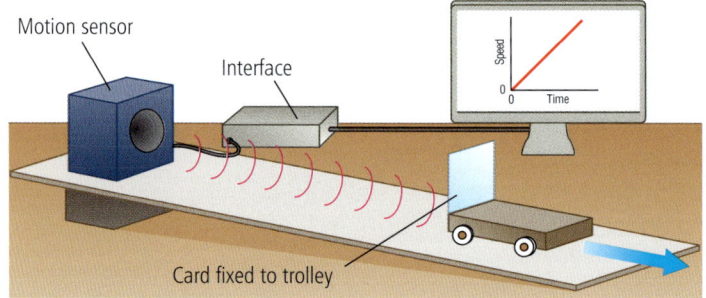

Figure 1.5.1 Investigating acceleration

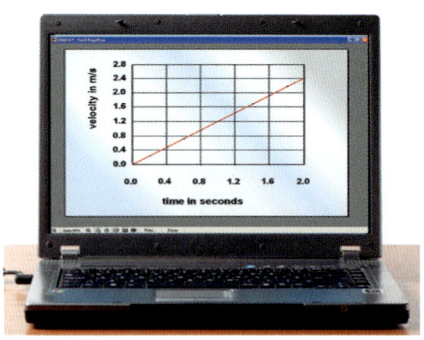

Measuring motion using a computer

Test A If we let the trolley accelerate down the runway, its speed increases with time. The computer on the left shows the speed–time graph from a test run.

- The line goes up because the speed increases with time. So it shows the trolley was accelerating as it ran down the runway.
- The line is straight which tells us that the increase of speed was the same every second. In other words, the acceleration of the trolley was constant (or *uniform*).

Test B If we make the runway steeper, the trolley accelerates faster. This would make the line on the graph steeper than for test A. So the acceleration in test B is greater.

The gradient of a graph is a measure of its steepness. The tests show that:

The gradient of a speed–time graph represents acceleration

Supplement

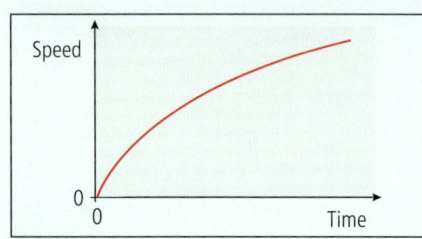

Figure 1.5.2 Decreasing acceleration

Changing acceleration

How is the acceleration affected in test A if the test is repeated with a 'windshield' attached to the front of the trolley? You should find that the gradient of the speed–time graph becomes less steep as the trolley gains speed, as shown by the graph in Figure 1.5.2. The graph shows that the acceleration is not constant and that it decreases as the speed increases. The air resistance due to the windshield causes this effect.

In the photograph on page 12:

- the gradient is given by the height divided by the base of the triangle under the graph
- the height of the triangle under the graph represents the change of velocity ($v - u$) and the base of the triangle represents the time taken t.

Therefore, the gradient represents the acceleration because **acceleration** $= \dfrac{\text{change of speed}}{\text{time taken}}$

Prove for yourself that the acceleration in Figure 1.5.3 in the last 5 seconds is $-4.0\,\text{m/s}^2$. Note that this is a **negative acceleration** because it decelerates.

More about speed–time graphs

Figure 1.5.3 shows the speed–time graph for a vehicle braking to a standstill at a set of traffic lights. We use the term **deceleration** for any situation where an object decelerates.

- Before the brakes are applied, the vehicle moves at a constant speed of 20 m/s for 10 s. It therefore travels 200 m in this time ($= 20$ m/s $\times$ 10 s). This distance is represented on the graph by the area under the graph from 0 to 10 s. This is the shaded rectangle in Figure 1.5.3.

- When the vehicle decelerates in Figure 1.5.3, its speed drops from 20 m/s to zero in 5 s. We can work out the distance travelled in this time from the area of the shaded triangle in Figure 1.5.3. This area is ½ × the height × the base of the triangle. So the vehicle must have travelled a distance of 50 m when it was decelerating.

The area under a speed–time graph represents distance travelled.

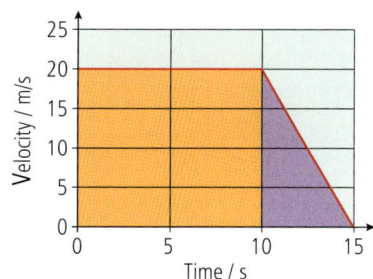

Figure 1.5.3 Braking

KEY POINTS

- The gradient of a graph represents acceleration.
- The area under a graph represents distance travelled.

SUMMARY QUESTIONS

1 Figure 1.5.4 shows four velocity–time graphs, labelled A, B, C and D.

a Match each of the following descriptions to one of the graphs.

i accelerated motion throughout

ii zero acceleration

iii accelerated motion then decelerated motion

iv deceleration.

b Which graph in Figure 1.5.4 represents the object that travelled:

i the furthest distance

ii the least distance?

2 Figure 1.5.5 shows the velocity–time graph of an object X moving with a constant acceleration.

a How can you tell from the graph that the acceleration of X is constant?

b Use the graph to calculate the distance moved by X in 10 s.

c Determine the acceleration of the object.

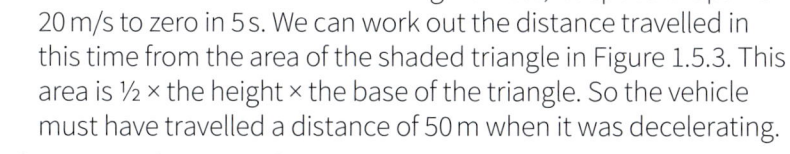

Figure 1.5.4

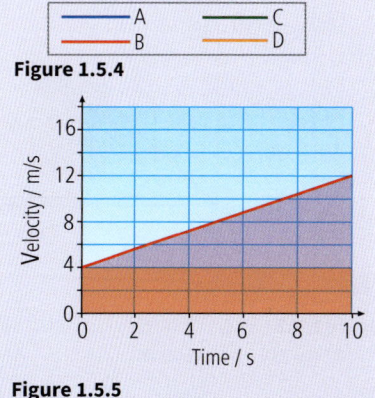

Figure 1.5.5

1.6 Free fall

- Recognise that objects in free fall accelerate at constant acceleration
- Know that the air resistance on objects in free fall is negligible
- **s** Recognise that where air resistance is not negligible, a falling object reaches a terminal speed

Investigating free fall

Does a falling object gain speed as it falls? We can investigate the motion of a falling ball by different methods, two of which are described below.

Using a camera and a light flashing at a constant rate, we can take a 'multiflash' photograph of a falling ball, as shown in Figure 1.6.1. With a suitable light source flashing at a constant rate, each time the light flashes, an image of the falling object is recorded by the camera. The photograph shows that the distance between successive images of the ball increases as it falls. This means that the ball's speed increases as it falls.

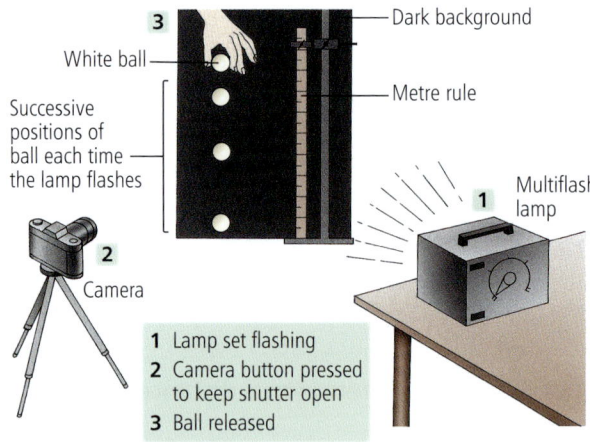

Figure 1.6.1 Multiflash photography

1 Lamp set flashing
2 Camera button pressed to keep shutter open
3 Ball released

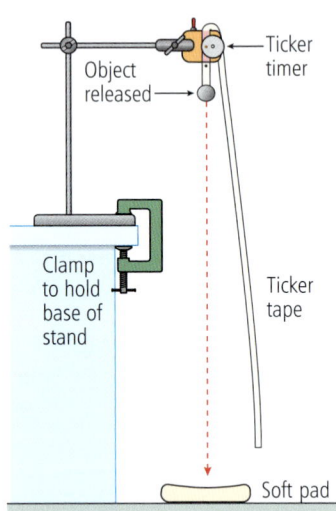

Using a ticker timer, as shown in Figure 1.6.2, we can make a tapechart to show the acceleration of a falling object is constant.

A paper tape attached to a suitable object is pulled through a ticker timer when the object is released. The ticker timer prints dots on the tape at a constant rate as the tape passes through it. The tape is then cut into single-dot lengths which are then stuck side-by-side on a sheet of paper to make a tape chart as shown in Figure 1.6.2.

- Each single-dot length is a measure of the speed of the object as the single-dot length passed through the ticker timer.
- The line through the tops of the tape lengths shows how the speed of the object changed as the object fell.
- The line on the tape chart in Figure 1.6.2 has a constant gradient. This shows that the speed of the falling object increased steadily. In other words, the object fell at constant acceleration.
- Any object released near the Earth's surface falls at a constant acceleration, provided air resistance is insignificant. Such motion is described as **free fall**.

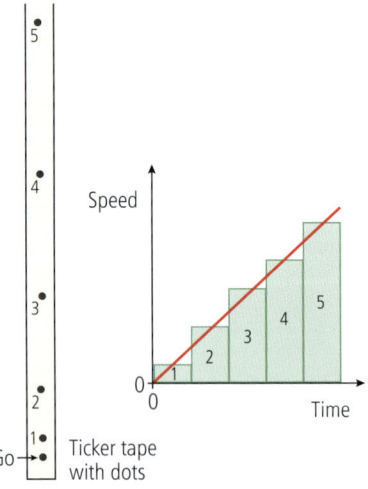

Figure 1.6.2 Using a ticker timer to show the acceleration of a falling object

The acceleration of a freely falling body near to the Earth's surface is constant.

Acceleration of free fall, *g*

The two methods described on page 14 can be used to measure the distance fallen by a falling object in different measured times. Such measurements may be used to show that a freely falling object has an acceleration of approximately 9.8 m/s². This acceleration is due to gravity and is referred to as the **acceleration of free fall, *g*.**

Supplement

The Earth's gravitational field decreases in strength with increased height above the Earth's surface. However, for heights which are very small compared with the Earth's radius, the field is effectively uniform (i.e. the same everywhere) with a value of *g* equal to approximately 9.8 m/s².

The effect of air resistance on falling objects

A parachutist who jumps out of a plane accelerates until the parachute opens. The parachutist descends to the ground at constant speed because the air resistance on the parachute (and the parachutist) opposes the force of gravity on them (i.e. their total weight) with an equal force. The air resistance is sometimes referred to as the **drag force.**

In general, the air resistance on any falling object increases as it gains speed. If it continues to fall, the increasing air resistance causes it to reach a constant speed vertically downwards, referred to as its **terminal velocity**. At this speed, the air resistance on the object opposes the force of gravity with an equal force. Figure 1.6.4 shows how the speed of such a falling object increases as it descends until it reaches its terminal speed.

Figure 1.6.3 Using a parachute

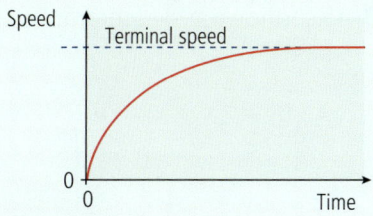

Figure 1.6.4 Terminal speed

SUMMARY QUESTIONS

1 Two objects, X and Y, are released simultaneously from different heights above the ground. X is released above Y, as shown in Figure 1.6.5.

Copy and complete the following sentences below using words from the list.

a greater   the same   a smaller

a Compared with X, Y has _____ acceleration as they fall.

b Compared with X, Y hits the ground with _____ speed.

c Compared with X, Y has _____ time of descent.

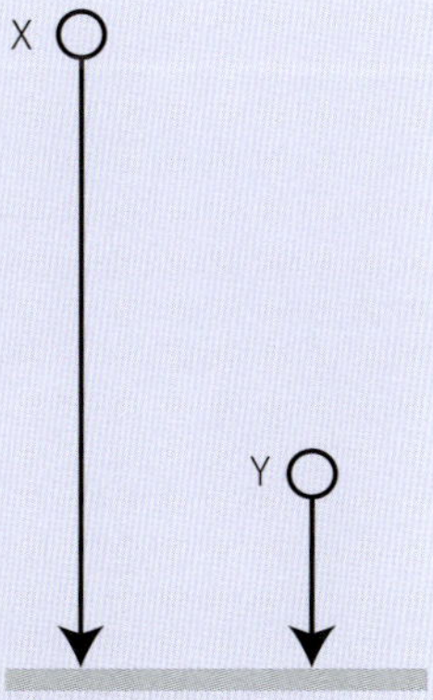

Figure 1.6.5

2 a Which feature in the graph in Figure 1.6.4 represents:

   i the distance fallen after the object was released?

   ii the acceleration of the object?

b Describe how the acceleration of the object changed after it was released.

KEY POINTS

- A freely falling object has a constant acceleration.

- The acceleration of a freely falling object near the Earth is 9.8 m/s².

- If air resistance is significant, a falling object reaches a terminal speed.

1 (a) Explain, with the aid of a diagram, how to avoid a 'line of sight' (or parallax) error when taking a measuring cylinder reading.

(a) The liquid surface in the measuring cylinder will not be entirely flat. Draw a diagram to show the shape of the meniscus and to show where to take the reading.

2 When using a pendulum, the experiment usually involves finding the period of the pendulum.

(a) Explain the meaning of the term 'period'. You may draw a diagram to help your description.

(b) A careful experimenter will measure the time for at least ten oscillations of the pendulum and then calculate the period. Explain why it is good practice to measure the time for at least ten oscillations.

3 The following sketch graphs show the motion of a toy car:

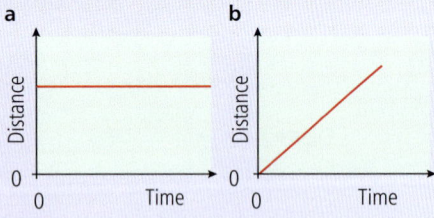

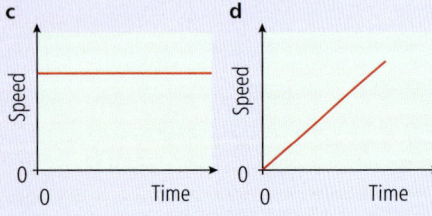

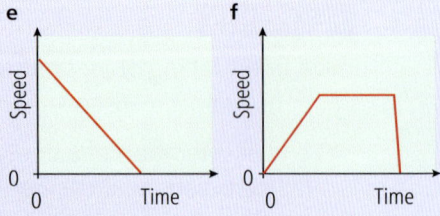

State briefly what each graph tells you about the motion.

4 The table (right) shows the results obtained by a student measuring the speed of a trolley travelling down a ramp.

Practice Questions

1 Which of these features of a distance–time graph shows the speed?
  A   The gradient
  B   The intercept on the distance axis
  C   The intercept on the time axis
  D   The area under the line of the graph

2 Which of these statements describes the motion of the motorcycle shown in the graph?

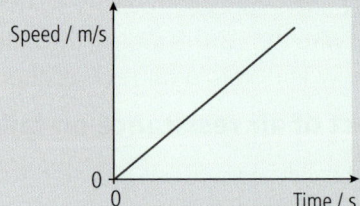

  A   moving forward at constant speed
  B   moving backwards at constant speed
  C   stationary   D   accelerating

3 A teacher uses a rule to measure the thickness of 500 sheets of paper as 5.5 cm. What is the thickness of a single sheet of paper?
  A 1.1×10^{-4} cm   C 1.1×10^{-4} m
  B 1.1×10^{-3} m   D 9.1×10^{-4} m

4 A car travelling at 3.5 m/s decelerates at 0.2 m/s^2 for 6.0 s as it approaches a junction. What is the new velocity of the car?
  A   4.7 m/s   C   0 m/s
  B   2.3 m/s   D   9.5 m/s

time / s	speed / m/s
0	0
0.2	1.4
0.4	3.0
0.6	4.6
0.8	6.1
1.0	7.4
1.2	9.0

(a) Use the information in the table to draw a speed–time graph of the motion of the trolley.

(b) Use the graph to determine the acceleration of the trolley.

5 Which of the following is the best definition of terminal velocity?

A The maximum speed an object will reach when falling through a vacuum

B The maximum speed an object will reach when falling through a fluid

C The velocity at the end of a fall

D The starting velocity of an object released in an experiment

6 A student is investigating the factors which affect the time it takes for a pendulum to make one swing. They find the pendulum is swinging too quickly to accurately measure the time for a single swing.

(a) What device should be used to measure the time of the swings? [1]

(b) Explain how the student can improve the reliability of the measurement of the swing time. [2]

7 The graph below shows the motion of an electric train between two stations.

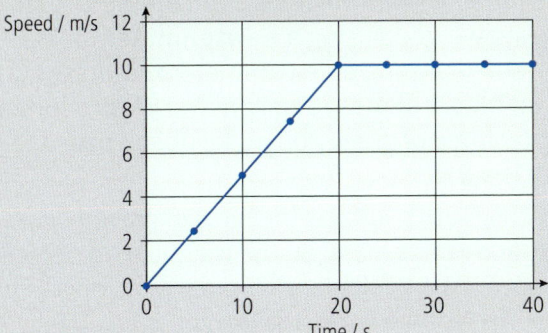

(a) Describe the motion of the train between 0 and 20 s. [1]

(b) Describe the motion of the train between 20 and 40 s. [2]

(c) Calculate the total distance travelled by the train in the first 40 s of motion. [4]

8 A student pushes a skateboard up a small slope and measures the speed travelled over time. The results are shown in the table.

time / s	speed / cm/s
0	20.0
1	15.0
2	11.0
3	8.0
4	6.0
5	6.0
6	6.0

(a) Use these results to draw a speed–time graph for the movement of the skateboard. [4]

(b) Describe what the graph shows about the movement of the skateboard during the first 4 seconds. [1]

(c) Describe what the graph shows about the movement of the skateboard in the final 2 seconds. [1]

9 A student is investigating how long it takes for an object to fall through a range of liquids in a measuring cylinder. They fill a set of identical measuring cylinders with 100 cm³ of different liquids and time how long it takes for a plastic ball to fall to the bottom of the liquids. They use their watch to time how long it takes the ball to fall through the liquids. The results are shown in this table.

Liquid	Water	Oil	Paste	Glue
Time / s	2	3	5	9

(a) Describe how the student should make sure that the volume of liquid is measured as precisely as possible. [2]

(b) Describe how the students could reduce random timing errors while using the watch. [3]

(c) The student decides to try to improve the experiment by using a pair of light gates to measure the time. Explain how these gates should be positioned and how they would reduce errors. [2]

10 A ball is dropped from the top of tall building and passes through a series of light gates as it falls. After falling for 0.5 s the ball is travelling at 4.90 m/s; after falling for 3.0 s the ball is travelling at 28.0 m/s.

(a) Calculate the average acceleration of the ball during the first 0.5 s. [3]

(b) Calculate the average acceleration of the ball during the first 3.0 s. [2]

(c) Explain why the acceleration for part (b) is less than the acceleration for part (a). [4]

2.1 Mass and weight

Using kilograms

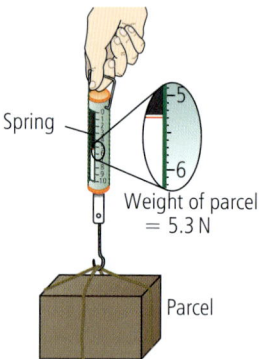

Spring

Weight of parcel = 5.3 N

Parcel

Figure 2.1.1 Using a newtonmeter to weigh an object

Mass and matter

The mass of an object depends on how much matter there is in the object. The amount of matter in an object determines its **mass**, regardless of whether the object is a solid or a liquid or a gas. Two objects of the same mass contain the same amount of matter. Two objects of different mass contain different amounts of matter.

The SI unit of mass is the **kilogram** (kg). We usually use this unit of mass in everyday life although we sometimes find it is more convenient to use the gram which is 0.001 kg.

The mass of a body is a measure of the amount of matter it contains.

Weight

The weight of an object depends on its mass. This is because weight is due to the downward pull of the Earth's gravity on the object and the force of gravity on an object depends on its mass.

The greater the mass of an object, the greater its weight.

We measure weight in newtons because the SI unit of force is the **newton** (abbreviated N) and weight is a force. Figure 2.1.1 shows an object being weighed using a **newtonmeter** marked in newtons. Measurements using a newtonmeter should show that the weight of an object of mass 1 kg near the Earth's surface is 9.8 N. Therefore the force of gravity on a 1 kg object near the surface of the Earth is 9.8 N.

For any object near the Earth's surface, the force of gravity on it is 9.8 N for every kilogram of its mass. So the weight of an object near the Earth's surface is 9.8 N for every kilogram of its mass. For example, near the surface of the Earth, the weight of an object:

- of mass 1 kg is 9.8 N
- of mass 20 kg is 196 N.

PRACTICAL

Using a newtonmeter

1 Check the pointer of the newtonmeter reads zero on the scale without any object suspended from the newtonmeter.

2 Suspend the object to be weighed from the newtonmeter hook. This causes the spring in the newtonmeter to stretch which makes the pointer move down the scale.

3 Read the position of the pointer on the scale to give the weight of the object in newtons.

The **gravitational field strength** at any point in a gravitational field is the force of gravity per unit mass on an object at that point. The symbol g is used for gravitational field strength because it is equivalent to the acceleration due to gravity. For example, we say g near the Earth's surface is 9.8 N/kg.

Weight is the gravitational force on an object that has mass. If we know the mass, m, and the weight, W, of an object, we can calculate *g* using the equation $g = \dfrac{W}{m}$

Rearranging this equation gives

weight W (in newtons) = mass m (in kilograms) × g (in N/kg)

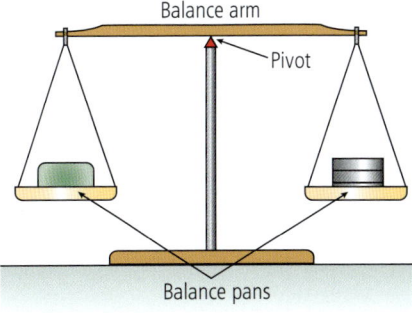

Figure 2.1.2 A balance to compare masses

Supplement

The weight of an object depends on its location. For example, the weight of a 50 kg person near the Earth's surface is 490 N (= 50 kg × 9.8 N/kg). However, the same person on the surface of the Moon where the gravitational field strength is 1.6 N/kg would weigh only 80 N (= 50 kg × 1.6 N/kg).

Comparing masses

We can compare the masses of two different objects using a balance as shown in Figure 2.1.2. For two objects of the same mass, the arm of the balance would be level. For two objects of different mass, the end of the arm supporting the heavier one would drop and the other end would rise.

We can find the mass of an object by placing it on one of the balance pans and placing 'standards' of known mass on the other pan until the arm is level.

SUMMARY QUESTIONS

1 Copy and complete the following sentences below using words from the list.

force   matter   mass   weight

a The _____ of an object is a measure of how much _____ it has.

b The _____ of an object is the _____ on it due to gravity.

c _____ is measured in newtons; _____ is measured in kilograms.

S 2 An object has a mass of 40 kg on the surface of the Earth.

a State whether **i** its mass, **ii** its weight would be smaller, the same or greater on the Moon.

b Calculate **i** the weight of the object on the Earth, **ii** its weight on the Moon.

Use the gravitational field strengths given in the worked example.

WORKED EXAMPLE

Calculate the weight in newtons of a person of mass 60 kg:

a near the Earth's surface

b on the surface of the Moon.

The gravitational field strength near the Earth's surface = 9.8 N/kg.

The gravitational field strength near the Moon's surface = 1.6 N/kg.

Solution

a Near the Earth's surface, the weight of the person = mass × gravitational field strength = 60 kg × 9.8 N/kg = 590 N.

b On the Moon's surface, the weight of the person = mass × gravitational field strength = 60 kg × 1.6 N/kg = 96 N.

KEY POINTS

- The greater the mass of an object, the greater is its weight.
- The weight of an object = its mass × *g*

WORKED EXAMPLE

A wooden post has a volume of 0.025 m³ and a mass of 20 kg. Calculate its density in kg/m³.

Solution

$$density = \frac{mass}{volume} = \frac{20\ kg}{0.25\ m^3}$$
$$= 800\ kg/m^3$$

Materials of different densities

Density comparisons

Any builder knows that a concrete post is much heavier than a wooden post of the same size. This is because the **density** of concrete is much greater than the density of wood. A volume of one cubic metre of wood has a mass of about 800 kg whereas a cubic metre of concrete has a mass of about 2400 kg. So the density of concrete is about three times the density of wood.

The density of two different materials can be compared by comparing the masses of same-size blocks of each material. We can do this using a balance as shown in Figure 2.1.2 on page 19 or we can use an electronic balance to measure the mass of each block. Each block has the same volume so the block with the greater mass has the greater density.

The density (ρ) of a substance is defined as its mass (m) per unit volume (V). We can use the equation below to calculate the density of a substance if we know the mass and the volume of a sample of it.

$$density = \frac{mass}{volume} \quad or \quad \rho = \frac{m}{V}$$

The SI unit of density is the kilogram per cubic metre (kg/m³) although the gram per cubic centimetre (g/cm³) is often used.

Density tests

For each of the tests below, measure the mass and the volume of the object as explained then use the formula, density = $\frac{mass}{volume}$ to calculate the density of the object.

1 Measuring the density of a regular solid object

- To measure the mass of the object, use a balance as shown in Figure 2.1.2 on page 19 or an electronic balance. Make sure the balance reads zero before placing the object on it.
- To find the volume of a regular solid such as a cube, a cuboid or a cylinder, measure its dimensions, using a millimetre ruler or a micrometer. Use the measurements and the correct formula shown in Figure 2.2.1 to calculate its volume.

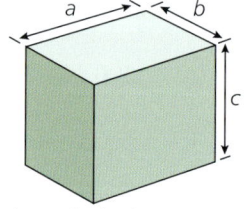

(i) Volume of cuboid = $a \times b \times c$

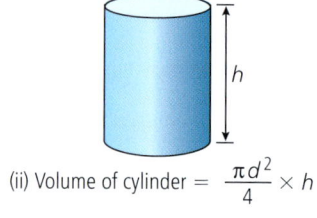

(ii) Volume of cylinder = $\frac{\pi d^2}{4} \times h$

Figure 2.2.1 Volume formulae

2 Measuring the density of a liquid

- Use a measuring cylinder to measure the volume of a certain amount of the liquid.

- Measure the mass of an empty beaker using a balance. Remove the beaker from the balance and pour the liquid from the measuring cylinder into the beaker. Use the balance again to measure the total mass of the beaker and the liquid. The mass of the liquid is worked out by subtracting the mass of the empty beaker from the total mass of the beaker and the liquid.

Use the results from your experiments to test the statements below.

1 An object will float in a liquid if the density of the object is less than the density of the liquid. If the density of the object is greater than the density of the liquid, the object will sink.

S 2 When two liquids that don't mix (e.g. oil and water) are in the same container, the less dense liquid floats on the other liquid.

WORKED EXAMPLE

A measuring cylinder contained a volume of 120 cm³ of a certain liquid. The liquid was then poured into an empty beaker of mass 51 g. The total mass of the beaker and the liquid was then found to be 145 g.

a Calculate the mass of the liquid in grams.

b Calculate the density of the liquid in g/cm³.

Solution

mass of liquid = 145 − 51 = 94 g; volume = 120 cm³

$$\text{density} = \frac{\text{mass}}{\text{volume}} = \frac{94\,g}{120\,cm^3} = 0.78\,g/cm^3$$

Measuring the density of an irregular solid

- Use a balance to measure the mass of the object.
- Determine the volume of the object using a beaker and a displacement can as shown in Figure 2.2.2. Water is the most suitable liquid to use provided the solid does not dissolve in it. Work out the density from the density equation on the opposite page.

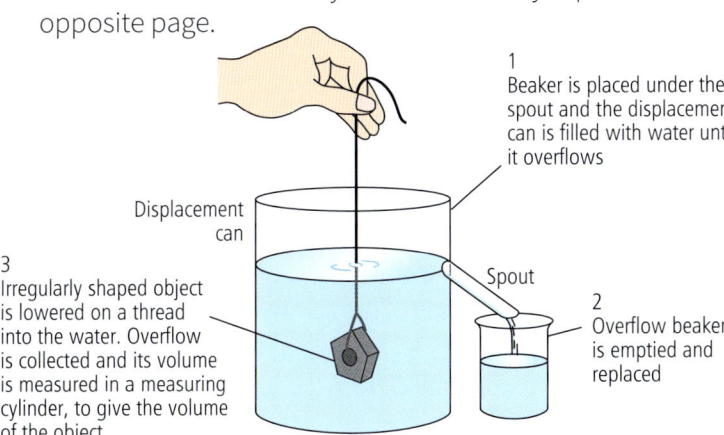

1
Beaker is placed under the spout and the displacement can is filled with water until it overflows

Displacement can

3
Irregularly shaped object is lowered on a thread into the water. Overflow is collected and its volume is measured in a measuring cylinder, to give the volume of the object

Spout

2
Overflow beaker is emptied and replaced

Figure 2.2.2 Measuring the volume of an irregular object in a measuring cylinder

SUMMARY QUESTIONS

1 A rectangular concrete slab is 0.80 m long, 0.60 m wide and 0.05 m thick.

   a Calculate its volume in m³.

   b The mass of the concrete slab is 60 kg. Calculate its density in kg/m³.

2 A measuring cylinder contains 80 cm³ of a certain liquid. The liquid is poured into an empty beaker of mass 48 g. The total mass of the beaker and the liquid was found to be 136 g.

   a Calculate the mass of the liquid in grams.

   b Calculate the density of the liquid in g/cm³.

3 A rectangular block of gold is 0.10 m in length, 0.08 m in width and 0.05 m in thickness.

   a i Calculate the volume of the block.

   ii If the mass of the block is 7.6 kg, calculate the density of gold.

   b A thin gold sheet has a length of 0.15 m and a width of 0.12 m. The mass of the sheet is 0.0015 kg. Use these measurements and the result of your density calculation in **a ii** to calculate the thickness of the sheet.

KEY POINTS

- $$\text{Density} = \frac{\text{mass}}{\text{volume}}$$
- The unit of density is kg/m³ or g/cm³

2.3 Force and shape

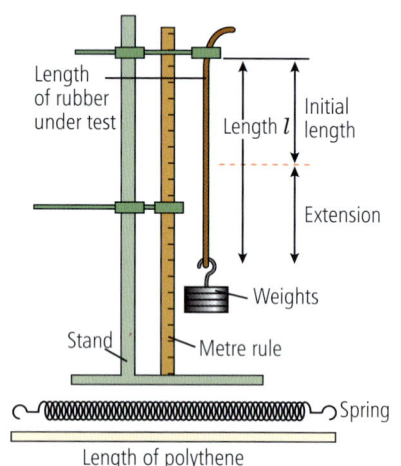

Figure 2.3.1 Investigating stretching

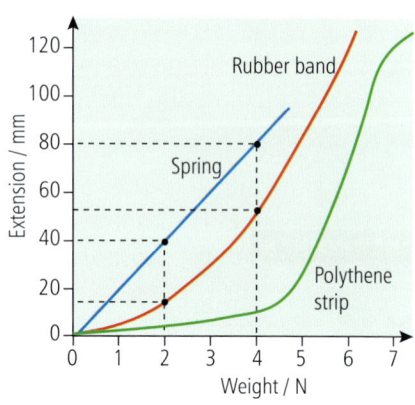

Figure 2.3.2 Extension versus weight for different materials

Stretching and squeezing

Squash players know that hitting a squash ball changes its shape briefly. An object is said to be **elastic** if it regains its original shape and size when the forces that deform it are removed. A squash ball is elastic because it regains its shape. So too is a rubber band, as it regains its original length after it is stretched and then released. Rubber is an example of an elastic material.

Stretch tests

We can investigate how easily a material stretches by hanging weights from it, as shown in Figure 2.3.1.

- The strip of material under test is clamped at its upper end and its initial length is measured using the metre rule. A small weight or a weight hanger attached to the material is used to keep it straight.
- The amount of weight hung from the material is then increased in stages. The strip stretches each time more weight is hung from it.
- At each stage, the total weight added is recorded in a table and the length of the strip is measured and also recorded in the table. The position of the upper end should stay the same throughout.

The change of length from the initial length is called the **extension**. This is calculated for each stage and recorded in the table, as shown below.

The extension of the strip of material at any stage = its length at that stage − its initial length

weight / N	length / mm	extension / mm
0	120	0
1.0	152	32
2.0	190	70
3.0	250	130

Force versus length measurements for a rubber strip

The measurements may be plotted on a graph of extension on the vertical axis against weight on the horizontal axis. Figure 2.3.2 shows the results for strips of different materials and a steel spring plotted on the same axes.

- The steel spring gives a straight line through the origin. This shows that the extension of the steel spring is directly proportional to the weight suspended on it. For example, doubling the weight from 2.0 to 4.0 N doubles the extension of the spring.
- The rubber band does not give a straight line. When the weight on the rubber band is doubled from 2.0 to 4.0 N, the extension more than doubles.
- The polythene strip does not give a straight line either. As the weight is increased from zero, the polythene strip stretches very little at first then it 'gives' and stretches easily.

Supplement

Hooke's law for springs states that the extension of a spring is directly proportional to the weight it supports.

Notes

1 Hooke's law applies up to a limit known as the **limit of proportionality**. The graphs in Figure 2.3.2 show that rubber and polythene have a low limit of proportionality. A steel spring has a much higher limit of proportionality.

2 Hooke's law may be written as an equation

$$F = kx$$

where F is the stretching force or tension, x is the extension and k is the **spring constant**. A graph of F against x gives a straight line through the origin (Figure 2.3.3).

3 As explained in Topic 2.1, a newtonmeter used to weigh an object on the Moon would read $0.16 \times$ its reading on the Earth. However, on the Moon, an object on a balance as in Topic 2.1 would be balanced by the same 'known' mass. The weaker gravity on the Moon has the same effect on both the known mass and the object on a balance but doesn't make the spring in a newtonmeter weaker.

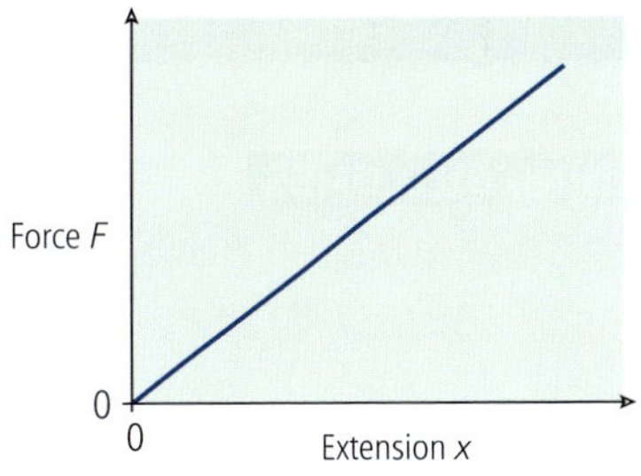

Figure 2.3.3 Graph to show Hooke's law

> **MATHS TIP**
>
> Rearranging the equation
>
> $F = kx$ gives $x = \dfrac{F}{k}$
>
> or $k = \dfrac{F}{x}$

SUMMARY QUESTIONS

1 Copy and complete the following sentences below using words from the list.

extension   length

a When a steel spring is stretched, its _____ is increased.

b When a strip of polythene is stretched too much, its _____ is permanent.

c When rubber is stretched and unstretched, its _____ afterwards is zero.

2 Describe how you would use the arrangement in Figure 2.3.1 to find out if a strip of material is an elastic material.

3 In a Hooke's law test on a spring, the following results were obtained.

weight / N	0	1.0	2.0	3.0	4.0	5.0	6.0
length / mm	245	285	324	366	405	446	484
extension / mm	0	40					

a Complete the bottom row of the table.

b Plot a graph of the extension on the vertical axis against the weight on the horizontal axis.

c If a weight of 7.0 N is suspended on the spring, what would be the extension of the spring?

d i Calculate the spring constant of the spring.

   ii An object suspended on the spring gives an extension of 140 mm. Calculate the weight of the object.

> **KEY POINTS**
>
> - The extension of a strip of stretched material is its extended length minus its initial length.
>
> - The extension of a spring is directly proportional to the weight it supports.

2.4 Force and motion

LEARNING OUTCOMES

- Describe how a force may change the motion of an object
- Recognise that a resultant force acts on an object when the object accelerates or decelerates
- S Recall and use the equation 'force = mass × acceleration'

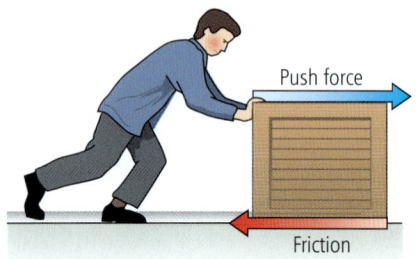

Figure 2.4.1 Overcoming friction

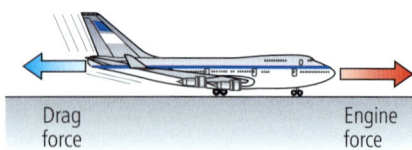

Figure 2.4.2 A passenger jet on take-off

Most objects around us are acted on by more than one force. We can work out the effect of the forces on the motion of an object by replacing them with a single force, the **resultant force**. This is a single force that has the same effect as all the forces acting on the object.

When the resultant force on an object is zero, the object:

- remains at rest if it was already at rest
- continues to move at the same speed and in the same direction if it was already moving.

For example, when a heavy crate is pushed across a rough floor, the crate moves at constant speed across the floor. The push force on the crate is equal and opposite to the force of friction of the floor on the crate. The resultant force on the crate is therefore zero. Frictional forces oppose the motion of any two surfaces that slide (or try to slide) across each other.

When the resultant force on an object is not zero, the movement of the object depends on the size and direction of the resultant force.

For example, when a jet plane is taking off, the force of its engines is greater than the force of air resistance on it. Air resistance is a form of friction called **drag** that acts on any object moving through a liquid or a gas. The resultant force on it is the difference between the thrust force and the force of air resistance on it. The resultant force is therefore not zero. The greater the resultant force, the sooner the plane reaches its take-off speed.

PRACTICAL

Investigating force and acceleration

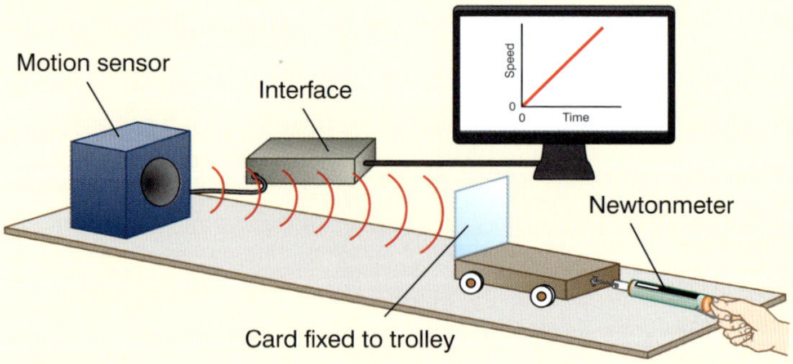

Figure 2.4.3 Investigating the link between force and motion

We can use the apparatus in Figure 2.4.3 to investigate how the acceleration of a trolley depends on the resultant force acting on it.

1. A newtonmeter is used to pull the trolley along with a constant force.
2. The total moving mass can be doubled or trebled by using double-deck and triple-deck trolleys.
3. A motion sensor and a computer are used to record the speed of the trolley as it accelerates. The results are displayed as a speed–time graph on the computer screen.

EXAM TIP

If an object is accelerating, there must be a resultant force acting on it in the same direction as the acceleration. This changes its velocity by changing its speed or its direction of motion or both.

Figure 2.4.4 shows speed–time graphs for different amounts of force. The gradient of each line gives the acceleration. These show that

- **for a given mass, the greater the force, the greater the acceleration**
- **for a given force, the greater the mass, the smaller the acceleration.**

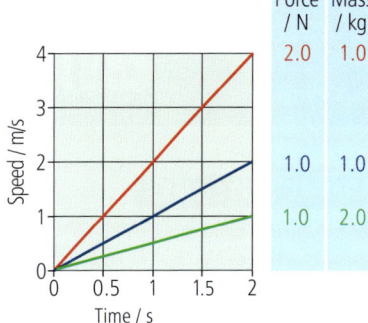

Force / N	Mass / kg
2.0	1.0
1.0	1.0
1.0	2.0

Figure 2.4.4 Speed–time graphs for different forces and masses

Supplement

An equation for force and acceleration

We can work out the acceleration from the gradient of the graph, as explained in Topic 1.4.

Some typical results are given in the table below.

resultant force / N	0.5	1.0	1.5	2.0	4.0	6.0
mass / kg	1.0	1.0	1.0	2.0	2.0	2.0
acceleration / m/s²	0.5	1.0	1.5	1.0	2.0	3.0
mass × acceleration / kg m/s²	0.5	1.0	1.5	2.0	4.0	6.0

The results show that the resultant force, the mass and the acceleration are linked by the equation:

$$\text{resultant force / N} = \text{mass / kg} \times \text{acceleration / m/s}^2$$

Mathematics notes

1 The word equation above may be written in the form:
   resultant force $F = ma$, where m = mass and a = acceleration

2 Rearranging this equation gives $a = \dfrac{F}{m}$ or $m = \dfrac{F}{a}$

WORKED EXAMPLE

Calculate the resultant force on an object of mass 6.0 kg when it has an acceleration of 3.0 m/s².

Solution

Resultant force $F = ma = 6.0 \text{ kg} \times 3.0 \text{ m/s}^2 = 18.0 \text{ N}$.

KEY POINTS

	object at the start	resultant force	effect on the object
1	at rest	zero	stays at rest
2	moving	zero	speed and direction of motion stay the same
3	moving	non-zero in the same direction as the direction of motion of the object	accelerates
4	moving	non-zero in the opposite direction to the direction of motion of the object	decelerates

S Resultant force / N = mass / kg × acceleration / m/s²

SUMMARY QUESTIONS

1 Copy and complete the sentences below using words from the list.

   acceleration   motion   mass   resultant force   speed

   a A moving object decelerates when a _____ acts on it in the opposite direction to its _____ .

   b The greater the _____ of an object, the less its acceleration when a _____ acts on it.

   c The _____ of a moving object increases when a _____ acts on it in the same direction as it is already moving.

2 A jet plane lands on a runway and stops.

   a What can you say about the direction of the resultant force on the plane as it lands?

   b What can you say about the resultant force on the plane when it has stopped?

S 3 Copy and complete the following table.

	force / N	mass / kg	acceleration / m/s²
a	?	20	0.80
b	200	?	5.0
c	840	70	?
d	?	0.40	6.0
e	5000	?	0.20

2.5 More about force and motion

A train on an incline

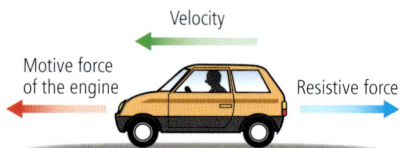

Figure 2.5.2 Terminal speed

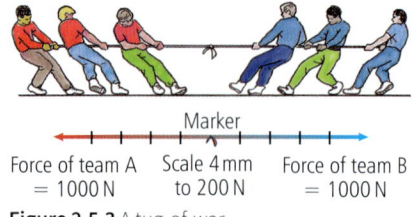

Figure 2.5.3 A tug-of-war

You will learn more about vectors in Topic 3.6 *More about vectors*

On the move

A railway engine pulling a train of carriages and wagons along a track needs to have enough power to pull the train up the steepest incline on the track. If the engine power is not enough, a second engine could be used to help. The force of the two engines on the train is equal to the sum of the force of each engine on the train.

A car stuck in mud can be difficult to shift. A tractor can be very useful here. Figure 2.5.1 shows the idea. One end of the rope is tied to the back of the tractor and the other end to the front of the car. To pull the car out of the mud, the force of the tractor on the car needs to be greater than the force of the mud on the car. If the force of the mud on the car is equal to the force of the tractor on the car, the car stays stuck in the mud.

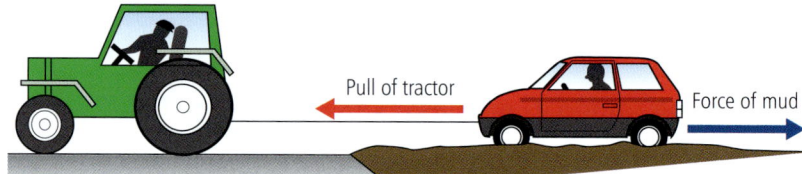

Figure 2.5.1 In the mud

A vehicle moving at its terminal speed on a flat road is pushed forward by the 'motive' force of its engine and opposed by a resistive force due to drag and friction. At terminal speed, the resistive force is equal and opposite to the engine force (Figure 2.5.2). Therefore, the resultant force is zero and so the acceleration is zero.

Supplement

Vectors

- The size and direction of a force can be represented by a **vector**. A vector is an arrow of length that represents the magnitude (size) of the force in the direction of the force. Any force has magnitude and direction and so can be drawn as a vector.

- Figure 2.5.3 shows the pull of two tug-of-war teams as vectors. A scale of 4 mm to 200 N is used here so the force of 1000 N is represented by a vector 2 cm long. In this example, the magnitudes of the two forces are the same so the two vectors are the same length. Because the forces are in opposite directions, the two vectors point in opposite directions.

- Suppose one team pulls with a force of 1000 N and the other team with a force of 750 N. The vector diagram for this situation is shown in Figure 2.5.4. The smaller force nearly cancels out the other force, but not quite. The stronger team exerts a force which is 250 N greater than the other team. So the resultant force (i.e. their combined effect) is 250 N.

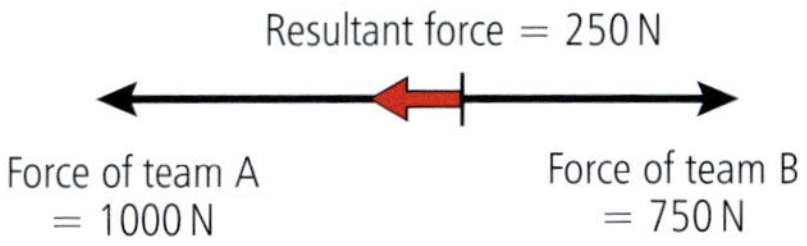

Supplement

Going around in circles

A satellite in a circular orbit above the Earth moves along its orbit at constant speed. The resultant force on the satellite is due only to the force of the Earth's gravitational field on it. This force acts towards the centre of the Earth, pulling on the satellite to prevent it from flying off 'at a tangent' into space. So it changes the direction of motion of the satellite without changing its speed. The same effect happens when an object is whirled around on the end of a string. The resultant force making the object go around in a circle is due to the pull of the string and the force of gravity on it.

Any object moving round a circular path is acted on by a resultant force which is directed towards the centre of the circle. The direction of motion of the object is always perpendicular to the resultant force. The resultant force depends on the mass and the speed of the object and the radius of the circle. For a satellite to stay in a particular orbit, it must move round the orbit at a particular speed that depends on the radius of the orbit. You will learn more about satellite orbits in Topic 16.3.

Resultant force = 250 N

Force of team A = 1000 N

Force of team B = 750 N

Figure 2.5.4 Unequal forces in opposite directions

In general, if an object is acted on by two forces:

- in the same direction, the resultant force is the sum of the two forces

- in opposite directions, the resultant force is the difference between the two forces.

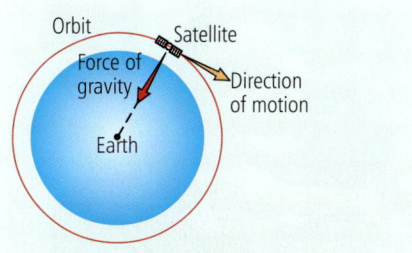

Figure 2.5.5 The effect of gravity

SUMMARY QUESTIONS

1 Copy and complete the following sentences using words from the list.

   equal to   greater than   less than   not zero   zero

   A car starts from rest and accelerates along a straight, flat road until it reaches a certain speed which it then travels at.

   a When it is accelerating, the force of its engine is _____ the resistive forces acting on it. The resultant force is _____ .

   b When it is travelling at constant speed, the force of air resistance on it is _____ the force of its engine. The resultant force on the vehicle is _____ .

2 **a** A lorry tows a car by means of a tow bar along a straight road at constant speed.

   i What can you say about the acceleration of the lorry?

   ii The force of the lorry on the car was 200 N. State the magnitude of resistive force on the car.

   b The lorry driver applies the brakes to the lorry, causing the lorry and the car to slow down and stop. In terms of the forces on the car, explain why the car comes to a stop.

KEY POINTS

- The resultant force due to two forces acting along the same line is given by:

  – the sum of the two forces if the forces act in the same direction

  – the difference between the two forces if they act in opposite directions.

- When the resultant force is non-zero, the object experiences an acceleration given by:

$$\text{acceleration} = \frac{\text{resultant force}}{\text{mass}}$$

- Recognise that momentum is defined as mass × velocity

- Recall that the unit of momentum is the kilogram metre per second (kg m/s)

- Use the equation 'momentum = mass × velocity'

A contact sport

MOMENTUM IN ACTION

If a vehicle crashes into the back of a line of cars, each car is shunted into the one in front. Momentum is transferred along the line of cars to the one at the front.

Momentum is important to anyone who plays a contact sport. In a game of rugby, a player with a lot of momentum is very difficult to stop.

The momentum of a moving object = its mass × its velocity.

So momentum has a magnitude and a direction.

The unit of momentum is the kilogram metre/second (kg m/s).

We can write the word equation using symbols: $p = m \times v$, where

p = momentum in kilogram metres/second, kg m/s

m = mass in kilograms, kg

v = speed in metres/second, m/s

PRACTICAL

Investigating collisions

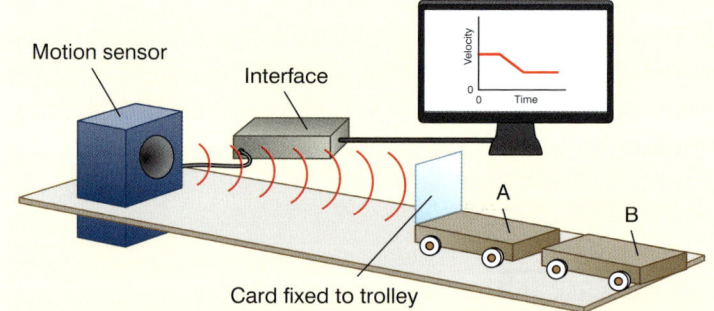

Motion sensor

Interface

Card fixed to trolley

Figure 2.6.1 Investigating collisions

When two objects collide, the momentum of each object changes. Figure 2.6.1 shows how to use a computer and a motion sensor to investigate a collision between two trolleys.

Trolley A is given a push so it collides with stationary trolley B. The two trolleys stick together after the collision. The computer gives the velocity of A before the collision and the velocity of both trolleys afterwards.

- What does each section of the velocity–time graph show?

1 **For two trolleys of the same mass,** the velocity of trolley A is halved by the impact. The combined mass after the collision is twice the moving mass before the collision. So the momentum (= mass × velocity) after the collision is the same as before the collision.

Figure 2.6.2 A 'shunt' collision

2 For a single trolley pushed into a double trolley, the velocity of A is reduced to one-third. The combined mass after the collision is three times the initial mass. So in this test as well, the momentum after the collision is the same as the momentum before the collision.

In both tests, the total momentum is unchanged (i.e. is conserved) by the collision. This is an example of the **conservation of momentum**. It applies to any system of objects as long as the system is closed, which means that no resultant force acts on it.

Safety: Use foam or an empty cardboard box to stop trolleys. Protect bench and feet from falling trolleys.

In general, the **law of conservation of momentum** states that:

In a closed system, the total momentum before an event is equal to the total momentum after the event.

We can use this law to predict what happens whenever objects collide or push each other apart in an 'explosion'. Momentum is conserved in any collision or explosion as long as no external forces act on the object.

SUMMARY QUESTIONS

1 a Define momentum and state its unit.

b Calculate the momentum of a 40 kg person running at 6 m/s.

2 a Calculate the momentum of an 80 kg rugby player running at a velocity of 5 m/s.

b An 800 kg car moves with the same momentum as the rugby player in **a**. Calculate the velocity of the car.

c Calculate the velocity of a 0.4 kg ball that has the same momentum as the rugby player in **a**.

3 A 1000 kg wagon moving at a velocity of 5.0 m/s on a level track collides with a stationary 1500 kg wagon. The two wagons move together after the collision.

a Calculate the momentum of the 1000 kg wagon before the collision.

b Calculate the velocity of the wagons after the collision.

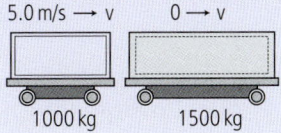

5.0 m/s → v   0 → v

1000 kg   1500 kg

A 3000 kg truck moving at a velocity of 16 m/s crashes into the back of a stationary 1000 kg car. The two vehicles move together immediately after the impact. Calculate their velocity.

Solution

Let v represent the velocity of the vehicles after the impact.

momentum of truck before impact = 48 000 kg m/s

momentum of the car before impact = 0 kg m/s

momentum of truck after impact = 3000 kg × v

momentum of car after impact = 1000 kg × v

$3000v + 1000v = 48\,000 + 0$

$4000v = 48\,000$; $v = \mathbf{12\,m/s}$

KEY POINTS

- Momentum = mass × velocity

- The unit of momentum is kg m/s.

- Momentum is conserved whenever objects interact, as long as objects are in a closed system so that no external forces act on them.

2.7 Explosions

Supplement

If you are skateboarder, you will know that the skateboard can shoot away from you when you jump off it. Its momentum is in the opposite direction to your own momentum. What can we say about the total momentum of objects when they fly apart from each other?

PRACTICAL

Investigating a controlled explosion

Figure 2.7.1 shows a controlled explosion using trolleys. When the trigger rod is tapped, a bolt springs out and the trolleys recoil (spring back) from each other.

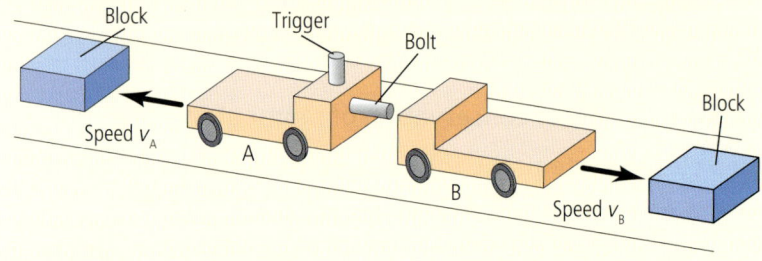

Figure 2.7.1 Investigating explosions

Using trial and error, we can place blocks on the runway so the trolleys reach them at the same time. This allows us to compare the speeds of the trolleys.

Some results are shown in Figure 2.7.2.

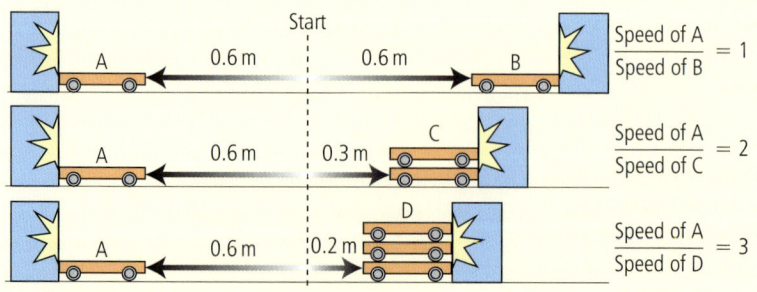

Figure 2.7.2 Using different masses

- Two single trolleys travel equal distances in the same time. This shows that they recoil at the same speed.

- A double trolley travels only half the distance that a single trolley does. Its speed is half that of the single trolley.

In each test:

1 The mass of the trolley × the speed of the trolley is the same.

2 They recoil in opposite directions.

So momentum has magnitude and direction. The results show that the trolleys recoil with equal and opposite momentum.

Conservation of momentum

In the trolley examples:

- momentum of A after the explosion = (mass of A × velocity of A)
- momentum of B after the explosion = (mass of B × velocity of B)
- total momentum before the explosion = 0 (because both trolleys were at rest).

Using conservation of momentum gives:

(mass of A × velocity of A) + (mass of B × velocity of B) = 0

Therefore

(mass of A × velocity of A) = − (mass of B × velocity of B)

The minus sign after the equals sign tells us that the momentum of B is in the opposite direction to the momentum of A. The equation tells us that A and B move apart with equal and opposite amounts of momentum. So, the total momentum after the explosion is the same as before it.

Momentum in action

When a shell is fired from a military gun, the gun barrel recoils backwards. The recoil of the gun barrel is slowed down by a spring. This lessens the backwards motion of the gun.

SUMMARY QUESTIONS

1 A 60 kg skater and an 80 kg skater standing in the middle of an ice rink, push each other away.

What can be said about:

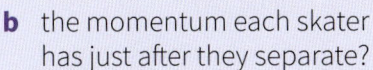
80 kg   60 kg

a the force they exert on each other when they push apart?

b the momentum each skater has just after they separate?

c each of their velocities after they separate?

d their total momentum just after they separate?

2 In Question 1, the 60 kg skater moves away at 2.0 m/s. Calculate:

a their momentum

b the velocity of the other skater.

3 A 600 kg cannon recoils at a speed of 0.5 m/s when a 12 kg cannonball is fired from it.

a Calculate the velocity of the cannonball when it leaves the cannon.

b How would the recoil velocity of the cannon have been different if a 4 kg cannonball had been used instead?

WORKED EXAMPLE

An artillery gun of mass 2000 kg fires a shell of mass 20 kg at a velocity of 120 m/s. Calculate the recoil velocity of the gun.

Solution

Applying the conservation of momentum gives:

mass of gun × recoil velocity of gun = − (mass of shell × velocity of shell)

If we let V represent the recoil velocity of the gun,

$$2000 \, kg \times V = -(20 \, kg \times 120 \, m/s)$$
$$V = -\frac{2400 \, kg \, m/s}{2000 \, kg} = -1.2 \, m/s$$

Figure 2.7.3 An artillery gun in action

KEY POINTS

- Momentum is mass × velocity and has direction.
- When two objects push each other apart, they move:
  - with different speeds if they have unequal masses
  - with equal and opposite momentum, so their total momentum is zero.

- Recognise that when objects collide, the force of the impact depends on how long the impact lasts for
- Recall that the impulse of a force F acting on an object for a time t is defined as $F \times t$
- Carry out calculations using the equation impulse = change of momentum

PRACTICAL

Investigating impacts

We can test an impact using a trolley and brick, as shown in Figure 2.8.1. When the trolley hits the brick, the plasticine flattens on impact, making the impact time longer. This is the key factor that reduces the impact force.

Safety: Protect bench and feet from bricks and trolley.

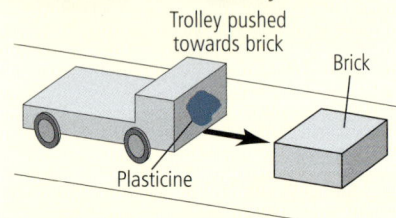

Figure 2.8.1 Investigating impacts

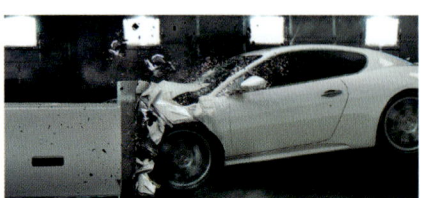

Figure 2.8.2 A crash test. Car makers test the design of a crumple zone by driving a remote control car into a wall.

Crumple zones at the front end and rear end of a car are designed to lessen the force of an impact. The force changes the momentum of the car.

- In a front-end impact, the momentum of the car is reduced.
- In a rear-end impact (where a vehicle is struck behind by another vehicle), the momentum of the car is increased.

In both cases the effect of a crumple zone is to increase the impact time and so lessen the impact force.

Impact time

Let's see why making the impact time longer reduces the impact force.

Suppose a moving trolley hits another object and stops. The impact force on the trolley acts for a certain time (the impact time) and causes it to stop. A soft pad on the front of the trolley would increase the impact time and would allow the trolley to travel further before it stops. The momentum of the trolley would be lost over a longer time and its kinetic energy would be transferred over a greater distance.

If we know the impact time t, we can calculate the impact force F as follows:

- From Topic 2.4, $F = ma$, where m is the mass of the object and a is its acceleration.
- From Topic 1.4, $a = \dfrac{v - u}{t}$, where u is the initial velocity, v is the final velocity and t is the impact time.

Therefore, $F = ma = m\dfrac{(v - u)}{t} = \dfrac{mv - mu}{t}$

So we can use the above equation to calculate F.

Also, multiplying both sides of the equation by t gives:

$$Ft = mv - mu$$

where the force $F \times$ the time t is called the **impulse** of the force. The equation above therefore tells us that:

The impulse of a force, Ft = the change of momentum ($mv - mu$).

The above equation shows that for a given change of momentum, the impact force can be reduced by increasing the impact time.

Rearranging the above equation gives $F = \dfrac{mv - mu}{t} = \dfrac{\text{change of momentum}}{\text{time taken}}$

Therefore

the resultant force can be defined as the change of momentum per unit time.

If a resultant force F acts on an object for a time Δt causing its momentum to change by Δp, then $F = \dfrac{\Delta p}{\Delta t}$

WORKED EXAMPLE

A bullet of mass 0.004 kg moving at a velocity of 90 m/s is stopped by a bulletproof vest in 0.0003 s. Calculate the impact force.

Solution

Initial momentum of bullet = mass × initial velocity = 0.004 kg × 90 m/s = 0.36 kg m/s

Final momentum of bullet = 0 (as the bullet is stopped)

Therefore the change of momentum
= final momentum – initial momentum = −0.36 kg m/s

Since Ft = change of momentum, where F is the impact force and t is the time taken, then

$F × 0.0003 \text{ s} = −0.36 \text{ kg m/s}$

Dividing both sides by 0.0003 s gives $F = \dfrac{−0.36 \text{ kg m/s}}{0.0003 \text{ s}} = −1200 \text{ N}$

The negative sign tells us that the impact force decelerates the bullet.

Two-vehicle collision

When two vehicles collide, they exert equal and opposite impact forces on each other at the same time. The change of momentum of one vehicle is therefore equal and opposite to the change of momentum of the other vehicle. The total momentum of the two vehicles is the same after impact as it was before the impact, so momentum is conserved – assuming no external forces act.

We sometimes express the effect of an impact on an object or person as a force-to-weight ratio. We call this **g-force**.

ABOUT g-FORCES

You would experience a g-force of:

- about 3 to 4g on a fairground ride that whirls you around
- about 10g in a low-speed car crash
- more than 50g in a high-speed car crash that you would be lucky to survive.

KEY POINTS

- When vehicles collide, the force of impact depends on mass, change of velocity and the duration of the impact.
- The longer the impact time, the more the impact force is reduced.
- When two vehicles collide:
  - they exert equal and opposite forces on each other
  - their total momentum is unchanged.

SUMMARY QUESTIONS

1 a In a car crash, when a passenger wears a seat belt, why does it reduce the impact force on him?

   b A ball of mass of 0.12 kg moving at a velocity of 18 m/s is caught by a person in 0.0003 s. Calculate the impact force.

2 a An 800 kg car travelling at 30 m/s is stopped safely when the brakes are applied. What braking force is required to stop it in 6.0 s?

   b If the vehicle in **a** had been stopped in a collision lasting less than a second, explain by referring to momentum why the force on it would have been much greater.

3 A 2000 kg van moving at a velocity of 12 m/s crashes into the back of a stationary truck of mass 10 000 kg. Immediately after the impact, the two vehicles move together.

   a Show that the velocity of the van and the truck immediately after the impact is 2 m/s.

   b The impact lasts for 0.3 seconds. Calculate:

      i the change of momentum of the van

      ii the force of the impact of the van.

1 The gravitational field strength at the Earth's surface is 10 N/kg. Calculate the weight of the following:

(a) a person of mass 80 kg

(b) a 2 kg bag of sugar

(c) a 125 g pack of tea

(d) a 70 g chocolate bar

2 (a) Write down the equation used for calculating density.

(b) Describe, with the aid of diagrams, a method for finding the density of a small, irregularly shaped object.

3 A rectangular block is 10 cm long, 4 cm wide and 3 cm high.

(a) Calculate the volume of the block.

(b) The block has a mass of 150 g. Calculate the density of the material of the block.

(c) Will this block float on water?

(d) Suggest a material from which the block might have been made.

4 (a) State the resultant force acting on a vehicle travelling at a constant speed of 120 km/h on a straight section of road.

(b)(i) A heavy truck is slowing down because the driver has applied the brakes. What can you conclude about the direction of the resultant force acting on the truck?

(ii) What is the resultant force acting on the truck when it has come to a halt?

5 (a) An object of mass 2.5 kg is acted on by a force of 5 N. Calculate the acceleration that this force produces.

(b) The same object is acted on by a different force and the acceleration is 3.2 m/s^2. Calculate the value of the force.

6 A stone is tied firmly to a length of string. The stone is then whirled around in a horizontal circle. The speed at which the stone is moving is constant.

Practice Questions

1 Which of the following is a definition of mass?

A   The mass of a body is a measure of the amount of matter it contains.

B   The mass of a body is the amount of force a body feels due to gravity.

C   The mass of a body is the number of atoms it has in it.

D   The mass of a body is a measure of how close its atoms are.

2 A steel ball has a mass of 2.2 kg. What is the weight of the ball if the gravitational field strength (g) is 10 N/kg?

A   0.22 N

B   2.2 N

C   22 N

D   4.5 N

3 Which of the following describes the motion of a moving object which has no resultant force acting on it?

A   The object will accelerate.

B   The object will decelerate.

C   The object will travel in a circle.

D   The object will continue to move at the same speed in a straight line.

(a) State whether or not the velocity of the stone is constant. Briefly explain your answer.

(b) The stone is accelerating. Explain how it can be accelerating whilst moving at constant speed.

4 What is the momentum of a remote controlled drone if it has a mass of 1.5 kg and is travelling at 4.0 m/s?

  A 5.5 kg m/s

  B 6.0 kg m/s

  C 2.7 kg m/s

  D 2.5 kg m/s

5 A student tries to measure the density of a liquid. They find that the volume is 32.0 cm³ and the mass is 66.0 g.

  (a) What instrument could be used to measure the volume of the liquid? [1]

  (b) What instrument could be used to measure the mass of the liquid? [1]

  (c) Calculate the density of the liquid in g/cm³. [2]

6 A spring is suspended from the stand and masses are added to the end as shown in the figure below. The spring has an unstretched length of 12.5 cm. When a 2.5 N weight is added, the spring stretches so that its length is 15.2 cm.

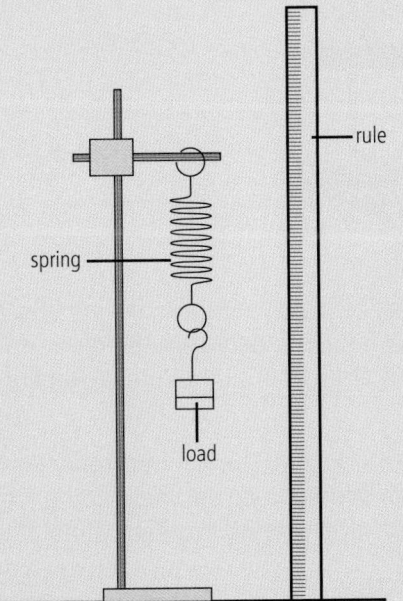

  (a) Calculate the extension when the spring has a 2.5 N weight attached. [1]

  (b) Describe how this equipment can be used to investigate the extension–load properties of the spring. [4]

7 At an airport check-in desk a conveyor belt is used to move cases and bags to the cargo area. The cases are placed on the belt, weighed and then transferred.

  Two suitcases are placed on the belt. One has a mass of 30 kg and the other a mass of 25 kg.

  (a) Calculate the weight of the 25 kg suitcase. [2]

  (b) Sketch a diagram showing the forces acting on a suitcase as it is accelerated away from the check-in desk on the conveyor belt. [3]

  (c) Explain why the conveyor belt needs to apply a larger force on the heavier suitcase to accelerate it away from the counter. [3]

  (d) Calculate the force required to produce an acceleration of 1.6 m/s² on a mass of 25 kg. [2]

8 A trolley (A) of mass 0.8 kg travelling at 1.2 m/s collides with a stationary trolley (B) of mass 1.6 kg as shown in the diagram below. The two trolleys stick together and move off at the same speed, v. The impact between the trolleys lasts for 0.2 s.

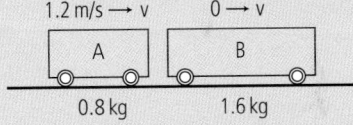

  (a) Calculate the momentum of trolley A before the collision. [2]

  (b) State the total momentum of the trolley combination after the collision. [1]

  (c) Calculate the speed of the trolleys after the impact. [3]

  (d) Calculate the average size of the force acting on trolley A during the collision. [4]

Using moments

A **spanner** is needed to undo a very tight wheel nut on a bicycle. The force you apply to the spanner has a turning effect on the nut. You couldn't undo a tight nut with your fingers but the spanner can turn it. The spanner exerts a much bigger force on the nut than the force you apply to the spanner.

If you had a choice between a long-handled spanner and a short-handled one, which would you choose? The longer the spanner's handle, the less force you need to exert on it to loosen the nut (Figure 3.1.1). In this example, the turning effect of the force, called the **moment** of the force, can be increased by:

- increasing the size of the force
- using a spanner with a longer handle.

Investigating the turning effect of a force

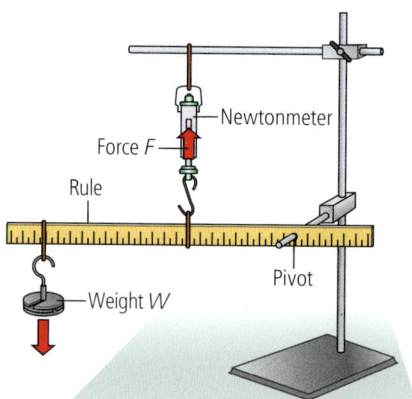

Figure 3.1.2 Investigating turning forces

Figure 3.1.1 A turning effect

Figure 3.1.2 shows one way to investigate the turning effect of a force. A known weight W is suspended from the metre rule as shown and then moved along it until the rule is exactly horizontal.

- How do you think the reading on the newtonmeter compares with the weight W? You should find that the reading (i.e. the force needed to support the weight) is greater than W, provided W is further from the pivot than the newtonmeter is. The rule is in equilibrium because the turning effect of the weight is equal and opposite to the turning effect of the newtonmeter.
- If the weight is moved further from the pivot, the free end of the rule drops down. This shows that the turning effect of the weight increases the further the weight is from the pivot.

The **moment of a force** is given by the equation:

moment	=	**force**	×	**perpendicular distance to pivot**
(in newton metres, N m)		(in newtons, N)		(in metres, m)

Note The moment of a force is the newton metre (N m).

The **claw hammer** in Figure 3.1.3 is being used to remove a nail from a wooden beam.

- The applied force F on the claw hammer tries to turn it clockwise about the pivot.
- The moment of force F about the pivot is $F \times d$, where d is the perpendicular distance from the pivot to the line of action of the force.
- The effect of the moment is to cause a much larger force to be exerted on the nail.

WORKED EXAMPLE

A force of 50 N is exerted on a claw hammer of length 0.30 m, as shown in Figure 3.1.3. Calculate the moment of the force.

Solution

Force = 50 N × 0.30 m = 15 N m

SUMMARY QUESTIONS

1 In Figure 3.1.1, a force is applied to a spanner to undo a nut. State whether the moment of the force is:

   a clockwise or anticlockwise

   b increased or decreased by **i** increasing the force, **ii** exerting the force nearer the nut.

2 Explain each of the following statements:

   a A claw hammer is easier to use to remove a nail if the hammer has a long handle.

   b A door with rusty hinges is more difficult to open than a door of the same size with lubricated hinges.

3 A spanner is used to tighten a nut as shown in Figure 3.1.1. A force of 50 N is exerted on the spanner at a distance of 0.24 m from the centre of the nut. Calculate the moment of the force.

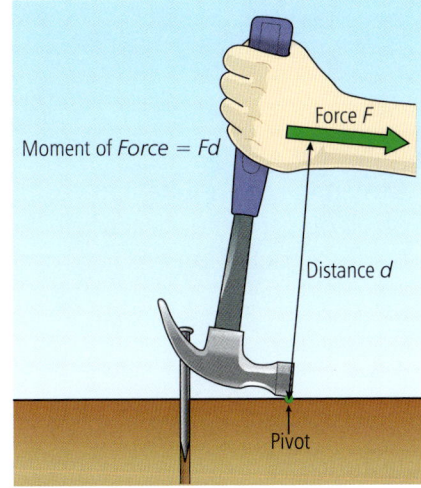

Moment of *Force* = *Fd*

Force *F*

Distance *d*

Pivot

Figure 3.1.3 Using a claw hammer

3.2 Moments in balance

LEARNING OUTCOMES

- Recognise that there is no resultant force or resultant turning effect on an object in equilibrium
- Use knowledge of forces and turning effects to explain why objects at rest don't move or turn

On a seesaw

Figure 3.2.1 Moments in action – the seesaw

A seesaw is an example in which clockwise and anticlockwise moments balance each other out. The girl in Figure 3.2.1 sits near the pivot to balance her younger brother at the far end of the seesaw. He is not as heavy as his big sister.

- His weight has a clockwise turning effect because it would make the seesaw turn clockwise if the girl were not on the seesaw.
- Her weight has an anticlockwise turning effect because it would make the seesaw turn anticlockwise if the boy were not on the seesaw.

Because the turning effect of each child depends on the child's weight and their distance to the pivot, the girl needs to sit nearer the pivot than her brother so that their turning effects about the pivot balance out. Therefore there is no resultant moment on the seesaw.

The pivot supports the weight of the two children and the weight of the seesaw beam. The support force acting upwards on the beam from the pivot must be equal and opposite to the total weight of the beam and the children. Therefore there is no resultant force on the seesaw beam.

On a building site

Figure 3.2.2 shows a horizontal plank supported near each end by steel scaffolding tubes X and Y. A builder stands on the plank between X and Y.

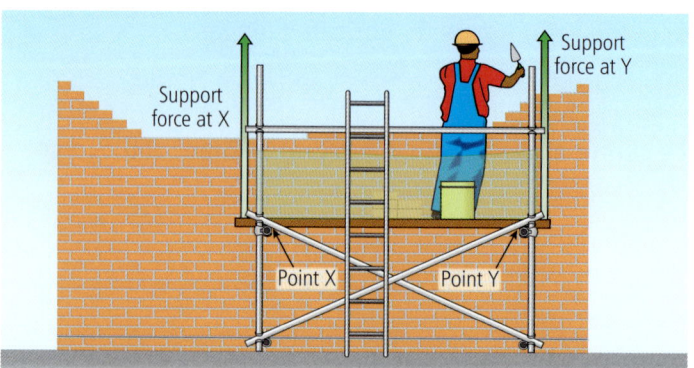

Figure 3.2.2 Moments at work – on a building site

EXAM TIP

Remember the resultant force on an object is the effect of <u>all</u> the forces acting on it.

KEY POINTS

There is no resultant force or resultant turning effect on an object in equilibrium.

- The tubes X and Y support the total weight of the plank and the builder. The support forces at X and Y:
  - act vertically upwards from the tubes X and Y
  - are equal to the total weight of the builder and the plank (which act downwards).

The resultant force on the plank is therefore zero.

- The support force at X is equal to the support force at Y if the builder is midway between X and Y, and the middle of the plank is also midway between X and Y.

The resultant turning effect on the plank is therefore zero.

These two examples show that, for any object in equilibrium:

- **there is no resultant moment on it**
- **there is no resultant force on it.**

PRACTICAL

Investigating a toy mobile

1 Measure the weight of a rule and of two small objects A and B, using a newtonmeter. Make a model of a toy mobile using the rule and the two objects as shown in Figure 3.2.3 and suspend it from a newtonmeter. Note that the point of suspension S of the rule is at its centre.

2 Adjust the positions of the two objects A and B so the rule is horizontal.

- You should find that the heavier of the two objects is nearer S than the other object. This is because the two objects cause equal and opposite moments on the rule when the rule is horizontal. So the resultant moment on the mobile is zero.

- You should also find the newtonmeter reading is equal to the sum of the weight of the rule and the weights of the two objects. This is because the support force from the newtonmeter on the rule acts vertically upwards and is equal and opposite to the total weight of the mobile. So there is no resultant force on the mobile.

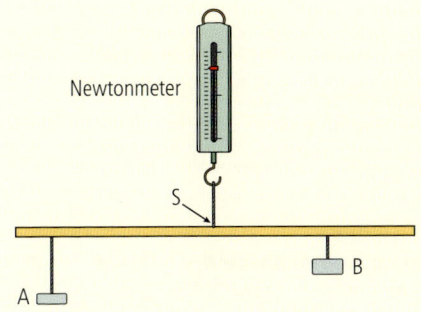

Figure 3.2.3 Investigating a toy mobile

1 a In the seesaw shown in Figure 3.2.1, state and explain whether the seesaw will turn clockwise or anticlockwise if the girl shifts away from the pivot.

b Figure 3.2.4 below shows a simple model of a seesaw in which a rule pivoted at its centre supports two weights A and B. The forces acting on the rule are shown in Figure 3.2.4.

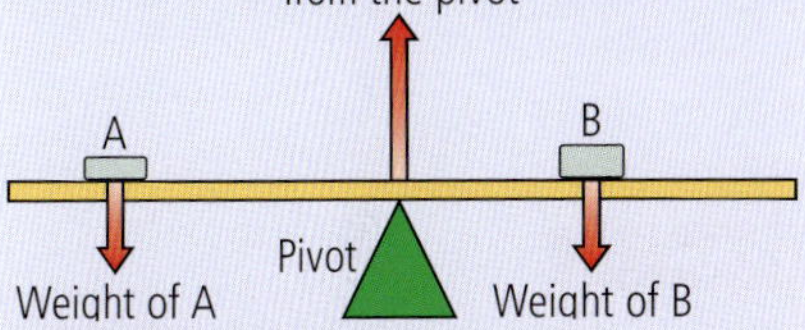

Support force from the pivot

A   B

Pivot

Weight of A   Weight of B

Figure 3.2.4

i When the rule is in equilibrium, A is further from its centre than B. What does this tell you about the weight of A compared with the weight of B?

ii What can you say about the support force on the rule from the pivot compared with the weights of A and B?

2 In Figure 3.2.2, the total weight of the plank and the builder is 860 N.

a The builder is at the centre of the plank which is midway between the supports X and Y. Calculate the support force from each support.

b State and explain how each support force changes when the builder moves nearer to Y.

3.3 The Principle of Moments

LEARNING OUTCOMES

- Describe and calculate the moment of a force
- Apply the principle of moments to a beam balanced about a pivot
- **S** Apply the principle of moments to different situations

EXAM TIP

There are many examples of turning effects – far more than can be given here. Use the examples here to make sure you understand the principles so you can apply them to other examples.

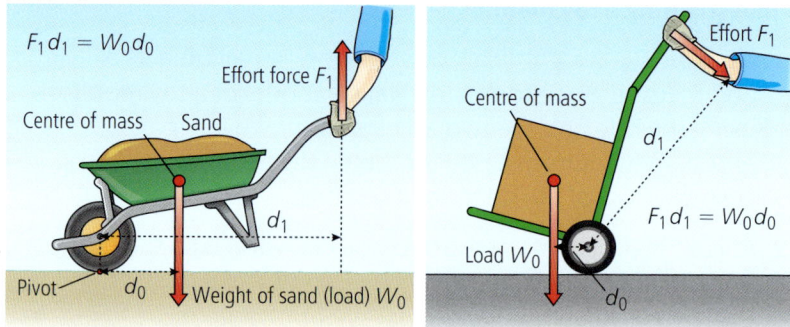

Turning effects at work

The next time you have to move a heavy load, think beforehand about how to make the job easier. Use the examples here to make sure you understand the principles so you can apply them to other examples. The load (weight W_0) is lifted and moved using a much smaller effort (force F_1). This is because:

- the pivot is the point where the wheel is in contact with the ground
- the turning effect of the effort about the pivot is equal and opposite to the turning effect of the load about the pivot
- the effort acts much further from the pivot than the load so the effort needed to raise the load is much smaller than the load.

$F_1 d_1 = W_0 d_0$

Effort force F_1

Centre of mass Sand

d_1

Pivot d_0 Weight of sand (load) W_0

a Using a wheelbarrow.

Effort F_1

Centre of mass

d_1

$F_1 d_1 = W_0 d_0$

Load W_0 d_0

b Using a trolley.

Figure 3.3.1 Moments at work

Balancing a beam

Figure 3.3.2 shows two weights W_1 and W_2 on a pivoted metre rule in equilibrium. Before placing the weights on the rule, the rule on its own was balanced horizontally on the pivot. With W_1 at a certain distance d_1 from the pivot, the distance d_2 of W_2 from the pivot is adjusted until the rule is horizontal. The turning effect of W_1 about the pivot is then equal and opposite to the turning effect of W_2 about the pivot.

$W_1 d_1 = W_2 d_2$

d_1 d_2

W_1 Pivot W_2

Figure 3.3.2 The Principle of Moments

The Principle of Moments

Measurements obtained using the arrangement in Figure 3.3.2 with different weights and distances are shown in the table below. Each row of measurements shows that, in each case, the moment of W_1 about the pivot ($= W_1 \times d_1$) is equal and opposite to the moment due to W_2 ($= W_2 \times d_2$).

weight W_1 / N	weight W_2 / N	distance d_1 / m	distance d_2 / m	$W_1 \times d_1$ / N m	$W_2 \times d_2$ / N m
1.0	2.0	0.48	0.24	0.48	0.48
1.0	3.0	0.48	0.16	0.48	0.48
1.0	4.0	0.48	0.12	0.48	0.48

The examples in the table demonstrate the **Principle of Moments**. This states that, **for an object in equilibrium**:

the sum of all the clockwise moments about any point = the sum of the all anticlockwise moments about that point.

Moments tests

1 Measuring an unknown weight

We can use the arrangement in Figure 3.3.2 to find an unknown weight, W_1, if we know the other weight, W_2, and we measure the distances d_1 and d_2. The unknown weight can then be calculated using the equation $W_1 d_1 = W_2 d_2$.

Supplement

2 Three weights on a beam

Check the rule on its own is balanced on the pivot. Place known weights W_1, W_2 and W_3 on the beam, as in Figure 3.3.3, with W_2 between W_1 and the pivot, and the pivot at the same position as before. Rebalance the beam by adjusting the positions of the weights. Measure the distances d_1, d_2 and d_3 from each weight to the pivot.

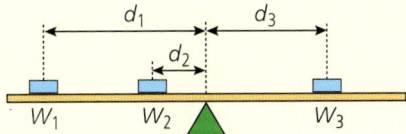

Figure 3.3.3

- The total anticlockwise moment about the pivot = the moment of W_1 + the moment of $W_2 = W_1 d_1 + W_2 d_2$.
- The total clockwise moment about the pivot = the moment of $W_3 = W_3 d_3$.

Applying the Principle of Moments gives $W_1 d_1 + W_2 d_2 = W_3 d_3$.

Use your measurements to confirm the above equation and explain why the resultant moment is zero.

Calculate W_1 in Figure 3.3.3 if $W_2 = 2.0\,N$, $W_3 = 4.0\,N$, $d_1 = 0.25\,m$, $d_2 = 0.10\,m$ and $d_3 = 0.20\,m$.

Solution

Using $W_1 d_1 + W_2 d_2 = W_3 d_3$ gives:
$(W_1 \times 0.25) + (2.0 \times 0.10) = (4.0 \times 0.20)$

Therefore $0.25 W_1 + 0.20 = 0.80$

$0.25 W_1 = 0.80 - 0.20 = 0.60$

$W_1 = \dfrac{0.60}{0.25} = 2.4\ N$

For an object in equilibrium:

the sum of the anticlockwise moments about any point = the sum of the clockwise moments about that point.

1 Dawn sits on a seesaw 2.50 m from the pivot. Jasmin balances the seesaw by sitting 2.00 m on the other side of the pivot.

  a Who is lighter, Dawn or Jasmin?

  b Dawn picks up her younger brother who weighs less than she does and sits him on the seesaw with her. Explain why Jasmin needs to move further away from the pivot to rebalance the seesaw.

2 a For the balanced beam in Figure 3.3.4, work out the unknown weight, W.

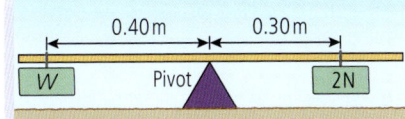

Figure 3.3.4

  b Figure 3.3.5 shows three weights on a beam that is balanced at its centre. Calculate the distance d from the 0.5 N weight to the pivot.

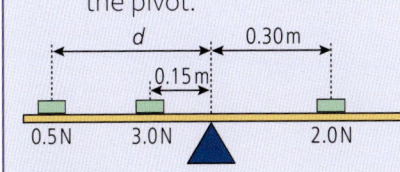

Figure 3.3.5

3.4 Centre of gravity

The design of racing cars has changed considerably since the first models, but one design feature that has not changed is the need to keep the car near the ground. The weight of the car must be as low as possible, otherwise the car will overturn when cornering at high speeds.

Modern racing car design

We can think of the weight of an object as if it acts at a single point. This point is called the **centre of gravity** (or the centre of mass) of the object.

Every object behaves as if its weight is concentrated at one point. This point is called the centre of gravity.

Suspended equilibrium

If you suspend an object and then release it, it will come to rest with its centre of gravity directly below the point of suspension, as shown in Figure 3.4.1a. The object is then in **equilibrium**. Its weight does not exert a turning effect on the object because its centre of gravity is directly below the point of suspension.

If the object is turned from this position and then released, it will swing back to its equilibrium position. This is because its weight has a turning effect that returns the object to equilibrium, as shown in Figure 3.4.1b.

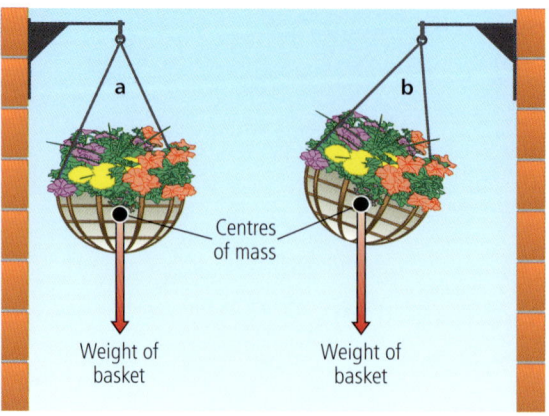

Figure 3.4.1 Suspension (a) in equilibrium, (b) non-equilibrium

Locating the centre of gravity of a flat object

Figure 3.4.2 shows how to find the centre of gravity of a flat card.

1 Suspend the card freely from a thin rod and allow it to come to rest. Its centre of gravity is then directly below the rod. Use a 'plumbline' to draw a vertical line on the card from the thin rod downwards. To do this, mark the bottom edge of card where the plumbline meets the edge. Remove the plumbline and card from the pivot. Draw a line joining the mark to the pivot hole.

2 Repeat the procedure with the card suspended from a second point to give another similar line. The centre of gravity of the card is where the two lines meet.

3 Test your result by suspending the card from a third point and using the plumbline to draw a vertical line through this third point. The third line should pass through the point where the other two lines meet.

See if you can balance the card at this point on the end of a pencil.

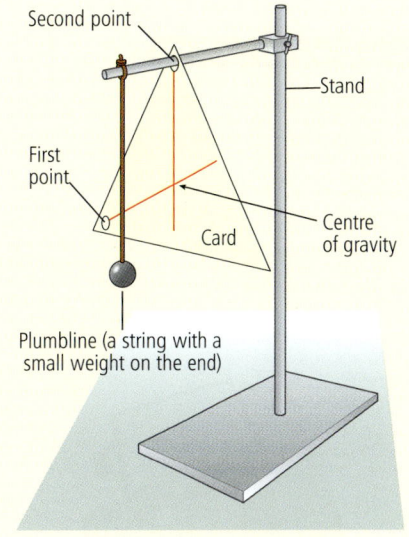

Figure 3.4.2 Finding the centre of gravity of a card

The centre of gravity of a symmetrical object

For a symmetrical object, its centre of gravity is along the axis of symmetry, as shown in Figure 3.4.3. If the object has more than one axis of symmetry, its centre of gravity is where the axes of symmetry meet.

- A rectangle has two axes of symmetry, as shown in Figure 3.4.3a. The centre of gravity is where the axes meet.

- The equilateral triangle in Figure 3.4.3b has three axes of symmetry, each bisecting one of the angles of the triangle. The three axes meet at the same point, which is the centre of gravity of the triangle.

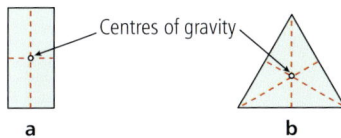

Figure 3.4.3 Symmetrical objects

Make sure you can describe all the steps in the above experiment.

- Every object behaves as if its weight is concentrated at one point. This point is called its centre of gravity.

- When a suspended object is in equilibrium, its centre of gravity is directly beneath the point of suspension.

- The centre of gravity of a symmetrical object is along the axis of symmetry.

SUMMARY QUESTIONS

1 a On a tightrope, a tightrope walker carrying a pole horizontally senses a slight movement to the left. Should the pole be shifted to the left or right? Give a reason for your answer.

b Explain why the centre of gravity of a spoon is not at its middle.

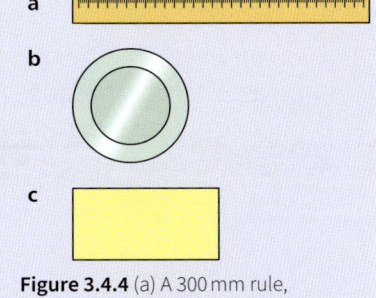

Figure 3.4.4 (a) A 300 mm rule, (b) a circular plate, (c) a rectangular card

2 Sketch each of the objects shown in Figure 3.4.4 and mark its centre of gravity.

3.5 Stability

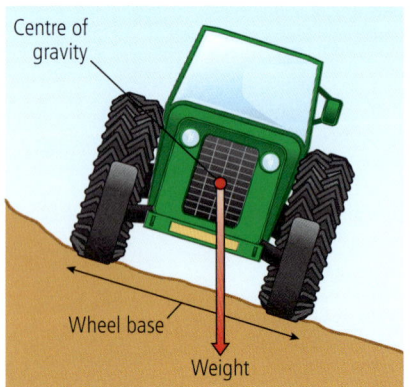
Figure 3.5.1 Forces on a tilting tractor

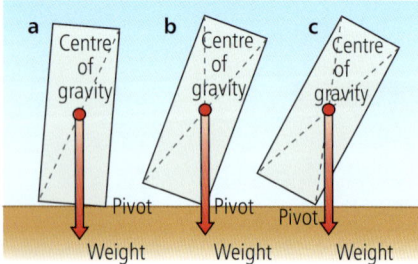
Figure 3.5.2 Tilting and toppling

A toppling test

Stability and safety

Look around you and see how many objects could topple over. Bottles, table lamps and floor-standing bookcases are just a few objects that can easily topple over. Lots of objects are designed for stability so they can't topple over easily.

1 Tractor safety

Figure 3.5.1 shows a tractor on a hillside. It doesn't topple over because the line of action of its weight acts within its wheelbase. If it is tilted more, it would topple over if the line of action of its weight acted outside its wheelbase. Its weight would then give a clockwise turning effect about the lower wheel.

2 Bus tests

The photo shows a double-decker bus being tested to see how much it can tilt without toppling over. Such tests are important to make sure buses are safe to travel on, especially when they go around bends and on hilly roads.

3 Ladders

If you climb a ladder, don't lean too far from it. If you do, you might shift your centre of gravity too far and topple over.

PRACTICAL

Tilting and toppling tests

How far can you tilt something before it topples over? Figure 3.5.2 shows how you can test your ideas using a tall box or a brick on its end.

1 If you tilt the brick slightly, as in **a**, and release it, the turning effect of its weight returns it to its upright position.

2 If you tilt the brick more, you can just about balance it on one edge, as in **b**. Its centre of gravity is then directly above the edge on which it balances. Its weight has no turning effect in this position.

3 If you tilt the brick even more, as in **c**, it will topple over if it is released. This is because the line of action of its weight is 'outside' its base. So its weight has a turning effect that makes it topple over.

Measuring the weight of a beam

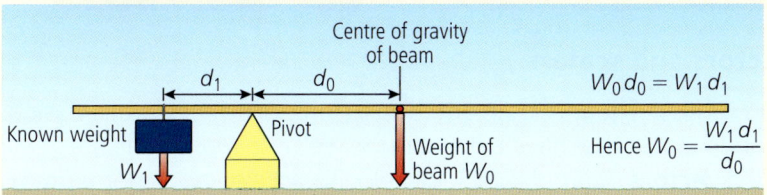

Figure 3.5.3 Finding the weight of a beam

1 A beam placed off-centre on a pivot will topple off the pivot because it is unstable.

2 Figure 3.5.3 shows how to balance a beam off-centre on a pivot using an object as a 'counterweight'. The weight of the beam acts at its centre of gravity which is at distance d_0 from the pivot.

• The moment of the beam about the pivot $= W_0 d_0$ clockwise, where W_0 is the weight of the beam.

• The moment of W_1 about the pivot $= W_1 d_1$ anticlockwise, where d_1 is the perpendicular distance from the pivot to the line of action of W_1.

Applying the principle of moments gives $W_1 d_1 = W_0 d_0$.

So we can calculate the beam's weight W_0 if we know W_1 and distances d_1 and d_0.

For example, if the weight $W_1 = 2.0\,\text{N}$, $d_1 = 0.15\,\text{m}$ and $d_0 = 0.25\,\text{m}$, prove for yourself that $W_0 = 1.2\,\text{N}$.

EXAM TIP

Make sure you understand the idea of stability so that you can explain even unfamiliar examples.

1 a Make a list of objects that are designed to be difficult to knock over.

   b A well-designed laboratory stool has a base that is wider than the seat. If the base was too narrow, why would the seat be unsafe?

   c Would a double-decker bus be more or less stable if everyone on it sat on the top deck?

2 The following measurements were made as shown in Figure 3.5.3 to find the weight of a beam.

   $W_1 = 2.4\,\text{N}$, $d_1 = 0.10\,\text{m}$, $d_0 = 0.30\,\text{m}$.

   Use these measurements to calculate:

   a the weight of the beam

   b the support force on the beam from the pivot.

KEY POINTS

• The stability of an object is increased by making its base as wide as possible and its centre of gravity as low as possible.

• An object will tend to topple over if the line of action of its weight is outside its base.

- Use the parallelogram of forces rule to find the resultant of two forces that do not act along the same line

- Describe an experiment to test the parallelogram of forces

- Recognise the difference between a vector and a scalar quantity

Under tow

Vectors and scalars

Many physical quantities in addition to force are directional. Physical quantities that that have a direction as well as a magnitude are called **vectors**. Examples include velocity, acceleration, momentum, weight and gravitational field strength.

Physical quantities that are not directional are called **scalars** because they only have a magnitude. Examples include distance, time, speed, mass, energy, temperature and power (which we will meet in the next chapter).

Force

We saw in Topic 2.5 how to calculate the resultant force of two forces that act along the same line. But what if the two forces do not act along the same line?

The photograph shows a ship being towed by cables from two tugboats. The tension force in each cable pulls on the ship. The combined effect of these tension forces is to pull the vessel as in Figure 3.6.1. This is the resultant force.

Figure 3.6.1 shows how the two tension forces T_1 and T_2 in the tow ropes can be represented as vectors and combined to produce the resultant force. The tension forces are drawn as adjacent sides of a parallelogram; the resultant force is the diagonal of the parallelogram from the origin of T_1 and T_2. This geometrical method is called the **parallelogram of forces**.

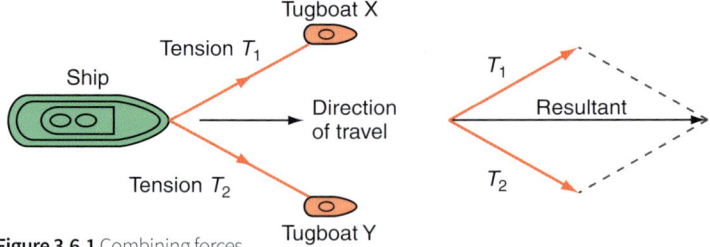

Figure 3.6.1 Combining forces

Investigating the parallelogram of forces

We can use weights and pulleys to demonstrate the parallelogram of forces, as shown in Figure 3.6.2. The tension in each string is equal to the weight it supports, either directly or over a pulley.

The point where the three strings meet is in equilibrium. The string supporting the middle weight (W_3) is vertical. The angles θ_1 and θ_2 between each of the other two strings and the vertical string are measured using a protractor.

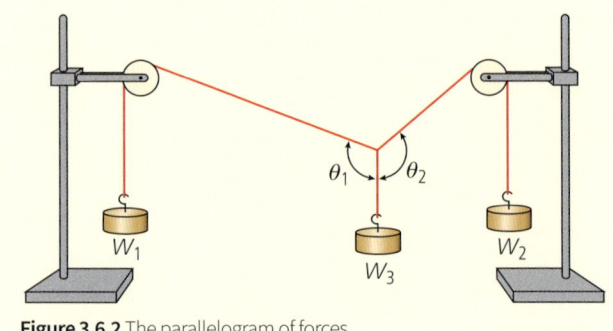

Figure 3.6.2 The parallelogram of forces

Using these measured angles, and the values of the three known weights, a scale diagram of a parallelogram is drawn, such that:

- the line down the centre of the diagram represents a vertical line
- adjacent sides of the parallelogram at angles θ_1 and θ_2 to the 'vertical' line represent the tensions in the strings supporting W_1 and W_2.
- The resultant of W_1 and W_2, represented by the diagonal, should be equal and opposite in direction to the vector representing W_3.

WORKED EXAMPLE

A tow rope is attached to a car at two points 0.8 m apart. The two sections of rope joined to the car are the same length and are at 30° to each other, as shown in Figure 3.6.3. The pull on each attachment should not exceed 3000 N. Use the parallelogram of forces to determine the maximum tension in the main tow rope.

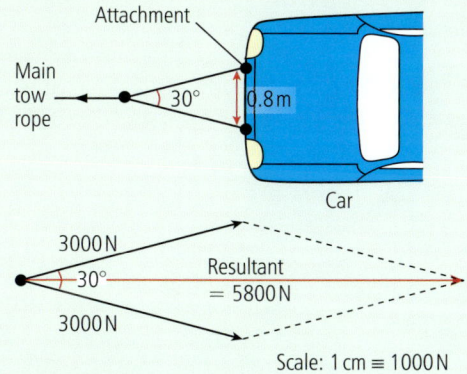

Figure 3.6.3 Using the parallelogram rule

Solution

The maximum tension T in the main tow rope is the resultant of the two 3000 N forces at 30° to each other. Drawing the parallelogram of forces as shown in Figure 3.6.3 gives $T = 5800\,\text{N}$.

SUMMARY QUESTIONS

1 Figure 3.6.4 shows two forces acting on an object X. Work out the magnitude and direction of the resultant force on X.

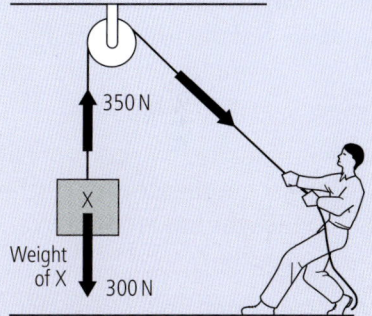

Figure 3.6.4

2 A force of 3.0 N and a force of 4.0 N act on a point object. Determine the magnitude of the resultant of these two forces if the angle between their lines of action is:

a 90°, b 60°, c 45°.

KEY POINTS

- The parallelogram of forces is used to find the resultant of two forces that do not act along the same line.
- A vector is a physical quantity that has a direction.
- A scalar is a physical quantity that does not have direction.

1 (a) Write the equation for the moment of a force.

(b) The diagram shows a force *F* being exerted by pulling on a rope to open a trap door. The weight of the trap door is 120 N. The trap door is square and the sides are of length 0.7 m. The centre of gravity is at the centre of the square.

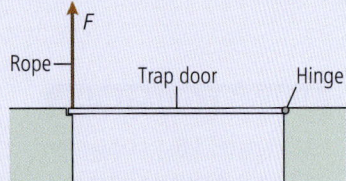

(i) Calculate the value of the force. Show your working.

(ii) What would be the value of the force if the rope were attached in the middle of the trap door instead of at the end?

2 A builder has two pairs of pliers. One has longer handles than the other. The builder applies the same force on each pair.

Explain which pair of pliers will exert a bigger force on the object in the jaw. Use the Principle of Moments in your explanation. It will probably help to include a labelled diagram in your answer.

3 (a) Suggest two household items that are designed to be stable.

(b) With the aid of suitable, labelled diagrams explain the features of the items that you have chosen that make them stable.

S 4 (a) Speed is a scalar quantity and velocity is a vector quantity. Explain briefly the terms scalar and vector.

(b) Write down two more examples of scalar quantities and two more examples of vector quantities.

(c) A large object is to be pulled along using two ropes that are attached at the same point. The angle between the two ropes is 30°. The force exerted in one rope is 500 N and in the other is 600 N. Calculate the resultant force acting on the object. Show your working.

Practice Questions

1 The diagram shows a wooden pole balanced at the centre on a pivot. Two weights are positioned as shown and the pole still balances.

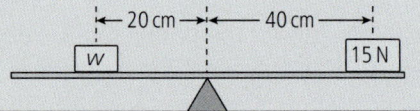

Which of these is the weight, *W*?

A 10 N

B 7.5 N

C 40 N

D 30 N

2 What is the moment produced by a force of 4.5 N acting 0.80 m from a pivot?

A 5.6 Nm

B 36 Nm

C 3.6 Nm

D 1.8 Nm

3 Which of the following is a statement of the principle of moments for an object in equilibrium?

A There are no moments acting on the object.

B The sum of the clockwise moments about any point = the sum of the anticlockwise moments about that point.

C The clockwise moments are greater than the anticlockwise moments.

D The anticlockwise moments are greater than the clockwise moments.

4 Two children are playing on a uniform seesaw. One of the children has a mass of 40 kg and sits 1.2 m from the pivot. The second child has a mass of 30 kg. How far should they sit from the pivot so that the seesaw balances?

A 1.6 m

B 0.9 m

C 1 440 m

D 1.2 m

5 Which of the following rows shows the properties of vectors and scalars correctly?

	Vector	Scalar
A	has magnitude only	has magnitude and direction
B	has magnitude and direction	has magnitude and direction
C	has magnitude only	has magnitude only
D	has magnitude and direction	has magnitude only

6

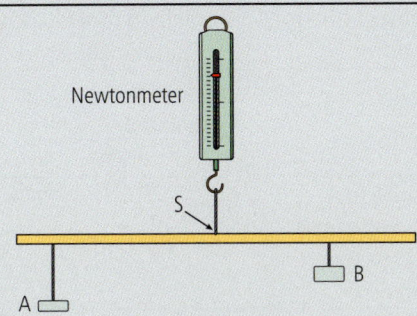

A pole with two weights is hung from a newtonmeter at the centre as shown in the figure above. The weight of the pole is 5.0 N, weight A is 3.0 N and weight B is 2.0 N. Weight A is positioned 12 cm from the centre. The system is in equilibrium.

(a) State the two conditions needed for a system to be in equilibrium. [2]

(b) State the reading on the newtonmeter. [1]

(c) Find the distance weight B must be from the centre. [3]

7 (a) What is meant by the term 'centre of gravity'? [1]

(b) Describe how to find the centre of gravity for a uniform sheet of plastic cut into an irregular shape. You can use diagrams to help your explanation. [4]

(c) How does the position of the centre of gravity affect an object's stability? [3]

8 A model flamingo stands on a single leg in a garden as shown in the figure.

(a) Copy the diagram and indicate the approximate centre of gravity of the flamingo with an X. [2]

(b) Explain why the flamingo standing like this may easily topple over when nudged. [3]

(c) Suggest how the plastic flamingo could be made more stable. [1]

9 A metal sculpture is suspended from a ceiling at points A and C using two strong wires as shown in the figure. The wires are separated by an angle of 60° and have a force of 75 newtons acting in each. The system is in equilibrium.

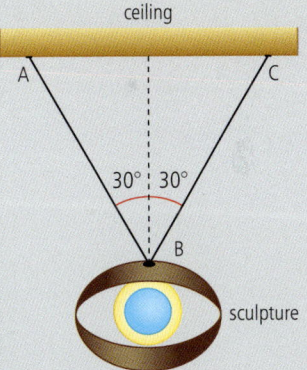

(a) Draw a scale diagram to find the resultant force acting on the sculpture due to the two wires. [4]

(b) State the weight of the sculpture. [1]

(c) Explain why the statue hangs with its centre of gravity directly below the point B which is halfway between points A and C. [3]

4.1 Energy transfers

The TGV electric train

On the move

Cars, buses, planes and ships all use energy from fuel. They carry their own fuel. Electric trains use energy from fuel in power stations. Electricity transfers energy from the power station to the train.

We describe energy stored or transferred in different ways. The table below shows some of the different ways in which energy can be stored or transferred.

energy stores	description	example
kinetic energy	energy of an object due to its motion	vehicle in motion has kinetic energy
gravitational potential energy	energy of an object due to its position	book lifted up gains gravitational potential energy
chemical energy	energy stored in a substance and released when chemical reactions take place	car battery
elastic strain energy	energy stored in an elastic object when we stretch or squash it	stretched spring
internal energy	energy of an object due to the internal motion and positions of its molecules	magnetised object or a hot object (Note: the energy of an object due to its temperature is sometimes referred to as *thermal energy*)
nuclear energy	energy released when the nucleus of an atom splits or disintegrates	a uranium fuel rod in a nuclear reactor

energy transfers	description	example
electrical energy	energy transferred by an electric current	electric heater switched on
thermal energy	energy transfer from a hot object to a cold object	radiation from burning coals
sound energy	energy transfer by sound waves	sound from a drum
light energy	energy transfer by light	light from a torch
work done	energy transfer by a force or by electricity	force in free fall of gravity, electric current in a circuit

Examples of energy transfers

An athlete performing a pole vault uses the force of their muscles to transfer energy from the chemical store in the muscles into kinetic energy and elastic strain energy of the pole. This energy is transferred by the force of gravity to the athlete's gravitational store as the athlete rises. Some energy is also transferred to the surroundings by air resistance and sound waves. The energy transfers are:

chemical energy → kinetic energy + elastic energy →
 gravitational potential energy (+ thermal energy + sound energy)

A pile-driver on a building site is used to make firm foundations for tall buildings. Engineers use a pile-driver to hammer steel girders end-on into the ground. The pile-driver lifts a heavy steel block above the top end of the girder then lets it crash down onto the girder. The energy transfers of the block from its release to when it hits the girder are:

gravitational potential energy → kinetic energy → thermal energy and sound energy on impact

A pole vaulter

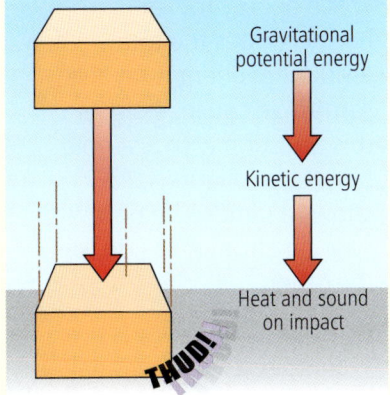

PRACTICAL

Energy transfers

1 Drop a small object onto a pad or cushion on the floor and observe the energy transfers. As the object falls, it gains kinetic energy because it speeds up as it falls. So its store of gravitational potential energy decreases and its store of kinetic energy increases as it falls. On impact, it stops so what happens to the kinetic energy it had just before it hit the pad?

Gravitational potential energy

Kinetic energy

Heat and sound on impact

THUD!

Figure 4.1.1 An energetic drop

2 Figure 4.1.1 shows a box that hits the floor with a thud. All its kinetic energy transfers to thermal energy and sound in the impact.

A pile-driver in action

SUMMARY QUESTIONS

1 Choose words from the list below for the spaces in the sentences. Copy them into your workbook.

electrical   kinetic   gravitational potential   thermal

  a When a ball falls in air, it loses _____ energy and gains _____ energy.

  b When an electric heater is switched on, it transfers _____ energy into _____ energy.

2 a List two different objects you could use to light a room in the event of a power cut. For each object, describe the energy transfers that happen when it produces light.

  b Which of the two objects in **a** is:
   i easier to obtain energy from? **ii** easier to use?

KEY POINTS

- Energy can be stored in objects in different ways and transferred in different ways.

- Energy transferred to an object can be stored in different ways to the ways in which it was stored before being transferred.

4.2 Conservation of energy

On a roller coaster

At the fairground

Fairgrounds are very exciting places because lots of energy transfers happen quickly. A roller coaster gains gravitational potential energy when it ascends and loses it when it descends.

As it descends the energy transfers are:

gravitational potential energy → kinetic energy + sound + thermal energy due to air resistance and friction

PRACTICAL

Investigating energy transfers

When energy transfers happen, does the total amount of energy stay the same? We can investigate this question with a simple pendulum. Figure 4.2.1 shows a pendulum bob swinging from one side to the other.

Maximum gravitational potential energy

Maximum gravitational potential energy

Maximum kinetic energy

Figure 4.2.1 A pendulum in motion

- As it moves towards the middle, its gravitational potential energy decreases and its kinetic energy increases.
- As it moves away from the middle, its kinetic energy decreases and its gravitational potential energy increases.

The principle of conservation of energy

Scientists have done lots of tests to find out if the total energy after a transfer is the same as before. All the tests so far show that it is the same.

The total amount of energy before and after a transfer is the same.

This important result is known as **conservation of energy**. It means that energy cannot be created or destroyed.

Supplement

But energy tends to spread out, for example where energy is transferred to the surroundings by heating or by sound waves. In such situations, we say energy is **dissipated**.

Bungee jumping

What energy transfers happen to a bungee jumper after jumping off the platform?

- The gravitational potential energy of the bungee jumper decreases and their kinetic energy increases as the jumper falls with the rope slack.

- Once the slack in the rope has been taken up, the rope slows the bungee jumper's fall. Most of the gravitational potential energy and kinetic energy of the jumper transfers into elastic energy in the rope as it stretches.

- After reaching the bottom, the rope pulls the jumper back up. As the jumper rises, **most of** the elastic energy of the rope transfers to gravitational potential energy and kinetic energy of the jumper.

Bungee jumping

The bungee jumper doesn't return to the same height as at the start. This is because some of the initial gravitational potential energy has been transferred to thermal energy as the rope stretched then shortened again.

PRACTICAL

Investigating a model bungee jump

1 You can try out the ideas about bungee jumping using the experiment shown in Figure 4.2.2.

2 Find out how much of the gravitational potential energy lost in the descent is regained at the end of the jump.

3 To do this, measure and compare the total height drop of the bungee jumper with the total height gain after the drop.

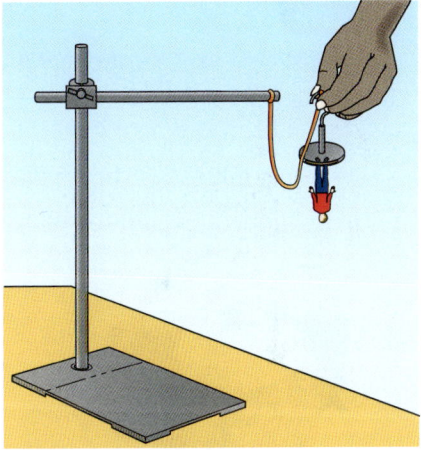
Figure 4.2.2 Testing a bungee jump

SUMMARY QUESTIONS

1 Copy and complete the sentences below using words from the list.

electrical   gravitational potential   thermal

A person going up in a lift has _____ energy. The lift is driven by electric motors. Some of the _____ energy supplied to the motors is changed to _____ instead of _____ energy.

2 A ball dropped onto a trampoline returns to the same height after the rebound.

   a Describe the energy transfers of the ball from the point of release to the top of the rebound.

   b What can you say about the energy of the ball at the point of release compared with at the top of the rebound?

3 One exciting fairground ride acts like a giant catapult. The capsule which the 'rider' is strapped in is fired high into the sky by rubber straps. Explain the energy transfers taking place in the ride.

KEY POINTS

- Energy can be transferred from one store to another or transferred from one place to another.

- Energy cannot be created or destroyed.

4.3 Fuel for electricity

Fuel for electricity

We can generate electricity by burning fuel and using the heat energy released to make an engine turn an electricity generator. The fuels include:

- fossil fuels such as oil (including petrol), natural gas and coal
- biofuels such as wood, straw, methane gas and ethanol.

Note that nuclear fuels (e.g. uranium) release energy as a result of the nuclei of atoms disintegrating, not as a result of burning fuel.

When electricity is generated as a result of burning fuel,

chemical energy $\longrightarrow$ *thermal energy (+ light energy)* $\longrightarrow$ *kinetic energy of the engine and generator (+ sound energy + thermal energy due to friction)* $\longrightarrow$ *electrical energy*

Power from electricity

Electricity generators supply electrical energy continuously from the energy released when fuel burns. The **power** of an electricity generator is the rate at which it supplies electrical energy.

Petrol-powered electricity generators usually consist of a petrol engine that turns an electricity generator. Petrol generators are used when or where mains electricity is not available.

Power stations generate the electricity used by most people except those in remote locations.

- **In a coal- or oil-fired power station or a biofuel power station**, the burning fuel heats water in a boiler to produce steam (Figure 4.3.1). The steam drives a turbine that turns an electricity generator.
- **In a gas-fired power station**, natural gas is burned directly in a gas turbine engine. This produces a powerful jet of hot gases and air that drives the turbine. A gas-fired turbine can be switched on very quickly.

A petrol-powered electricity generator

In a gas-fired power station

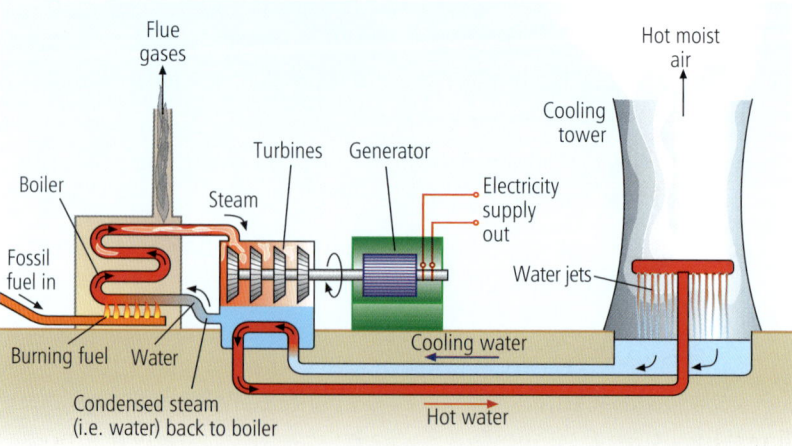

Figure 4.3.1 Inside a fossil fuel power station

Electricity generator tests

1. A cycle dynamo is used to light a torch bulb. The dynamo consists of a magnet that spins inside a coil of wire to which the torch bulb is connected. Turn the dynamo steadily and light the bulb.

2. A clockwork radio stores energy in a clockwork spring which is used to drive a small electricity generator in the radio. Wind the clockwork spring and listen to the radio.

Energy and efficiency

Energy is measured in joules (J). As explained in detail in Topic 4.8, 1 J of energy is the energy transferred to an object of weight 1 N when it is raised by a vertical distance of 1 m.

When energy is supplied to a device, not all the energy supplied to it is used for the intended purpose. For example, when an electric motor is used to raise a weight, some of the electrical energy supplied to the motor is wasted as thermal energy due to friction and as sound energy. The energy transferred by the motor to raise the load is **useful energy** because it is used for the intended purpose of raising the weight.

The energy supplied to it = the useful energy output of the device + the energy wasted by the device

Supplement

We can define the **efficiency** of a device as:

$$\frac{\text{the useful energy output of the device}}{\text{the energy supplied to the device}} \times 100\% \quad \text{or}$$

$$\frac{\text{the useful power output}}{\text{the power input}} \times 100\%$$

For example, suppose an electric motor used to raise a weight transfers 30 J of energy to the weight as gravitational potential energy for every 100 J of energy supplied to the motor. The other 70 J is transferred to the surroundings, mostly as thermal energy and also as sound energy.

The percentage efficiency of the motor is 30% $\left(\frac{30\,\text{J}}{100\,\text{J}} \times 100\%\right)$.

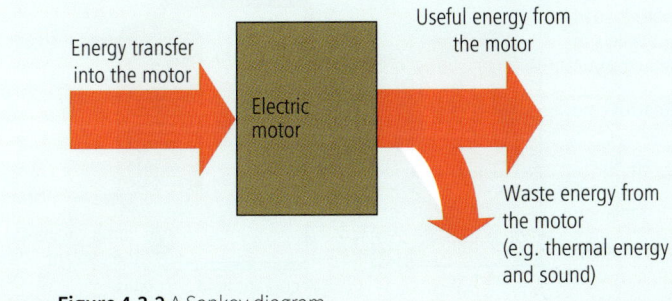

Figure 4.3.2 A Sankey diagram

The energy transfers through a system are shown in a **Sankey diagram**. Figure 4.3.2 shows a Sankey diagram for an electric motor.

- A petrol generator consists of a petrol engine and an electricity generator.
- Electricity generators in power stations are driven by turbines.
- The unit of energy is the joule.
- Efficiency is a measure of how much of the energy supplied to a device is usefully used.

1. Copy and complete the sentences below using words from the list.

   coal   gas   oil   uranium   wood

   a. _____ and _____ are not fossil fuels.

   b. Power stations that use _____ as the fuel can be switched on very quickly.

   c. Steam is used to make the turbines rotate in a power station that uses coal, _____ or _____ as fuel.

2. a. Name the device used to turn an electricity generator in:
      i. a petrol generator
      ii. a gas-fired power station.

   b. State one advantage and one disadvantage of a gas-fired power station compared with a coal-fired power station.

   c. Draw an energy flow diagram to show the energy transfers in a clockwork radio when the radio is on.

4.4 Nuclear energy

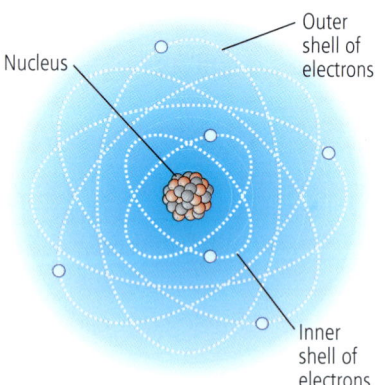

Figure 4.4.1 The structure of an atom

Inside the atom

Figure 4.4.1 shows the particles inside an atom.

- Every atom contains a positively charged nucleus surrounded by electrons which are negatively charged.
- The nucleus is composed of neutrons and protons.
- Atoms of the same element can have different numbers of neutrons in the nucleus.

Nuclear fission

The fuel in a nuclear power station is uranium. It releases about 10 000 times as much energy per kilogram as fossil fuel or biofuel. The uranium fuel is contained in sealed cans in the core of the reactor.

The nucleus of a uranium atom is unstable and can split in two when a neutron from outside the atom collides with it. This process is called **nuclear fission**. Energy is released in the process.

Nuclear power stations do not produce greenhouse gases whereas fossil fuel power stations do. However, they produce radioactive waste which needs to be stored safely for many years.

Inside a fission reactor

S When a uranium nucleus undergoes fission, two or three neutrons may be released, as well as energy. These neutrons may cause other uranium nuclei to undergo fission. When this happens, a chain reaction can occur in the core of the reactor.

As there are lots of uranium atoms in the core, it becomes very hot. The thermal energy of the core is taken away by a fluid (called the 'coolant') that is pumped through the core. The coolant is very hot when it leaves the core. It flows through a pipe to a 'heat exchanger' then back to the reactor core. The thermal energy of the coolant is used to turn water into steam in the heat exchanger. The steam drives turbines which turn electricity generators (Figure 4.4.2).

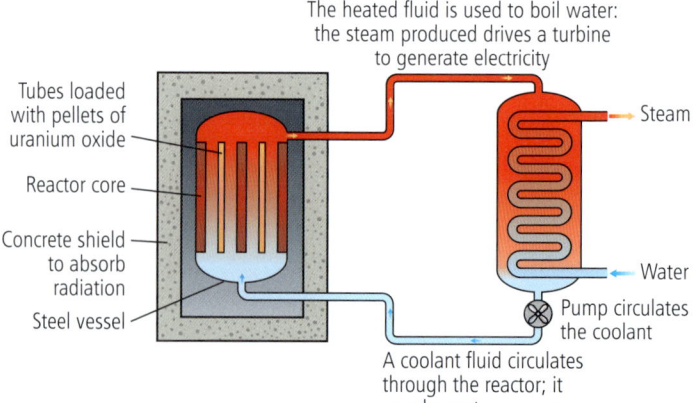

Figure 4.4.2 Left: a nuclear power station. Right: components of a nuclear reactor

Supplement

Nuclear fusion

The Sun and the stars release energy as a result of fusing small nuclei, such as hydrogen, to form larger nuclei. Two small nuclei release energy when they are fused together to form a single larger nucleus (Figure 4.4.3). The process is called **nuclear fusion**.

Fusion reactors

Fusion reactors release energy by fusing hydrogen nuclei and other light nuclei. The fuel needed, such as hydrogen, is available in abundance and the fusion products are not radioactive. However, there are enormous technical difficulties in sustaining fusion in a fusion reactor. A collection of unbound nuclei and electrons, as in a fusion reactor, is called a plasma. The plasma must be heated to very high temperatures before any of the nuclei will fuse with each other. This is because two nuclei approaching each other will repel each other due to their positive charge. If the nuclei are moving fast enough, they can overcome the force of repulsion and fuse together.

In a fusion reactor:

- the plasma is heated by passing a very large electric current through it

- the plasma is contained by a magnetic field so it doesn't touch the reactor walls. If it did, it would go cold and fusion would stop.

Scientists have not yet developed a successful fusion reactor which would release more energy than it uses to heat the plasma. At the present time, scientists working on experimental fusion reactors are able to do this by fusing hydrogen nuclei to form helium nuclei – but only for a few minutes!

EXAM TIP

'Fission' means splitting. Don't confuse nuclear fission with nuclear fusion.

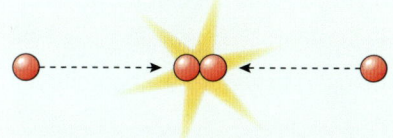

Figure 4.4.3 A fusion reaction. The fusion of two nuclei to form a new nucleus which is lighter than iron's generally releases energy

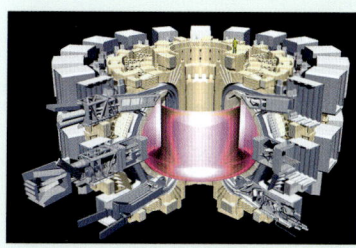

An experimental fusion reactor

SUMMARY QUESTIONS

1 Copy and complete the sentences below using the following words.

large   small   stable

a When nuclear fission occurs, a _____ nucleus splits into two _____ nuclei.

b Energy is released in nuclear fusion if the product nucleus is not as _____ as an iron nucleus.

c When two _____ nuclei moving at high speed collide, they form a _____ nucleus.

2 a Explain what is meant by a chain reaction in a nuclear fission reactor.

b Why does the plasma in a fusion reactor need to be very hot?

KEY POINTS

- Energy is released in nuclear fission.

- Energy is also released in nuclear fusion.

- Nuclear fission occurs when a uranium nucleus splits as a result of being struck by a neutron.

- Nuclear fusion occurs when two nuclei are forced close enough together so they form a single larger nucleus.

4.5 Energy from wind and water

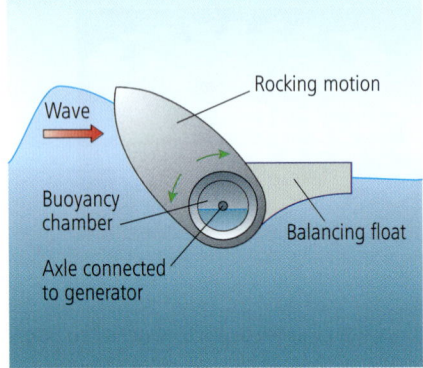

Figure 4.5.1 Energy from waves

A hydroelectric scheme

Wind energy

A wind turbine is an electricity generator at the top of a narrow tower. The generator is driven by the force of the wind on its blades. The power generated increases as the wind speed increases. A wind farm consisting of 40 wind turbines could generate enough electricity to supply over 10 000 homes.

Wind turbines do not emit greenhouse gases and do not cause airborne pollution. However, if the wind does not blow, they cannot produce electricity – so they are unreliable.

A wind farm

Wave energy

A wave generator at a coastal site uses the motion of waves to make a floating section move up and down (Figure 4.5.1). This motion drives a turbine which turns a generator. A cable between the generator and the shore delivers electricity to local users or via a network of cables (referred to as the grid system) to distant users.

Wave generators need to withstand storms and they don't produce a constant supply of electricity. Furthermore, lots of cables (and buildings) would be needed along the coast to connect the wave generators to the electricity grid. This would spoil the views of the coastline. In addition, tidal flow patterns might be changed, affecting the habitats of marine life and birds.

Hydroelectricity

Hydroelectricity is generated when rain water collected in an uphill reservoir flows downhill. The water flow drives turbines that turn electricity generators at the foot of the hill. Large hydroelectric power stations can supply enough electricity for a city. The 'Three Gorges' hydroelectric scheme in China traps fast-flowing river water behind a massive dam and channels the water through generators that supply electricity to millions of homes.

Hydroelectric schemes do not produce greenhouse gases, although they can have a considerable environmental impact if they require the construction of large dams and the flooding of large areas of land.

Tidal energy

A tidal power station in an estuary traps each high tide behind a barrage. The high tide is then released into the sea through turbines that drive generators in the barrage.

Tidal energy is more reliable than wind energy because ocean tides occur roughly twice every day. They happen because the Moon's gravity pulls on the Earth's oceans.

The table below shows that a tidal power station can generate much more power than a wind turbine or a hydroelectric power station.

A tidal power station at a suitable location can generate as much power as a large fossil fuel power station or a nuclear power station, enough to supply a city. An estuary that becomes narrower as you move 'upriver' away from the open sea is a good site for a tidal power station. This is because it 'funnels' the incoming tide and makes it higher than elsewhere.

power station (or devices)	typical output	location	cost[a] per MW h
coal-fired	1000 MW	Any	1
hydroelectric	500 MW	Upland	0.8
nuclear	5000 MW	Coastal	1
solar cells	1 kW per m^2	Any	1.3
tidal	2000 MW	Estuary	2
wave generators	20 MW per km	Coast	0.9
wind turbines	2 MW per turbine	Windy site	0.8[b] 2.1[c]

*Notes   **1** 1 MW = 1 million watts = 1 million joules per second*

*   **2 a** Costs are given relative to coal; **b** onshore; **c** offshore*

The energy sources described in this topic and solar energy in the next topic are called renewable energy sources because they never run out – unlike non-renewable sources such as fossil fuels, nuclear fuels and geothermal energy sources.

KEY POINTS

- A wind turbine is an electricity generator on top of a tall tower.
- A wave generator is a floating generator turned by the waves.
- Hydroelectricity generators are turned by water running downhill.
- A tidal power station traps each high tide and uses it to turn generators.

SUMMARY QUESTIONS

1 Copy and complete the following sentences using words from the list below.

hydroelectric   tidal   wave   wind

a _____ energy does not need water.

b _____ energy does not need energy from the Sun.

c _____ energy is obtained from water running downhill.

d _____ energy is obtained from water moving up and down.

2 a Use the table above for this question.

   i How many wind turbines would give the same total output as a tidal power station?

   ii How many kilometres of wave generators would give the same total output as a hydroelectric power station?

b i How many 2 MW wind turbines would give the same power output as a 5000 MW nuclear power station?

   ii State one advantage and one disadvantage of installing wind turbines instead of building a nuclear power station.

4.6 Energy from the Sun and the Earth

A solar powered vehicle

Solar energy

Solar radiation transfers energy to you from the Sun – sometimes more than you want if you get sunburnt. We can use it to generate electricity using **solar cells** and we can also use it to heat water directly in **solar heating panels**.

1 **Solar cells** in use now convert about 15% of the solar energy they absorb into electrical energy. We connect them together to make solar cell panels.

- They are useful where only small amounts of electricity are needed (e.g. watches and calculators) or in remote locations (e.g. satellites).
- They are very expensive to buy even though they cost nothing to run.
- You need lots of them to generate enough power to be useful – and plenty of sunshine!
- They don't work at night and aren't very effective on a cloudy day.

2 **A solar heating panel** heats water flowing through it. Even on a cloudy day, a solar heating panel on a house roof can supply plenty of hot water (Figure 4.6.1).

- The panel has a black cover which absorbs sunlight better than other colours.
- The back of the panel is insulated so water flowing through the panel doesn't lose thermal energy.
- The panel is filled with oil so it heats up faster than water.
- Water passes through copper pipes in the oil. Copper is a good conductor of thermal energy so thermal energy is readily transferred from the hot oil to the water in the pipes.
- Cold water enters the pipes and is heated by the oil. The hot water from the outflow pipe is collected in an insulated tank.

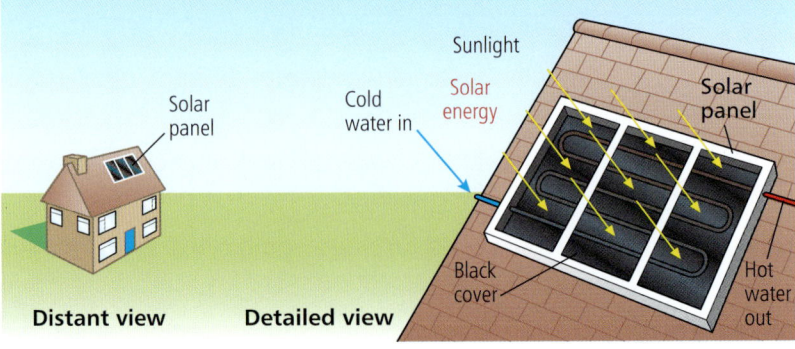

Figure 4.6.1 Solar water heating

Investigating solar cells

1 Use a solar cell to drive a small motor (Figure 4.6.2).

2 Observe the change of motor speed when you use a card to cover the solar cell partially then totally.

3 Carry out the test on a sunny day and on a cloudy day to see what difference cloud cover makes.

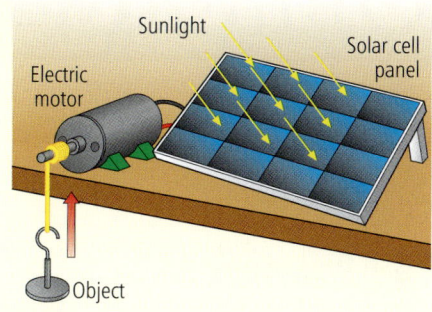

Figure 4.6.2 Solar cells at work

The Sun is the source of energy for all our energy resources except nuclear, tidal and geothermal energy. Fossil fuels formed from the remains of plants and animals that grew using solar energy long ago. The Sun heats our atmosphere, creating winds, waves and rain as a result.

Geothermal energy

Geothermal energy is released by radioactive substances deep within the Earth.

- The energy released by these substances heats the surrounding matter.

- As a result, thermal energy transfers outward towards the Earth's surface.

- Rocks below the surface become very hot because they gain thermal energy from the flow of thermal energy.

Geothermal power stations are built where there are hot rocks deep below the surface. Water is pumped down to these rocks to produce steam. Electricity turbines at ground level are driven by the steam.

Geothermal power stations do not need energy from the Sun because the energy they use comes from radioactive substances deep within the Earth.

Geothermal energy

1 Copy and complete the following sentences using words from the list.

geothermal energy   solar energy   radiation radioactivity

a   The best energy resource to use in a calculator is _____.

b   _____ inside the Earth releases _____ energy.

c   _____ from the Sun generates electricity in a solar cell.

2 a   A satellite in space uses a solar cell panel for electricity. The satellite carries batteries that are charged by electricity from the solar cell panels. Why are batteries carried as well as solar cell panels?

b   If the water stopped flowing through a solar heating panel, what would happen to the temperature of the water in the panel? Give a reason for your answer.

- Solar energy can be converted into electricity using **solar cells** or used to heat water directly in **solar heating panels**.

- Geothermal energy is released by radioactive substances deep inside the Earth.

- The Sun is the source of energy for all energy resources except geothermal energy, nuclear energy and tidal energy.

4.7 Energy and the environment

Can you get energy without creating any problems? Figure 4.7.1 shows the energy sources people use today to generate electricity. What effect does each one have on your environment?

Fossil fuel problems

When coal, oil, or gas is burnt, **greenhouse gases** such as carbon dioxide are released. The amount of these gases in the atmosphere is increasing, and most scientists believe that this is causing more global warming and climate change. Some electricity comes from oil-fired power stations. People use much more oil to produce fuels for transport.

Burning fossil fuels can also produce sulfur dioxide. This gas causes **acid rain**. The sulfur can be removed from a fuel before burning it, to stop acid rain. For example, natural gas has its sulfur impurities removed before it is used.

Fossil fuels are non-renewable. Sooner or later, people will have used up the Earth's reserves of fossil fuels. Alternative sources of energy will then have to be found. But how soon? Oil and gas reserves could be used up within the next 50 years. Coal reserves will last much longer.

Carbon capture and storage (CCS) technology could be used to stop carbon dioxide emissions into the atmosphere from fossil fuel power stations. Old oil and gas fields could be used for storage.

Nuclear versus renewable

People need to use less fossil fuels to stop global warming. Should people rely on nuclear power or on renewable energy in the future?

Nuclear power

Advantages

- No greenhouse gases (unlike fossil fuel)
- Much more energy is released from each kilogram of uranium (or plutonium) fuel than from fossil fuel.

Disadvantages

- Used fuel rods contain radioactive waste, which has to be stored safely for centuries
- Nuclear reactors are safe in normal operation. However, an explosion in a reactor could release radioactive material over a wide area. This would affect this area, and the people living there, for many years.

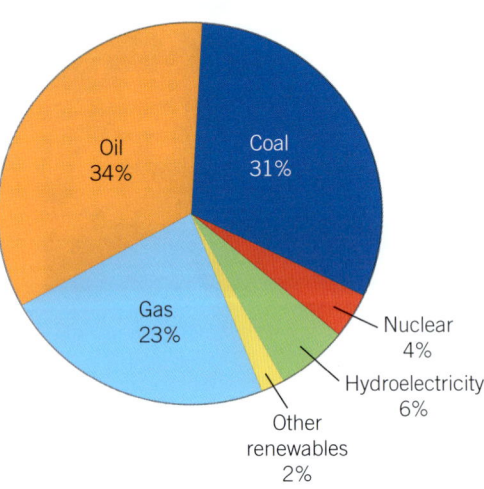

Figure 4.7.1 Energy sources for electricity

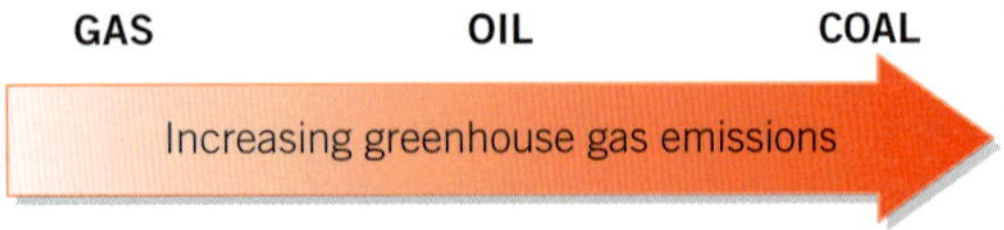

GAS   **OIL**   **COAL**

Increasing greenhouse gas emissions

Figure 4.7.2 Greenhouse gases from fossil fuels

Renewable energy sources and the environment

Advantages

- They will never run out because they are always being replenished by natural processes.
- They do not produce greenhouse gases or acid rain.
- They do not create radioactive waste products.
- They can be used where connection to the National Grid is uneconomic. For example, solar cells can be used for road signs and to provide people with electricity in remote areas.

Disadvantages

- Renewable energy resources are not currently able to meet the world demand. So fossil fuels are still needed to provide some of the energy demand.
- Wind turbines create a whining noise that can upset people nearby, and some people consider them unsightly.
- Tidal barrages affect river estuaries and the habitats of creatures and plants there.
- Hydroelectric schemes need large reservoirs of water, which can affect nearby plant and animal life. Habitats are often flooded to create dams.
- Solar cells need to cover large areas to generate large amounts of power.
- Some renewable energy resources are not available all the time or can be unreliable. For example, solar power is not produced at night and is affected by cloudy weather. Wind power is reduced when there is little or no wind, and hydroelectricity is affected by droughts if reservoirs dry up.

SUMMARY QUESTIONS

1 Match each energy source with a problem it causes:

Energy source	Problem
a Coal	A Noise
b Hydroelectricity	B Acid rain
c Uranium	C Radioactive waste
d Wind power	D Takes up land

2 a Name three possible renewable energy resources that could be used to generate electricity for people on a remote flat island in a hot climate.

   b Name three types of power stations that do not release greenhouse gases into the atmosphere.

3 A tidal power station, a nuclear power station, or 1000 wind turbines can each supply enough power to meet the electricity needs of a large city on an estuary. Describe the advantages and disadvantages of each type of power station for this purpose.

Climate change

Climate change is happening because of global warming caused by burning fossil fuels to meet the demand for energy. To stop global warming, all countries that rely on fossil fuels need to reduce their use of fossil fuels drastically and implement carbon capture and storage technology. They will need to switch to renewable energy sources and/or nuclear energy. To do this, every country needs to assess factors such as the scale of energy demand, the available energy sources, their reliability and their environmental impact.

See Topic 6.8 for more about how greenhouse gases cause global warming.

KEY POINTS

- Fossil fuels produce increased levels of greenhouse gases, which could cause global warming.
- Nuclear fuels produce radioactive waste.
- Renewable energy resources will never run out, they do not produce harmful waste products (e.g., greenhouse gases or radioactive waste), and they can be used in remote places. But they cover large areas, and they can disturb natural habitats.
- Different energy resources can be evaluated in terms of reliability, environmental effects, pollution, and waste.

4.8 Energy and work

- Recognise that when a force moves an object, increasing the force or the distance moved increases the work done
- Recall and use equations to calculate:
  - the work done by a force
  - the change of gravitational potential energy when an object is raised or lowered
  - the kinetic energy of a moving object

S

EXAM TIP

This is a simple equation but one that students often seem to forget!

Working out

EXAM TIP

This is another important equation to learn.

Working out

In a fitness centre or a gym, you have to work hard to keep fit. Raising weights and using a running machine are just two ways to keep fit. Whichever way you choose to keep fit, you have to apply a force to move something. The work you do causes transfer of energy. For example, if you raise an object and increase its gravitational potential energy by 20 J, the work you do on the object is 20 J.

work done = energy transferred

The work done by a force depends on the force and the distance moved and is defined as the force (in newtons) × distance moved in the direction of the force (in metres).

Thus the work done, ΔW, (in joules) when a force F moves an object by a distance d in the direction of the force is given by the equation:

$$\Delta W = F \times d$$

where the force is in newtons and the distance moved is in metres.

Supplement

Gravitational potential energy

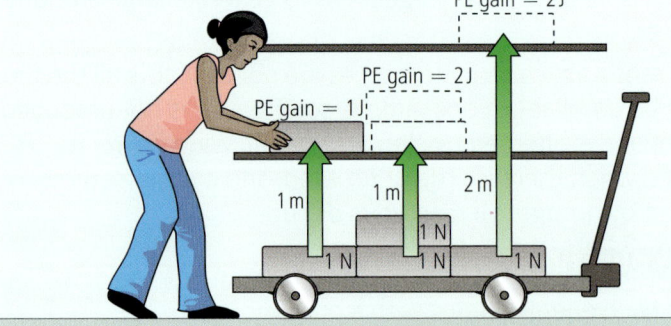

Figure 4.8.1 Using joules

Gravitational potential energy is the energy of an object due to its position. Consider an object of mass m raised through a vertical distance h:

- the weight of the object = mg, where g is the gravitational field strength at the Earth's surface,
- the force needed to raise the object is equal and opposite to its weight.

Therefore:

the gain of gravitational potential energy of the object = work done on the object to raise it

= force × distance moved = $(mg)h = mgh$

In general, when an object of mass m is raised or lowered by a vertical distance, its

change of gravitational potential energy = mgh

Kinetic energy

The kinetic energy of an object is its energy due to its motion. It can be shown that the kinetic energy of an object of mass m moving at speed v is given by the equation:

kinetic energy $= \frac{1}{2}mv^2$, where $m =$ the mass of the object in kilograms (in joules, J) and $v =$ the speed of the object in m/s.

WORKED EXAMPLE

Calculate the kinetic energy of a vehicle of mass 500 kg moving at a speed of 12 m/s.

Solution

Kinetic energy $= \frac{1}{2}mv^2 = 0.5 \times 500\,\text{kg} \times (12\,\text{m/s})^2 = 36\,000\,\text{J}$

A ride on a roller coaster at a fairground causes rapid changes in the kinetic energy and gravitational potential energy of the riders and the train. Suppose the highest point of the track is followed by a steep descent. In the descent, the train loses gravitational potential energy and gains kinetic energy. Assuming the loss of gravitational potential energy of the train is equal to its gain of kinetic energy, the speed, v, at the bottom of the descent is given by:

$\frac{1}{2}mv^2 = mgh$

where h is the vertical distance between the highest point of the track and the bottom of the descent.

In practice, due to air resistance and friction, the speed given by the above equation is not reached.

WORKED EXAMPLE

A 20 N weight is raised through a height of 0.4 m. Calculate: **a** the work done, **b** the gain of gravitational potential energy of the object.

Solution

a The force needed to lift the weight = 20 N

Work done
= force × distance moved in the direction of the force
= 20 N × 0.4 m = 8.0 J.

b Gain of gravitational potential energy = work done = 8.0 J.

The top speed on a roller coaster of height 60 m is about 35 m/s (which is about 125 km/h).

SUMMARY QUESTIONS

Use g = 9.8 N/kg

1 a Calculate the work done when:

   i a force of 20 N makes an object move 4.8 m in the direction of the force,

   ii an object of weight 80 N is raised through a height of 1.2 m.

S 2 a An object of weight 2.0 N fired vertically upwards from a catapult reaches a maximum height of 5.0 m. Calculate:

   i the gain of gravitational potential energy of the object,

   ii the kinetic energy of the object when it left the catapult.

b A roller coaster train of mass 1800 kg descends through a vertical distance of 50 m from rest to the bottom of the track. Calculate:

   i the loss of gravitational potential energy of the train at the bottom of the descent,

   ii the maximum possible speed of the train at the bottom of the descent.

KEY POINTS

- Work done ΔW by a force depends on the force F and distance d moved in the direction of the force:

  $\Delta W = Fd$

- For an object of mass m, its change of gravitational potential energy $= mgh$ where $h =$ vertical distance moved, its

  kinetic energy $= \frac{1}{2}mv^2$,

  where $v =$ its speed.

4.9 Power

LEARNING OUTCOMES

- Recognise that power is rate of transfer of energy
- Recall that power is measured in watts
- Recall and use the equation:

$$\text{power} = \frac{\text{energy transferred}}{\text{time taken}}$$

Rocket power for the first moon landing

EXAM TIP

Note that all types of power are measured in watts (W).

Powerful machines

When you use a lift to go up, a powerful electric motor pulls you and the lift up. The work done by the lift motor transfers energy from electricity to gravitational potential energy. Thermal energy is also produced and some sound energy.

- The work done per second by the motor is the output **power** of the motor.
- The more powerful the lift motor is, the faster it takes you up.

We measure the power of an appliance in watts (W) or kilowatts (kW) or megawatts (millions of watts, MW). One watt is a rate of transfer of energy of 1 joule per second (J/s). For example:

- a 5 W electric torch would transfer 5 J every second as light energy and thermal energy to its surroundings.
- a lift motor with an output power of 6000 W would transfer 6000 J to the lift as gravitational potential energy every second.

Here are typical values of power levels for different energy transfer 'mechanisms':

- A torch   1 W
- A filament light bulb   100 W
- An electric cooker   10 000 W
                             = 10 kW (where 1 kW = 1000 watts)
- A railway engine   1 000 000 W
                             = 1 megawatt (MW) = 1 million watts
- A Saturn V rocket   100 MW
- A very large power station 10 000 MW
- The Sun   100 000 000 000 000 000 000 MW

Energy and power

Machines are labour-saving devices that do work for us. The faster a machine can do work, the more powerful it is.

Whenever a machine does work on an object, energy is transferred to the object. The useful energy transferred is equal to the work done.

The output power of a machine is the rate at which it does work. This is the same as the rate at which it transfers useful energy.

$$\text{power } P \text{ (in watts)} = \frac{\text{work done (in joules)}}{\text{time taken (in seconds)}}$$
$$= \frac{\text{useful energy transferred (in joules)}}{\text{time taken (in seconds)}}$$

If energy ΔE is transferred in time t, **power** $P = \dfrac{\Delta E}{t}$

WORKED EXAMPLE

1 A crane lifts an object of weight 4000 N through a vertical distance of 2.5 m in 5.0 s. Calculate:

   a the gain of gravitational potential energy of the object

   b the output power of the crane.

Solution

a Gain of gravitational potential energy =
mgh = 4000 N × 2.5 m = 10 000 J

b Work done by the crane = gain of gravitational potential energy = 10 000 J

Output power = $\dfrac{\text{work done}}{\text{time taken}} = \dfrac{10\,000\,\text{J}}{5.0\,\text{s}} = 2000\,\text{W}$

A crane at work

PRACTICAL

Measure your personal power

1 Ask a friend to time how long it takes you to step on and off a suitable platform ten times. If you are medically unfit, you do the timing and your friend can do the steps.

2 For each step, your increase of gravitational potential energy = your weight × the height of the platform. This is the work you do each step.

3 Calculate the work you do for ten steps. Your personal power = work done for ten steps ÷ time taken.

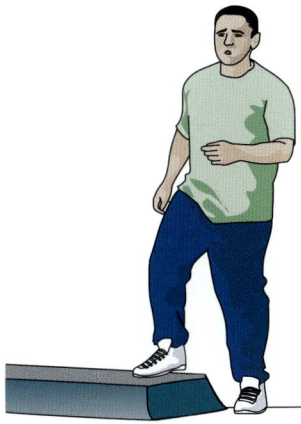
Figure 4.9.1 Steps

HOW POWERFUL IS A WEIGHTLIFTER?

A 30 kg dumb-bell has a weight of 300 N. Raising it by 1 m would give it 300 J of gravitational potential energy. A weight lifter could lift it in about 0.5 s. The rate of transfer of energy would be 600 J/s (= 300 J / 0.5 s). So a weightlifter's output power would be about 600 W in total!

SUMMARY QUESTIONS

1 **a** Which is more powerful?

   i a torch bulb or a mains filament lamp?

   ii a 3 kW electric kettle or a 10 000 W electric cooker?

   b There are about 2 million homes in a certain city. If a 3 kW electric kettle was switched on in 1 in 10 homes in the city at the same time, how much power would need to be supplied?

2 An electric motor raises an object of weight of 200 N through a height of 4.0 m in 5.0 s. Calculate:

   a the gain of gravitational potential energy of the object

   b the work done by the motor on the object

   c the output power of the motor.

KEY POINTS

The unit of power is the watt (W), equal to 1 J/s.

1 kilowatt = 1000 watts

$$\text{power (in watts)} = \dfrac{\text{energy transferred}}{\text{time taken}}$$

1 Describe the following forms of energy:

   (a) kinetic energy

   (b) gravitational potential energy

   (c) nuclear energy

   (d) chemical energy.

2 The school laboratory has a small electric motor that runs from two 1.5 V cells. A pulley wheel is mounted on the axle and this can be connected to turn another pulley wheel that is mounted on the axle of a small dynamo (dc generator). The dynamo is connected to a lamp. When the circuit is switched on, the motor turns. This makes the dynamo turn and this lights the lamp.

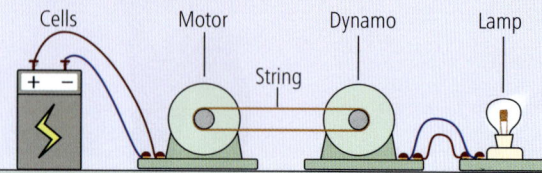

Cells   Motor   Dynamo   Lamp   String

   (a) Show with an energy flow diagram the energy changes at each stage from the cells to the lamp.

   (b) Explain why the lamp will be lit more brightly if connected directly to the cells.

3 An electric motor is used to lift a load. When 6000 J of electrical energy are supplied to the motor, the load gains 2400 J of gravitational potential energy.

   (a) Calculate the amount of energy wasted by the motor.

   (b) Suggest in which form(s) the energy is wasted.

4 Suggest one advantage and one disadvantage of each of the following types of energy resource when used as a method of electricity generation.

   (a) tidal energy   **(c)** geothermal energy

   (b) wave energy   **(d)** solar energy

5 Using the information in question **3**:

   Calculate the efficiency of the motor.

Practice Questions

1 Which of these is the S.I. unit for power?

   A J

   B kg

   C N

   D W

S 2 A pulley system is used to lift a 250 kg mass through a height of 40 m. What is the gain in gravitational potential energy for the mass?

   A 10 kJ

   B 100 kJ

   C 6.25 J

   D 290 J

3 Which of the following energy resources relies on the radioactive decay of materials in the Earth's core?

   A Wind turbines

   B Solar panels

   C Geothermal energy

6 A student has a mass of 65 kg. He climbs up a flight of 12 stairs. Each stair has a height of 16 cm. The time taken to reach the top is 2.2 s ($g = 9.8$ N/kg). **S**

   (a) Calculate the student's weight.

   (b) Calculate the gravitational energy gained by the student between the bottom and the top of the flight of stairs.

   (c) Calculate the student's power.

16 cm

D Hydroelectricity

4 Which of the following uses a non-renewable energy resource?

  A Nuclear power

  B Geothermal energy

  C Solar power

  D Tidal power

5 A stadium floodlight transfers 500 kJ of energy in one hour. What is the power rating of the floodlight?

  A 500 W

  B 140 W

  C 140 kW

  D 8.3 kW

6 A hydroelectric dam can be used to generate electricity.

  (a) Describe the advantages of using a hydroelectric dam when compared to a fossil fuel based power station. [3]

  (b) Describe the negative impact building a hydroelectric dam can have on the local environment. [2]

  (c) Explain why it is not possible to use hydroelectric dams in some areas. [2]

7 A model steam engine is used to lift a load from the floor to the bench. The fuel released 1.5 kJ of energy and the load gained 0.7 kJ of energy as it was lifted.

  (a) Describe the energy transfers which are happening as the load is lifted from the floor. [3]

  (b) Calculate how much energy was transferred to the surroundings during the lifting of the weight. [1]

  (c) Explain how this energy is transferred to the surroundings. [1]

S 8 A motorcycle and rider of combined mass 300 kg travel in a straight line and accelerate from 5 m/s to 7 m/s taking 5.0 s to do this.

  (a) Calculate the gain in kinetic energy for the motorcycle and rider as they accelerated. [3]

  (b) State the minimum amount of work done by the engine as the motorcycle accelerated. [1]

  (c) Explain why the work done by the engine must be more than your answer to part b. [2]

  (d) Calculate the effective power of the motorcycle engine. [2]

9 An electric motor is used to lift a metal girder of mass 450 kg through a height of 30.0 m. The motor has a power rating of 4000 W and operates for 60 s during the lift.

  g = 9.8 N/kg

  (a) Calculate the gain in gravitational potential energy for the girder. [3]

  (b) Calculate the energy transferred electrically to the motor. [2]

  (c) Calculate the efficiency of the motor. [3]

10 A hydroelectric power station has an upland reservoir that is 400 m above the power station. The power station is designed to produce 96 MW of electrical power with an efficiency of 60%.

  (a) Estimate the loss of gravitational potential energy per second when the hydroelectric power station generates 96 MW of power. [3]

  (b) Use your estimate to calculate the volume of water per second that flows from the reservoir through the power station generators when 96 MW of power is generated. [4]

  The density of water is 1000 kg/m^3.

5.1 Under pressure

What is pressure?

If you stand barefoot on a sharp object, you will find out about pressure in a very painful way. All your weight acts on the tip of the object so there is huge pressure on your foot at the area of contact.

Pressure is caused when objects exert forces on each other. The pressure caused by any force depends on the area of contact where the force acts, as well as on the size of the force.

Caterpillar tracks fitted to vehicles are essential on sandy or muddy ground or on snow-covered ground. The reason is that the contact area of the tracks on the ground is much larger than it would be if the vehicle had wheels instead. The tracks therefore reduce the pressure of the vehicle on the ground as its weight is spread over a much larger contact area. It is therefore less likely to sink into the sand, mud or snow.

Pressure is defined as force per unit area. The unit of pressure is the pascal (Pa) which is equal to one newton per square metre (N/m^2).

For a force F acting evenly on a surface of area A, at right angles to the surface, the pressure p on the surface is given by the equation:

$$p = \frac{F}{A}$$

Note Rearranging this equation gives $F = p \times A$ or $A = \dfrac{F}{p}$.

WORKED EXAMPLE

A caterpillar vehicle of weight 12 000 N is fitted with tracks that have an area of 3.0 m² in contact with the ground. Calculate the pressure of the vehicle on the ground.

Solution

$$\text{Pressure} = \frac{\text{force}}{\text{area}} = \frac{12\,000\,\text{N}}{3.0\,\text{m}^2} = 4000\,\text{Pa}$$

PRACTICAL

Measure your foot pressure

1 Draw around your shoes using centimetre squared paper. Count the number of centimetre squares in each footprint (ignoring any square that is less than half-filled) to find the area of contact in square centimetres. Convert this to square metres using the conversion $1\,\text{m}^2 = 10\,000\,\text{cm}^2$.

2 Use suitable scales (e.g. bathroom scales) to measure your weight. If the scales read mass in kilograms, find your weight in newtons using $g = 10\,\text{N/kg}$.

Work out your pressure using: $\text{pressure} = \dfrac{\text{weight}}{\text{area}}$

Caterpillar tracks

In hospital

A sharp knife cuts more easily than a blunt knife. Surgical knives used in operating theatres need to be very sharp. A sharp knife has a much smaller area of contact when it is used than a blunt knife. So the pressure (= force ÷ contact area) of a sharp knife is much greater than the pressure of a blunt knife for the same force. Therefore the same force applied to a sharp knife has a much greater effect than it would with a blunt knife.

Bed sores are a problem for patients confined to bed for long periods of time. Such sores occur where the body presses on the bed for a long time. The skin in contact with the bed is not as tough as the skin under your feet. As a result, the skin in the contact area is easily rubbed away, causing bed sores.

A pressure test

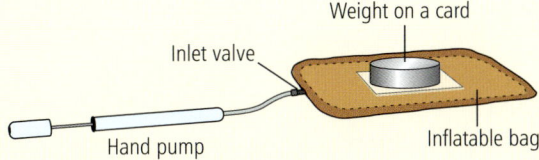

Weight on a card
Inlet valve
Hand pump
Inflatable bag

1 Use the arrangement shown to measure the pressure inside an inflatable bag.

2 With the bag deflated, place a weight on a card on the bag. Use the hand pump to inflate the bag until the card and weight are raised. The pressure of the air in the bag is then equal to the pressure of the weight on the card.

3 Measure the area of the card and calculate the pressure of the air in the bag (= weight in newtons ÷ area of the card in square metres).

KEY POINTS

- Pressure is force per unit area.
- The unit of pressure is the pascal (Pa) which is equal to $1 \, N/m^2$.
- For a force F acting at right angles on an area A, the pressure $p = \dfrac{F}{A}$.

SUMMARY QUESTIONS

1 Explain each of the following:

   a When you do a handstand, the pressure on your hands is greater than the pressure on your feet when you stand upright.

   b Snowshoes like those shown in the photograph are useful for walking across soft snow.

2 A rectangular concrete paving slab of weight 1200 N has sides of length 0.60 m and 0.40 m and a thickness of 0.05 m. Calculate the pressure of the paving slab on the ground when it is:

   a laid flat on a bed of sand

   b standing upright on its short side.

5.2 Pressure at work

- Recognise that pressure can be transmitted through a fluid
- Explain how a hydraulic system works
- Recall that the force exerted by a hydraulic system depends on the pressure and the area of the cylinder that exerts the force

a

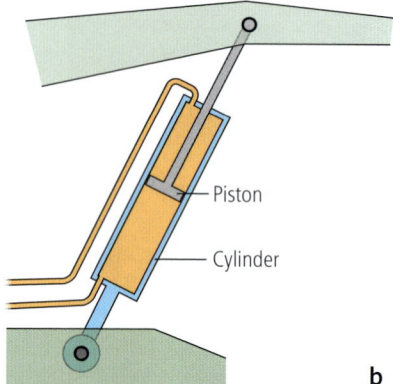

Piston

Cylinder

b

Figure 5.2.1 (a) A mechanical digger
(b) A hydraulic system

A remote-controlled robot handling a dangerous object

Mechanical diggers are used to remove large quantities of earth, for example when soil has to be removed above an underground pipe to reach the pipe. The 'grab' of the digger is operated by a **hydraulic pressure** system. The hydraulic system of a machine is its 'muscle power'.

In the hydraulic systems shown in Figure 5.2.1, oil is pumped into the upper or lower part of the cylinder to make the piston move in or out of the cylinder.

Robots use hydraulics for muscle power. Robotic machines in factories are fixed machines that operate such things as welding gear or paint sprays on assembly lines. Robot muscles use compressed air or oil and they work non-stop without the need for a human operator. Remote-controlled robots are used for dangerous tasks such as bomb disposal and handling suspect packages.

Vehicle brakes use pressure. When the driver presses on the brake pedal of the car, pressure is exerted on the brake fluid in the main cylinder. This pressure is transmitted along the brake pipes to wider cylinders at the wheels. The fluid pressure forces the piston in each wheel cylinder to push the brake disc pads on to the wheel disc.

Power-assisted brakes fitted to heavy goods vehicles and coaches use compressed air. When the driver applies the brakes, compressed air at very high pressure is released to push on the piston in the main cylinder. The compressed air is used instead of the brake pedal to exert the force on the main cylinder. This is why such vehicles hiss when the brakes are released.

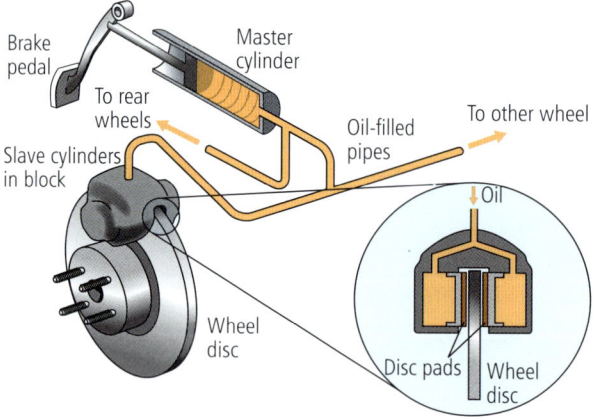

Brake pedal

Master cylinder

To rear wheels

Oil-filled pipes

To other wheel

Slave cylinders in block

Oil

Wheel disc

Disc pads

Wheel disc

Figure 5.2.2 Disc brakes in a car

A hydraulic car jack can be used to lift a car, as shown in Figure 5.2.3. When the handle is pressed down, the piston in the narrow cylinder is forced into the oil-filled cylinder. Oil is forced out of this cylinder, through the pipe and into a wider cylinder. The pressure of the oil on the piston in the wider cylinder forces this piston outwards which forces the pivoted lever to raise the car.

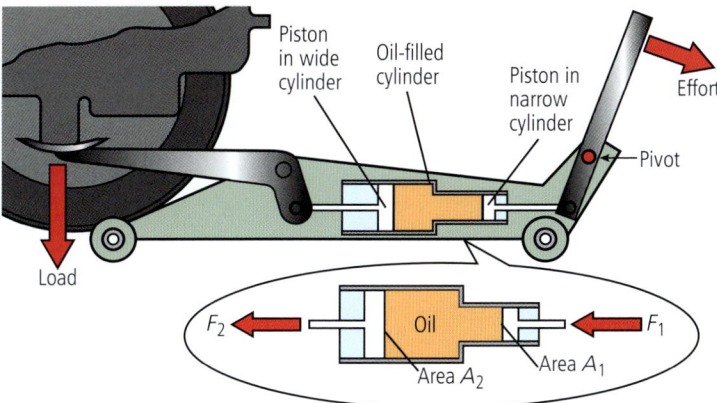

Figure 5.2.3 A hydraulic car jack

Supplement

In Figure 5.2.3:

- The force F_1 created by moving the lever acts on the narrow cylinder and creates a pressure $p = \dfrac{F_1}{A_1}$ on the fluid in the cylinder, where A_1 is the area of the narrow cylinder.

- This pressure is transmitted through the fluid in the pipe to the wider cylinder.

- The force on the larger piston, $F_2 = p \times A_2$ where A_2 is the area of the wide cylinder. Therefore $F_2 = \dfrac{F_1}{A_1} \times A_2$

The force F_2 is therefore much greater than F_1 because area A_2 is much greater than area A_1.

SUMMARY QUESTIONS

1 a Write down as many machines as you can think of that are operated hydraulically.

b The photograph shows the arm of a mechanical digger. It is controlled by three hydraulic pistons called 'rams', labelled X, Y and Z.

i Explain why the arm is raised when compressed air is released into ram X so it extends.

ii State and explain what happens to the 'bucket' on the end of the arm when rams Y and Z are both extended.

2 The hydraulic lift shown in Figure 5.2.4 is used to raise a vehicle so its underside can be inspected.

The lift has four pistons, each of area $0.01\,\text{m}^2$ to lift the platform. The pressure in the system must not be greater than $5.0 \times 10^5\,\text{Pa}$. The platform weight is $2000\,\text{N}$. Calculate the maximum load that can be lifted on the platform.

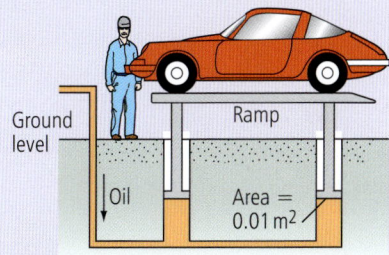

Figure 5.2.4

An underwater swimmer using a snorkel tube can breathe safely provided the top of the tube is above the water. However, a deep-sea diver could not breathe through a very long snorkel tube because the pressure of the water increases with depth. At depths of more than a few metres, the diver's chest muscles would not be strong enough to expand his or her chest muscles against the water pressure on the body.

The pressure of a liquid increases with depth. Figure 5.3.1a shows water jets from the holes down the side of an open plastic bottle filled with water. The further the hole is below the level of water in the bottle, the greater the pressure of the jet.

Using a snorkel tube

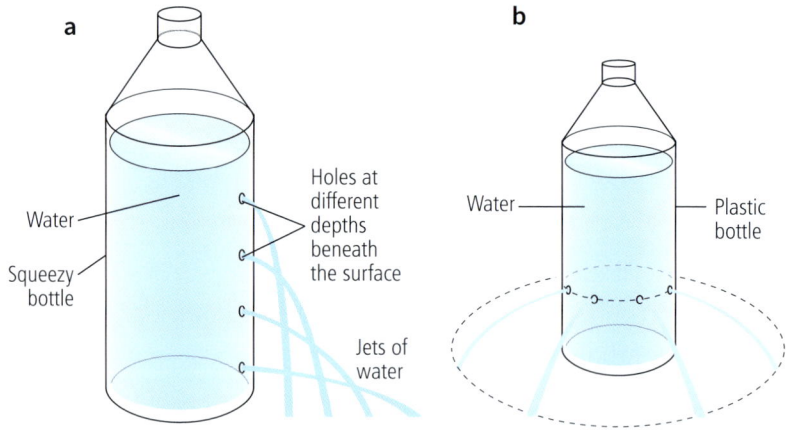
Figure 5.3.1 Pressure in a liquid at rest: (a) pressure increases with depth (b) same pressure at the same depth

The pressure along a horizontal line in a liquid is constant. This can be demonstrated by making several holes around the bottle at the same level as in Figure 5.3.1b. The jets from these holes are at the same pressure.

Water levels

Figure 5.3.2 shows several containers joined to a sealed pipe. If a liquid is poured into one of the containers, some of the liquid flows into the other containers. The flow stops when the level of the liquid in each container is the same. This is because the pressure in each container along the same horizontal line has become equal. The unit of pressure, the pascal (Pa), is named after the 17th century French scientist, Blaise Pascal, who first demonstrated this effect and made many other discoveries about pressure.

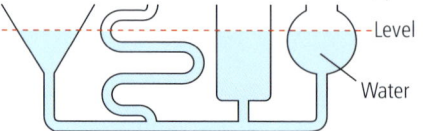
Figure 5.3.2 Pascal's vases

The pressure in a liquid depends on the density of the liquid.
Suppose water is poured into one side of a U-shaped tube, then oil is carefully poured into the other side, as in Figure 5.3.3. When the liquids settle, the oil level is higher than the water level on the other side. This is because oil is less dense than water so a greater depth of oil is needed to create the same pressure at the same level.

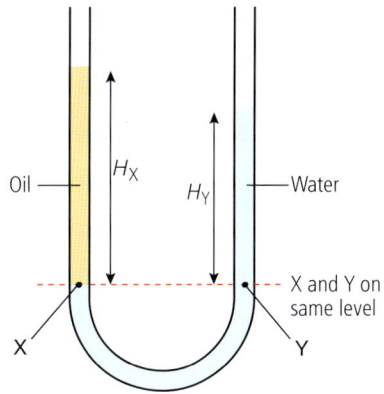

Figure 5.3.3 Comparing densities

Supplement

The pressure of a liquid column

Consider the column of liquid in the container shown in Figure 5.3.4. The pressure caused by the liquid column on the bottom of the container is due to the weight of the liquid.

For a column of height Δh and area of cross-section A, the volume of liquid in the container $= A\,\Delta h$.

The mass of liquid = its density × volume of the liquid = $\rho A\,\Delta h$, where ρ is the density of the liquid.

The weight of the liquid = mass × g = $\rho A\,\Delta h \times g$, where the gravitational field strength $g = 9.8\ \text{N/kg}$.

The pressure Δp due to the liquid at the base of the liquid column

$$= \frac{\text{weight}}{\text{area of cross-section}} = \frac{\rho A\,\Delta h\, g}{A}.$$

Therefore, cancelling A gives: $\boldsymbol{\Delta p = \rho g \Delta h}$

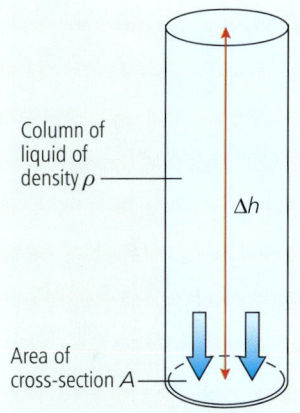

Figure 5.3.4 Calculating liquid pressure

Column of liquid of density ρ

Δh

Area of cross-section A

> **WORKED EXAMPLE**
>
> Use $g = 9.8\ \text{N/kg}$
>
> Calculate the pressure due to sea water of density $1050\ \text{kg/m}^3$ at a depth in the sea of $200\ \text{m}$.
>
> **Solution**
>
> Pressure $p = \rho g \Delta h = 1050\ \text{kg/m}^3 \times 9.8 \times 200\ \text{m} = 2.1 \times 10^6\ \text{Pa}$.

> **SUMMARY QUESTIONS**
>
> Use $g = 9.8\ \text{N/kg}$
>
> **1 a** Explain why the wall of a dam needs to be thicker at the base than at the top.
>
> **b** A water tank at the top of a tall building supplies water to taps in the building. Explain why the pressure of the water from a tap on the ground floor is greater than the pressure from a tap on a higher floor.
>
> **2** A sink plug has an area of $0.0006\ \text{m}^2$. It is used to block the outlet of a sink filled with water to a depth of $0.090\ \text{m}$. Calculate:
>
> **a** the pressure on the plug due to the water
>
> **b** the force needed to remove the plug from the outlet; use density of water = $1000\ \text{kg/m}^3$.

> **KEY POINTS**
>
> • The pressure in a liquid increases with increase of depth and of density.
>
> • The pressure Δp due to the column of height Δh of liquid of density ρ is given by the equation $\Delta p = \rho g \Delta h$.

5.4 Solids, liquids and gases

Spot the three states of matter

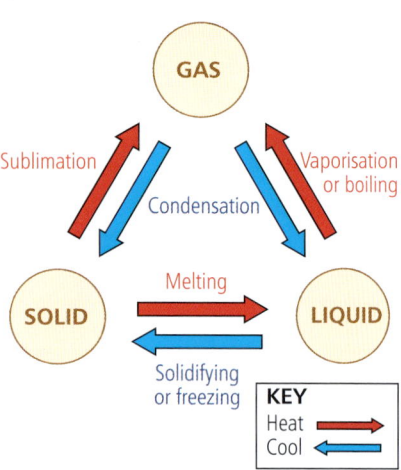

Figure 5.4.1 Changes of state

States of matter

Everything around us is made of matter, mostly in either a solid or a liquid or a gaseous state. Each state of matter has its own characteristic properties.

- Solids have a fixed shape and a fixed volume. In general, solid objects do not lose their shape unless they are deformed by force beyond their **elastic limit**.

- Liquids have a fixed volume and they flow. They change their shape to fit the shape of the container. If a liquid is stirred, it moves around internally at and below its surface. A stirred liquid gradually slows down and stops because the container 'drags' on the liquid where they are in contact.

- Gases have neither a fixed volume nor a fixed shape. Any gas can be compressed or expanded to change its volume. A gas in a container takes the shape of the container. Gases flow and they are much less dense than solids or liquids.

This table summarises the main properties of solids, liquids and gases.

	flow	shape	volume	density
solid	no	fixed	fixed	much higher than a gas
liquid	yes	fits container shape	fixed	much higher than a gas
gas	yes	fills container	can be changed	low compared with a solid or liquid

Change of state

Matter can change from one state to another. Figure 5.4.1 shows the changes between different states. The changes are brought about by heating or cooling the substance. For example:

- when liquid water in a beaker is heated sufficiently, it boils and turns to steam, which is water in the gaseous state

- when solid carbon dioxide warms up, it sublimes because it turns to vapour directly

- when liquid water freezes, it solidifies and turns to ice, which is water in the solid state

- when steam touches a cold surface, it condenses and turns to liquid water.

Molecules and matter

The smallest particle in a pure substance that can be identified as belonging to the substance is called a **molecule**. Depending on the substance, a molecule may be a single atom (in certain elements) or it may be a group of atoms. Figure 5.4.2 shows how the molecules of a substance in the solid, liquid or gaseous state are arranged.

- The molecules in a solid hold each other together in fixed positions to form a rigid structure with its own shape.

- The molecules in a liquid move about in contact with each other but they don't stay in fixed positions. This is why a liquid doesn't have its own shape and can flow.

- The molecules in a gas are far apart compared with molecules in a liquid or solid. This is why the density of a gas is much less than that of a solid or liquid. They move very rapidly throughout their container and make contact only when they collide.

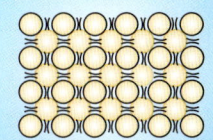 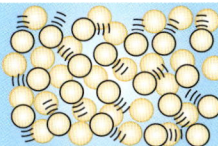 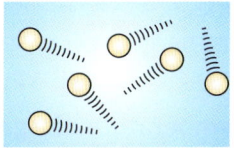

Figure 5.4.2 The arrangement of molecules in a solid, a liquid and a gas

KEY POINTS

- Molecules in a solid are in a rigid structure.
- Molecules in a liquid move about in contact with each other.
- Molecules in a gas are far apart and move about at high speed.

SUMMARY QUESTIONS

1 Copy and complete the following sentences using words from the list

   gas   liquid   solid

   a A _____ has a fixed shape and volume.

   b A _____ has a fixed volume but no shape.

   c A _____ and a _____ can flow.

   d A _____ does not have a fixed volume.

2 State the scientific word for each of the following changes.

   a A mist appears on the inside of a window on a bus full of people.

   b Steam is produced from the surface of the water in a pan when the water is heated.

   c Food taken from a freezer thaws out.

   d An ice cube in a beaker of warm water gradually disappears.

PRACTICAL

Changing state

1 Place a beaker of water on a tripod and use a Bunsen burner to heat the water. Observe that evaporation takes place at the water surface before it boils. When water boils, bubbles of steam form throughout the water and rise to the surface to release the steam.

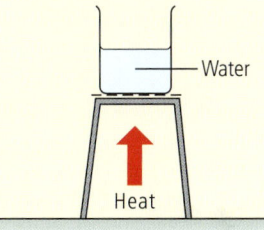

Figure 5.4.3 Changing state

2 Switch the gas burner off and hold a cold object above the beaker. Observe condensation of steam from the beaker on the object.

EXAM TIP

The next time you see ice melt or water boil, think about what is happening to the water molecules.

5.5 More about solids, liquids and gases

EXAM TIP

See page 50 for internal energy.

Molecules in a gas

The pressure of a gas on a surface is due to the impacts of the gas molecules with the surface. The gas molecules move about very fast in random directions. Imagine a squash court with lots of squash balls flying about, rebounding from the walls, the floor and the ceiling. Now imagine the squash court and the balls scaled down to a thousandth of a millionth of their size. On this scale, the balls would be like gas molecules moving about at random.

The molecules in a gas collide repeatedly with each other and with the surface of their container, rebounding after each collision. They move at different speeds which can change each time they collide with each other. Each impact with the surface exerts a tiny force on the surface, as shown in Figure 5.5.1. Millions of millions of such impacts happen every second and their overall effect is to cause a steady pressure on the surface.

The temperature of a gas affects the speed of the gas molecules. If some gas in a sealed container is heated, its temperature increases because its molecules gain more kinetic energy. Therefore, the average speed of the gas molecules increases. The temperature of a gas is a measure of the average kinetic energy of the gas molecules. The molecules still move about at random but their average speed is greater. Cooling a gas lowers the average speed of the molecules.

Figure 5.5.1 Gas molecules in a box

EXAM TIP

'Random' means unpredictable or haphazard. The speed and direction of motion of a gas molecule is unpredictable because its collisions with other molecules cannot be predicted.

Supplement

Comparison of molecules in solids, liquids and gases

In a solid, the molecules are arranged in a three-dimensional structure.

- There are strong forces of attraction between the molecules.

- Each molecule vibrates about a mean position that is fixed.

- When a solid is heated, its internal energy increases as the molecules gain energy and vibrate more. If heated sufficiently, the solid melts (or sublimes) because the molecules have gained enough energy to break away from the structure.

Molecular model of ice. Each water molecule consists of two hydrogen atoms (white) and one oxygen atom (red)

In a liquid, the molecules are in contact with each and they move around each other.

- There are still forces of attraction between the molecules but they are not strong enough to hold the molecules in a rigid structure. So a liquid can flow and has no fixed shape.

- The forces of attraction are strong enough to stop the molecules moving away from each other completely at the surface.

- When a liquid is heated, its internal energy increases because the molecules move about faster. Some of the molecules gain enough energy to break away from the other molecules. The molecules that escape from the liquid become a gas.

In a gas, the molecules are much further apart on average than in a solid or a liquid. This is why the density of a gas is much less than that of a liquid.

- The forces of attraction between the molecules are negligible. So a gas can flow and has no fixed shape or volume.

- The molecules move about at high speed, colliding with each other and the internal surface of their container.

- When each molecule hits the container surface, its momentum changes and it exerts a force on the container. The pressure of the gas on the container surface is due to all the molecules striking the surface.

- When a gas is heated, its internal energy increases because its molecules gain kinetic energy and move faster on average. This causes the pressure of the gas to increase because the molecules collide with the container surface with more force.

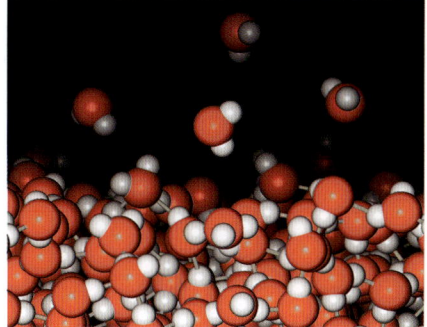

Model of molecules in water. Note the water molecules above the water surface that have broken away from the surface

SUMMARY QUESTIONS

1 Explain the following statements in terms of molecules.

   a A gas exerts a pressure on any surface it is in contact with.

   b Heating a solid makes it melt.

2 The table below lists the properties of the molecules in four different substances. State with a reason whether each substance is a solid, a liquid or a gas or does not exist.

substance	distance between the molecules	particle arrangement	movement of the molecules
A	close together	not fixed	move about
B	far apart	not fixed	move about
C	close together	fixed	vibrate
D	far apart	fixed	vibrate

KEY POINTS

- Increasing the temperature of a gas increases the average speed of its molecules.

- The pressure of a gas on a surface is caused by its molecules repeatedly hitting the surface.

5.6 Gas pressure and temperature

- Describe how the pressure of a gas in a sealed container is affected by changing the temperature of the gas

- Explain why raising the temperature of a gas in a sealed container increases its pressure

- Describe evidence that gas molecules move about at random

The link between gas temperature and the kinetic energy of the molecules is important to know.

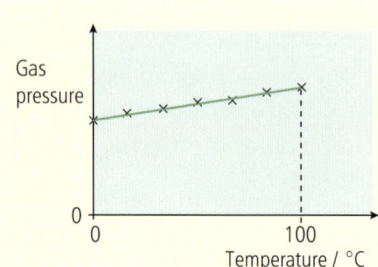

Figure 5.6.2 Graph of gas pressure v temperature

In the kitchen

Never heat food in a sealed can. The can is likely to explode because the pressure of gas inside it increases with an increase in temperature. Increasing the temperature of any sealed container increases the pressure of the gas inside it. This is because:

- the gas molecules move about very fast in random directions – the pressure of the gas is due to the impacts of the gas molecules with the container surface

- the thermal energy supplied to the gas to raise its temperature is supplied to the gas molecules as kinetic energy.

Therefore, the average kinetic energy of the gas molecules increases when the gas is heated. As a result, the average speed of the molecules therefore increases.

When the temperature of a gas is increased, the molecules on average move about faster inside the container. They hit the surfaces harder and more often, so, if the gas is in a sealed container, the force per unit area of the impacts increases so its pressure increases.

Investigating how the pressure of gas varies with temperature

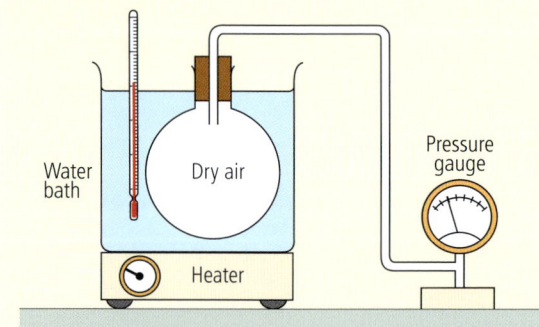

Figure 5.6.1 Measuring gas pressure at different temperatures

1 Figure 5.6.1 shows dry air in a sealed flask connected to a pressure gauge. The flask is in a large beaker of water which is heated to raise the temperature of the gas. The water is heated in stages to raise the temperature in stages. At each stage, the water is stirred to ensure its temperature is the same throughout. The temperature of the water is measured using the thermometer. The pressure is read off the pressure gauge.

2 If the measurements are plotted on a graph of pressure against temperature in °C, the results give a straight line graph as shown in Figure 5.6.2. This shows that the increase of pressure is the same for equal increases of temperature.

Observing random motion

Individual molecules are too small to see directly. However, we can see their direct effects by observing the motion of microscopic smoke particles in air. Figure 5.6.3 shows how we can do this using a smoke cell and a microscope. The smoke particles move about haphazardly and follow erratic (irregular and unpredictable) paths.

1 A small glass cell is filled with smoke

2 Light is shone through the cell

3 The smoke is viewed through a microscope

4 You see the smoke particles constantly moving and changing direction. The path taken by one smoke particle will look something like this

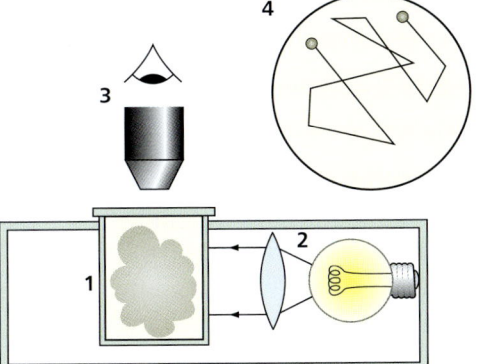

Figure 5.6.3 A smoke cell

The microscopic smoke particle is much larger than the air molecules

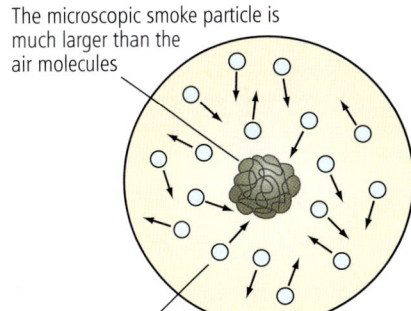

The cell contains air molecules which are in constant erratic motion. As they collide with the smoke particle they give it a push. The direction of the push changes at random

Figure 5.6.4 Brownian motion

This kind of motion is called **Brownian motion**. It is named after the botanist Robert Brown who first observed it in 1785. He used a microscope to observe pollen grains floating on water. He was amazed to see that the pollen grains were constantly moving about and changing direction as if they had a life of their own. Brown could not explain what he saw. Brownian motion puzzled scientists until the molecular theory of matter provided an explanation.

Figure 5.6.4 shows how the Brownian motion of smoke particles in air is caused. Air molecules repeatedly collide at random with each smoke particle. What we observe is the erratic motion of the smoke particles caused by the random impacts of gas molecules on each one.

S The air molecules must be moving very fast to cause this effect because they are far too small to see and the smoke particles are massive compared with the gas molecules.

SUMMARY QUESTIONS

1 Copy and complete the following sentences using words from the list.

impacts   kinetic energy   pressure   temperature

a The _____ of a gas can be increased by increasing its _____.

b Reducing the _____ of a gas reduces the average _____ of its molecules.

c The _____ of a gas is caused by repeated _____ of its molecules on the surface of its container.

2 a Explain why smoke particles in air move about erratically.

b Explain why smoke particles in air move about faster if the temperature of the air is increased.

KEY POINTS

- The pressure of a gas in a sealed container increases if the gas temperature is increased.

- Brownian motion is the erratic motion of microscopic particles due to random impacts of gas molecules on each particle.

5.7 Evaporation

LEARNING OUTCOMES

- Explain the difference between an unsaturated vapour and a saturated vapour
- Recognise that a liquid can be cooled by evaporation
- **s** Explain cooling by evaporation of a liquid
- Explain the factors that can increase evaporation from a liquid

Unsaturated and saturated vapour

If a saucer of water is left in a room, the water gradually disappears. This is because water molecules move from the water in the saucer into the air space above the water. The molecules that have left the liquid spread throughout the air in the room. The water in the saucer is said to **evaporate** as it turns to water vapour. If all the water in the bowl evaporates, and the air in the room can still 'hold' more water vapour, the air is said to be **unsaturated** (Figure 5.7.1).

If a bowl containing sufficient water is left in a room that is not ventilated, evaporation occurs until the air in the room is **saturated** with water vapour. Water vapour will then condense at the same rate that water evaporates from the bowl or from condensation elsewhere. When this happens, water molecules from the water vapour in the air return to water in the liquid state at the same rate as water molecules leave the water. Figure 5.7.2 represents condensation in a room which is saturated with water vapour.

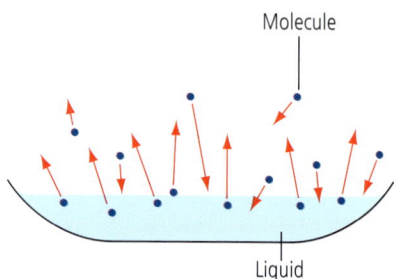

Figure 5.7.1 An unsaturated vapour

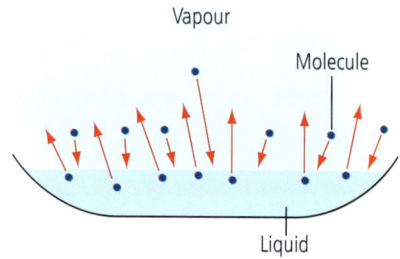

Figure 5.7.2 A saturated vapour

Cooling by evaporation

When a liquid evaporates, it becomes cooler. Figure 5.7.3 shows a demonstration of this effect. It can be explained in terms of molecules and their movement, as shown in Figure 5.7.4.

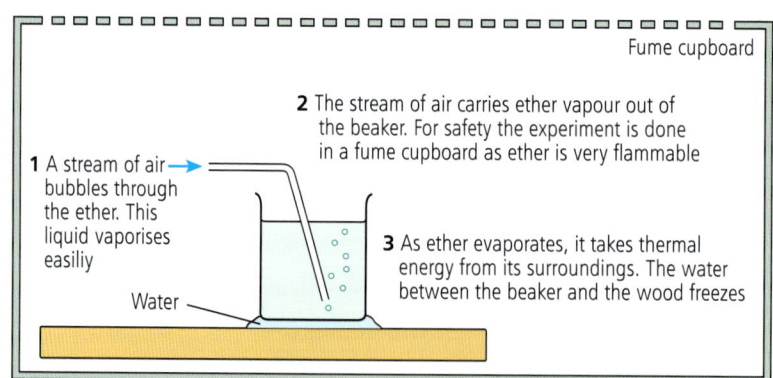

Fume cupboard

1 A stream of air bubbles through the ether. This liquid vaporises easiliy

2 The stream of air carries ether vapour out of the beaker. For safety the experiment is done in a fume cupboard as ether is very flammable

3 As ether evaporates, it takes thermal energy from its surroundings. The water between the beaker and the wood freezes

Water

Figure 5.7.3 Cooling by evaporation

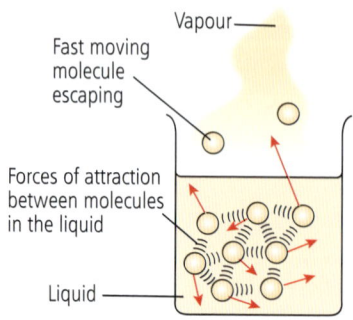

Vapour

Fast moving molecule escaping

Forces of attraction between molecules in the liquid

Liquid

Figure 5.7.4 Explanation of cooling by evaporation

- Weak attractive forces exist between the molecules in the liquid.
- Molecules in the liquid have a range of kinetic energies. Near the surface, the more energetic molecules can break away from the attraction of the other molecules and escape from the liquid.
- After they escape, the average kinetic energy of the remaining molecules in the liquid decreases so the liquid becomes cooler.
- An object in contact with an evaporating liquid is cooled by the liquid because the liquid becomes colder than the object. **s**

Using evaporation

1 **Salt can be obtained by allowing water to evaporate from salt water.** In hot countries, heating by the Sun causes evaporation from salt water in large, enclosed, shallow areas known as 'salt pans'. Dry salt is left in each salt pan when all the water has evaporated.

2 **An air cooling system** transfers thermal energy from inside a hot building to the outside. A sealed circuit of pipes carries a volatile substance (one that easily evaporates) through an air-conditioning unit in the building and a heat exchanger outside.

- The substance evaporates and cools in the pipes in the air-conditioning unit. Air blown over the pipes cools down and is circulated throughout the building.

- The vapour in the pipes is pumped into a heat exchanger where it condenses and releases thermal energy to the surroundings.

- The condensed vapour is then forced through valves and pipes back to the air-conditioning unit.

An air-conditioning unit

Supplement

Factors affecting evaporation

The examples above show that evaporation from a liquid is increased as a result of:

- increasing the surface area or the temperature of the liquid,
- creating a draught of air across the liquid surface.

1 The salt pans need to be shallow so the salt water heats up easily and they need to cover a large area so evaporation readily occurs.

2 If the building heats up, more liquid evaporates in the air-conditioning unit, resulting in more condensation in the heat exchanger and more energy transfer from it to the outside.

SUMMARY QUESTIONS

1 Copy and complete the following sentences using words from the list.

equal to   greater than   less than

a In an unsaturated liquid, the rate of transfer of molecules from the liquid to the vapour is _____ the rate of transfer in the opposite direction.

b In a saturated liquid, the rate of transfer of molecules from the liquid to the vapour is _____ the rate of transfer in the opposite direction.

c When the temperature of a liquid in an open container is cooled, the rate of transfer of molecules from the liquid to the vapour becomes _____ the rate of transfer in the opposite direction.

S 2 Explain the following statements:

a Wet clothes on a washing line dry faster on a hot day than on a cold day.

b A person wearing wet clothes on a cold windy day is likely to feel much colder than someone wearing dry clothes.

5.8 Gas pressure and volume

The volume of a fixed mass of gas depends on its pressure and on its temperature. For example, suppose the air in a tube is trapped by a piston in the tube, as shown in Figure 5.8.1.

- If the piston is forced into the tube, the volume of the air in the tube decreases as the air is compressed. As a result, the pressure of the air in the tube increases. Provided the compression is carried out slowly, the temperature of the air in the tube does not change.

- If the tube is heated, for example by placing it in hot water, the air expands and forces the piston towards the open end of the tube. In this situation, the volume of the air in the tube increases. The air pressure in the tube stays the same as the air expands.

Investigating the pressure and volume of a fixed mass of air at constant temperature

Figure 5.8.2 shows a gas trapped by oil in an inverted glass tube. The pressure of the air in the tube is measured using a pressure gauge. The volume of the air is measured in cubic centimetres using the vertical scale alongside the tube. The pressure of the gas is increased using a foot pump. The tube has a thick wall so that it can withstand the high internal pressure of the gas.

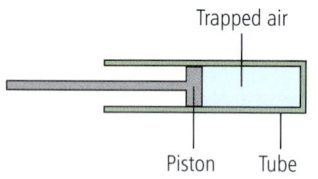

Trapped air

Piston   Tube

Figure 5.8.1 Changing the volume of a gas

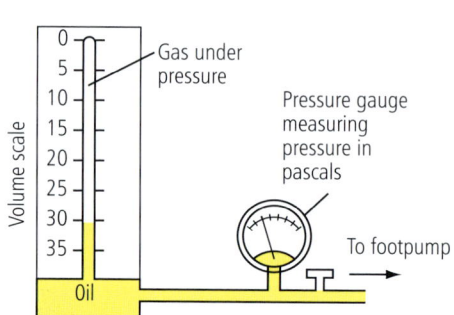

Figure 5.8.2 Testing the variation of pressure and volume of a fixed mass of air

The volume of the gas in the tube is measured at different pressures as the pressure is increased slowly from atmospheric pressure in stages. At each stage, the tap is closed so the gas in the tube does not leak out and the pressure and volume of the gas are measured when they have stopped changing. The measurements are recorded in the following table and plotted in the graph in Figure 5.8.3. The graph shows that

- the pressure increases as the volume decreases

- the pressure decreases as the volume increases.

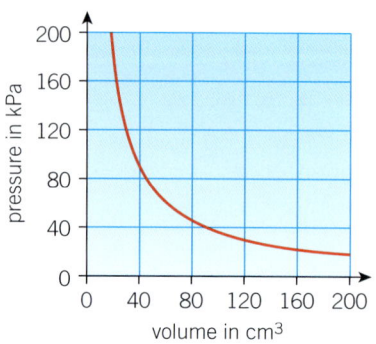

Figure 5.8.3 The inversely proportional relationship between volume and pressure

pressure / kPa	100	120	150	180
volume / cm³	36	30	24	20
pressure × volume / kPa cm³	3600	3600	3600	3600

Supplement

The measurements in the table on the previous page show that the product, pressure × volume, is constant. This is known as **Boyle's law** after Robert Boyle who first discovered the law in 1662.

For a fixed mass of gas at constant temperature:

pV = constant where p = the gas pressure and V = the gas volume.

Note Rearranging the above equation gives $p = \dfrac{\text{constant}}{V}$, which means that the pressure varies in inverse proportion to the volume of the gas.

Explanation of the variation of pressure with volume

For a fixed mass of gas, the number of gas molecules is constant. If the temperature is constant, the average speed of the molecules is constant.

If the volume of a fixed mass of gas at constant temperature is reduced, the gas pressure increases because:

- the space the molecules move in is smaller so they do not travel as far between successive impacts on the surface
- the molecules hit the surfaces more often so the average force of impact increases.

SUMMARY QUESTIONS

1 Copy and complete the following sentences using words from the list.

   decreases   increases   stays the same

   a When a fixed mass of gas expands at constant temperature, its volume _____ and its pressure _____.

   b When a fixed mass of gas is compressed, its pressure _____, its volume _____ and its density _____.

   c When the volume of a gas is reduced at constant temperature, the average kinetic energy of its molecules _____ and the number of impacts per second of gas molecules on the surfaces in contact with the gas _____.

2 Calculate the unknown quantity in each of the following changes involving a fixed mass of gas at constant pressure.

	initial pressure / Pa	initial volume / m³	final pressure / Pa	final volume / m³
a	100 000	0.000 20	50 000	?
b	100 000	0.000 30	?	0.000 15
c	120 000	?	100 000	0.000 60
d	?	0.000 15	60 000	0.000 45

WORKED EXAMPLE

In a chemistry experiment, 0.000 20 m³ (= 200 cm³) of gas was collected in a flask at a pressure of 125 kPa. Calculate the volume of this mass of gas at a pressure of 100 kPa and the same temperature.

Solution

Let p_1 = 125 kPa = 125 000 Pa and V_1 = 0.000 20 m³, $p_1 V_1$ = 125 000 Pa × 0.00020 m³ = 25 Pa m³.

Let p_2 = 100 kPa = 100 000 Pa, where V_2 is the volume to be calculated.

Applying $p_2 V_2 = p_1 V_1$

therefore gives 100 000 Pa × V_2 = 25 Pa m³

Hence $V_2 = \dfrac{25 \text{ Pa m}^3}{100\,000 \text{ Pa}} =$ 0.000 25 m³ (= 250 cm³)

KEY POINTS

For a fixed mass of gas at constant temperature, if its volume is decreased:

- its molecules hit the surfaces of its container more often
- its pressure increases
- its pressure × its volume is constant.

1 Describe an experiment to study the relationship between the pressure and volume of a fixed mass of gas at constant temperature.

Include a labelled diagram and suggest any precautions that you would take to make the experiment as reliable as possible.

2 Explain the reason for the following.

(a) A tractor has large, wide rear wheels.

(b) A sharp knife cuts more easily than a blunt knife.

(c) A small stone trapped in your shoe, under your foot, can be very painful.

3 The diagram shows gas molecules in a box.

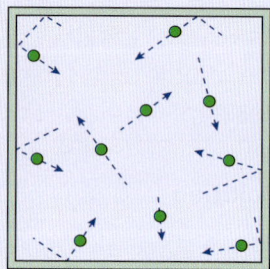

(a) Describe the motion of the gas particles.

(b) Comment on the forces between the particles.

(c) The size of the container is fixed. Explain why increasing the temperature of the gas will increase its pressure.

4 (a) When a liquid evaporates quickly, there is a noticeable cooling effect. Explain why this cooling occurs.

(b) State and explain the factors that affect the rate of evaporation of a liquid.

S 5 In a hydraulic machine a force of 40 N is applied to a piston of area $0.40\,m^2$. The area of the other piston is $4.0\,m^2$.

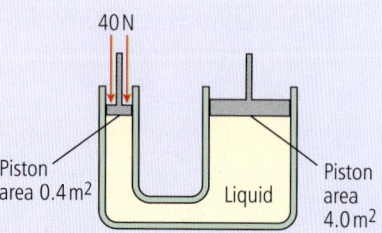

40 N

Piston area $0.4\,m^2$ Liquid   Piston area $4.0\,m^2$

(a) Calculate the pressure transmitted through the liquid.

(b) Calculate the force on the other piston.

Practice Questions

1 Which of these is a correct unit for pressure?

A N

B Pa

C m

D Nm

2 The block in the diagram has a weight of 3000 N. What is the maximum pressure this object can exert on a surface when resting on one of its sides?

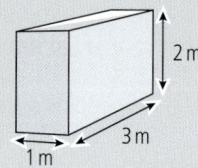

2 m

3 m

1 m

A 1500 Pa

B 500 Pa

C 1000 Pa

D 1200 Pa

3 Which of the following is the correct term for the random motion of microscopic smoke particles caused by collisions with air molecules?

A Brownian motion

B Diffusion

C Evaporation

D Sublimation

4 In which of the following are the particles furthest apart?

A Solids

B Liquids

C Gases

D Solutions

5 A hydraulic cylinder is used to change the size of the force as shown in the diagram.

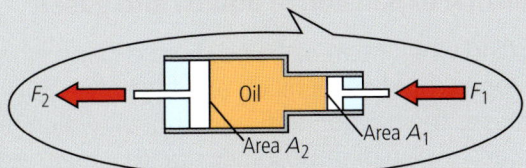

The force F_1 is 40 N, Area A_1 0.25 m² and area A_2 0.60 m².

Calculate the size of the force F_2.

A 6.0 N

B 167

C 267 N

D 96 N

6 The figure below shows a simple mercury barometer. At standard atmospheric pressure the height of the mercury column is 760 mm.

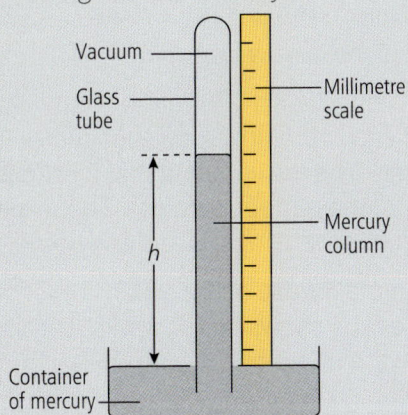

(a) Explain why the mercury in the glass tube does not flow out of the glass tube. [3]

(b) Describe what happens to the height of the mercury column as the atmospheric pressure increases. [2]

(c) Suggest why mercury is used in a barometer instead of a liquid such as coloured water. [2]

(d) The density of mercury is 13600 kg/m³. Calculate the pressure due to a column of mercury of length 760 mm. Use $g = 9.8$ N/kg. [3]

7 A metal pan containing hot water is left in a room. Over time the mass of water in the pan decreases slightly and its temperature falls.

(a) Why does the mass of water in the pan decrease? [1]

(b) Describe how evaporation helps to cool the water in the pan. [3]

(c) Give one application which uses evaporation. [1]

8 A cylinder is used to compress air. Initially the cylinder has a volume of 200 cm³ at a pressure of 1.0 kPa.

The cylinder is compressed so that the volume decreases to 40 cm³.

(a) Describe how and why the pressure in the container changes as the air is compressed. [3]

(b) Calculate the pressure of the air when it has been compressed. [3]

9 A cylindrical metal barrel has a height of 0.80 m and a diameter of 0.90 m. When filled, the barrel contains 506 kg of oil.
Use $g = 9.8$ N/kg.

(a) Calculate the volume of oil in the barrel. [3]

(b) Calculate the density of the oil in the barrel. [2]

(c) Calculate the pressure at the bottom of the barrel caused by the oil. [3]

10 A bicycle pump contains 25 cm³ of air at a pressure of 105 kPa. The air is then pumped in a single stroke through a valve into a tyre of volume 95 cm³ which contains air at the same pressure. Calculate the pressure of the air in the tyre after the stroke. Assume the volume of the inflated tyre does not change. [4]

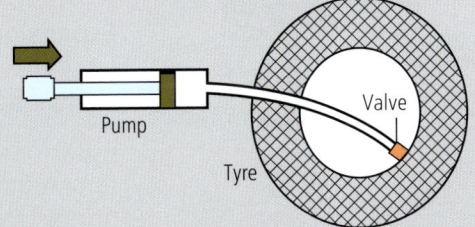

6.1 Thermal expansion

LEARNING OUTCOMES

- Recognise that solids, liquids and gases expand when heated
- Describe and explain some applications of thermal expansion
- Describe and explain some consequences of thermal expansion
- **S** Appreciate that gases expand much more than solids and liquids

Filling a hot-air balloon

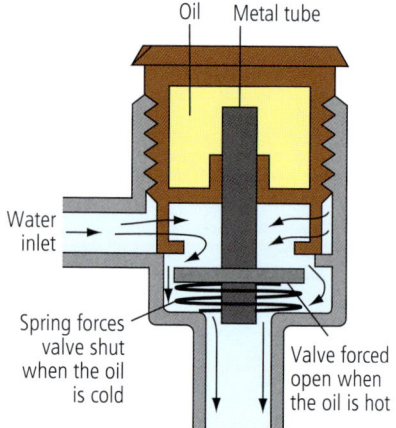

Figure 6.1.2 A radiator thermostat

Oil | Metal tube

Water inlet

Spring forces valve shut when the oil is cold

Valve forced open when the oil is hot

Comparing the thermal expansion of liquids and gases

Most substances expand when heated. If air did not expand when heated, a hot-air balloon would never fill up and take off. A burner under the balloon causes the balloon to fill with hot air which then lifts the balloon. The expansion of a substance due to increasing its temperature is called **thermal expansion.**

A simple example of the thermal expansion of air in a test tube is shown in Figure 6.1.1. If an identical test tube is filled completely with water, as shown in Figure 6.1.1, we can compare the thermal expansion of air with that of water. The air in the air-filled test tube is trapped by a 'thread' of water in the narrow tube. In each case, expansion of the air or water in the test tube causes the level of the water in the narrow tube to move towards the open end of the narrow tube.

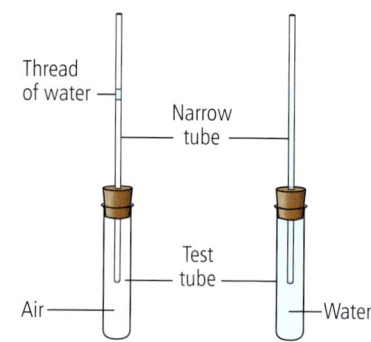

Figure 6.1.1 Comparing the thermal expansion of air and water

Thread of water

Narrow tube

Test tube

Air | Water

To ensure both test tubes are heated exactly the same, both test tubes could be placed next to each other in strong sunlight (or in the same beaker of warm water). The results show that air expands much more than water for the same temperature rise. The same result applies to a comparison between any gas and any liquid.

A radiator thermostat in a car makes use of the expansion of oil when it becomes warm. Figure 6.1.2 shows how it works. The valve stays closed until it has been warmed up by thermal energy from the engine. The metal tube is forced out of the oil chamber as the oil becomes hotter and expands. The movement of the metal tube opens the valve, allowing the hot water to flow to the radiator.

Thermal expansion of solids

A jar with a tight-fitting lid is sometimes difficult to open. Warming the lid in warm water may help to open the jar. The increase of temperature of the lid causes it to expand just enough to help you unscrew the lid.

Expansion gaps are necessary in buildings, bridges and railway tracks to allow for thermal expansion. Outdoor temperatures can change by as much as 50 °C between summer and winter.

A 100 m concrete bridge span would expand by about 5 cm if its temperature increased by 50 °C. Without expansion gaps, the bridge would buckle. The gaps are usually filled with soft material such as rubber to prevent rubble falling in.

Steel tyres are fitted on train wheels by heating the tyre so it expands, then fitting it on the wheel. As the tyre cools, it contracts so it fits very tightly on the wheel.

Bimetallic strips are used in thermostats in devices fitted with thermal cut-out switches or valves. A bimetallic switch consists of a strip of two different metals such as brass and steel stuck together. When the temperature of the strip rises, one metal expands more than the other (e.g. brass expands more than steel) so the strip bends. This can be used to switch on or off an electrical device. Figure 6.1.3 shows a bimetallic strip in a fire alarm. When heated, the strip bends towards the contact screw and when it touches it, the circuit is completed and the bell rings.

In a heater thermostat, when the strip becomes hot, it bends away from the contact screw and loses contact with it. As a result, the heater is switched off.

S In general, the volume of a solid increases by no more than about 0.01% for a temperature increase of 1°C. Most liquids expand slightly more.

Gases at constant pressure expand about 30 times more. This is because there are attractive forces between the molecules in any solid or liquid whereas intermolecular forces are negligible in a gas.

Expansion gaps

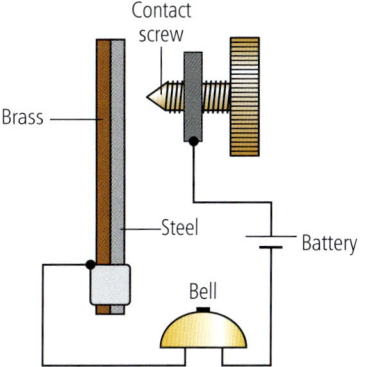
Figure 6.1.3 A bimetallic strip in an alarm circuit

SUMMARY QUESTIONS

1 Figure 6.1.4 shows a thermostat in a gas oven. When the oven overheats, the brass tube expands more than the invar rod. Explain why this change reduces the flow of gas through the valve.

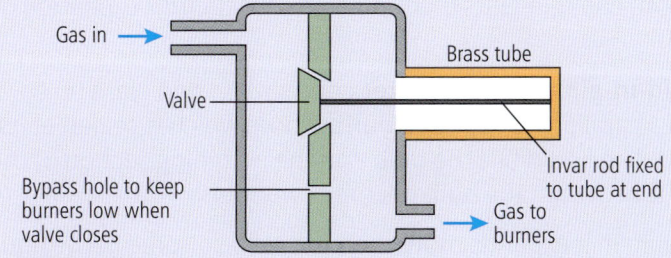
Figure 6.1.4

2 **a** Explain why expansion gaps are **i** necessary between concrete sections in buildings, **ii** usually filled with soft material such as rubber.

b i Explain why a bimetallic strip bends when it is heated.

ii A bimetallic thermostat is used to switch off the electricity supply to a heater if the heater overheats. With the aid of a diagram, describe such a thermostat and explain how it works.

KEY POINTS

● Applications of thermal expansion include liquid-in-glass thermometers and thermostats.

● Thermal expansion must be allowed for in buildings and bridges.

● Gases expand much more than solids and liquids. **S**

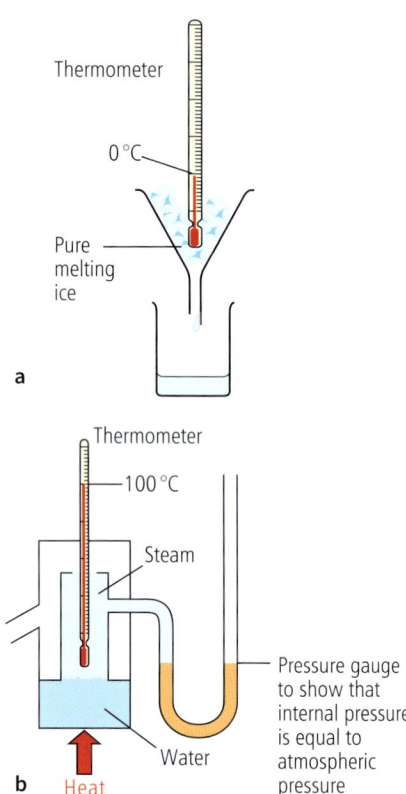

Figure 6.2.1 Calibrating a liquid-in-glass thermometer: (a) ice point, (b) steam point

Temperature

Temperature is a measure of hotness. When a weather forecast tells you that tomorrow's outdoor temperature could be 10 °C lower than today, you can expect a much cooler day. **Fixed points** are used to define a scale of temperature. These points are 'degrees of hotness' that can be reproduced precisely. They are usually **melting** points or **boiling** points of pure substances.

The Celsius scale of temperature, denoted by °C, is defined by two fixed points which are:

- Ice point at 0 °C, the temperature at which pure ice melts, and
- Steam point at 100 °C, the temperature at which pure water boils at standard atmospheric pressure.

Figure 6.2.1 shows how a liquid-in-glass thermometer is calibrated (i.e. marked) at ice point and at steam point.

The length between the two calibrations can then be marked with 100 equal intervals, each corresponding to a temperature change of 1 degree Celsius. The temperature according to the thermometer is then read from the position of the liquid meniscus in the narrow section of the thermometer.

Types of thermometer

Every thermometer makes use of a physical property that varies with temperature. This property is referred to as the **thermometric property** of the thermometer. For example, the thermometric property of a liquid-in-glass thermometer is the thermal expansion of the liquid.

The liquid-in-glass thermometer consists of a thin glass bulb joined to a capillary tube with a narrow bore which is sealed at its other end. The liquid fills the bulb and the adjoining section of the capillary tube. When the bulb becomes warmer:

- the liquid in it expands more than the bulb so some of the liquid in the bulb is forced into the capillary tube
- the thread of liquid in the capillary tube increases in length
- the thinner the bulb wall is, the faster the response of the thermometer will be when the temperature changes.

The liquid used usually contains mercury or coloured alcohol. Alcohol has a lower freezing point than mercury so it is more suitable for low-temperature measurements.

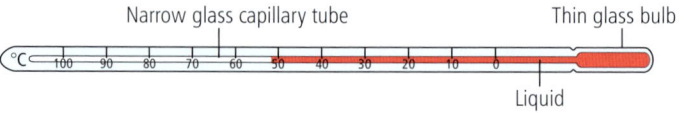

Figure 6.2.2 A liquid-in-glass thermometer

Absolute zero

In Topic 5.6, you learned that when a gas is heated in a sealed container, the gas pressure increases because its temperature increases. Suppose the gas had been cooled down instead by putting ice in the water bath instead of heating it. In this situation, the gas pressure would have decreased. Instead of using a water bath, suppose the flask containing the gas had been put in a very cold refrigerator – the gas pressure would have decreased even more. The graph in Figure 6.2.3 shows how the pressure of a gas decreases as it is cooled more and more. Notice where the line in the graph is drawn back (extrapolated) to where it cuts the temperature axis at –273°C. At this temperature, the pressure of the gas would be zero if it could be cooled more and more. This temperature is called **the absolute zero of temperature** because it is the same whichever gas is tested in the flask.

When the temperature of any substance is reduced, the kinetic energy of its molecules decreases. At absolute zero, the internal energy of the substance would be at its lowest because the molecules would have no kinetic energy.

The Kelvin scale of temperature, denoted by **K**, is obtained by adding 273 on the Celsius temperature. On the kelvin scale, the absolute zero is therefore zero kelvins (written 0 K). This reflects the fact that the molecules have no kinetic energy at the absolute zero of temperature.

Temperature T in kelvins = temperature θ in °C + 273

For example,

Absolute zero = 0 K = –273°C

Ice point = 273 K = 0°C

Steam point = 373 K = 100 °C

Notes

1   The Kelvin scale of temperatures is named after the UK scientist Lord Kelvin who made many discoveries in physics in the nineteenth century.

2   The scientific symbol for temperature in kelvins (K) is T and the symbol θ (pronounced 'theta') is used for the temperature in °C.

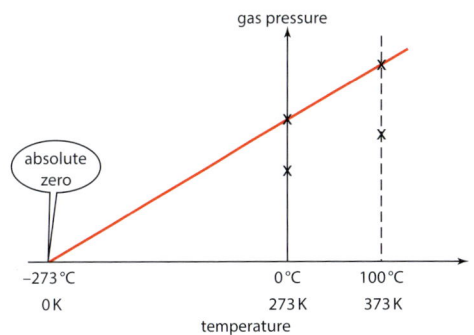

Figure 6.2.3 Towards absolute zero

CLINICAL THERMOMETERS

Body temperature for a healthy person is about 37 °C. Clinical thermometers are designed to measure body temperature so they are graduated in intervals of 0.1 °C from about 32 °C to 42 °C. A liquid-in-glass clinical thermometer has a constriction (i.e. a very narrow section) near the bulb to prevent the liquid returning to the bulb after the thermometer is removed from the patient.

KEY POINTS

- The temperature of an object is a measure of its degree of hotness.

- The fixed points of the Celsius scale are ice point (0 °C) and steam point (100 °C).

- Temperature T in K = temperature θ in °C + 273

SUMMARY QUESTIONS

1   Copy and complete the sentences below using words from the following list.

expansion   temperature   pressure

  a   The degree of hotness of an object is a measure of its _____.

  b   The liquid-in-glass thermometer makes use of the _____ of the liquid when its _____ changes.

2   a   Explain what is meant by the absolute zero of temperature

  b   Calculate the kelvin temperature for
i 20°C, ii –77°C

  c   Calculate the celsius temperature for
i 300 K, ii 10 K

6.3 Specific heat capacity

LEARNING OUTCOMES

- Explain what is meant by specific heat capacity
- **S** Recall and use $E = mc\,\Delta T$ in specific heat calculations
- Describe an experiment to measure the specific heat capacity of a body

EXAM TIP

Increasing an object's temperature increases its internal energy. See page 50 for internal energy.

A piece of metal in strong sunlight can become very hot. A stone of equal mass in the sunlight for the same time would not become as hot even though it gains the same amount of internal energy. The stone needs more energy to give it the same temperature rise as the piece of metal. When any substance is heated or cooled, its temperature change depends on the nature of the substance as well as its mass and how much energy is transferred to or from it.

The energy transferred

- to an object when its temperature rises increases its store of internal energy
- from an object when its temperature falls decreases its store of internal energy.

PRACTICAL

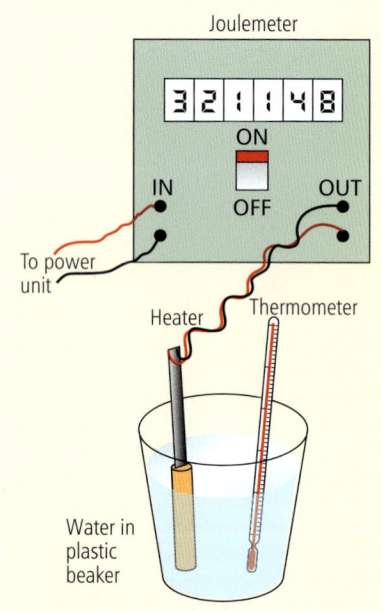

Joulemeter

$3\ 2\ 1\ 1\ 4\ 8$

ON

IN   OUT

OFF

To power unit

Heater   Thermometer

Water in plastic beaker

Figure 6.3.1 Heating water

mass of water (in kg)	0.10	0.20
energy supplied (in J)	6000	6000
temperature rise (in °C)	12	6

Investigating heating

1 Use a low voltage electric heater connected to a power supply and a joulemeter to heat some water in a plastic cup, as shown in Figure 6.3.1. Measure the temperature rise of the water using a thermometer and measure the energy supplied using the joulemeter.

2 Repeat the test using twice as much water. Typical results are shown in the table below. They show that the larger the mass of water, the smaller the temperature rise.

Supplement

Specific heat capacity

When a substance is heated, its temperature rise depends on:

1 **the amount of energy supplied**

2 **the mass of the substance**

3 **the nature of the substance.**

The specific heat capacity, c, of a substance is defined as the energy needed to raise the temperature of 1 kg of the substance by 1 °C.

The unit of specific heat capacity is the joule per kilogram per °C or J/(kg °C) in symbols.

If mass m of a substance is supplied with energy E, its specific heat capacity, $c = \dfrac{E}{m(\Delta T)}$ where ΔT is the temperature increase.

Rearranging this equation gives: energy supplied $\boldsymbol{E = mc(\Delta T)}$.

Measuring the specific heat capacity of a metal

1. A block of the metal of known mass m in an insulated container is used. A 12 V electrical heater is used to heat the metal by supplying measured amount of electrical energy as shown in Figure 6.3.2. The initial temperature T_1 of the block is measured using the thermometer.

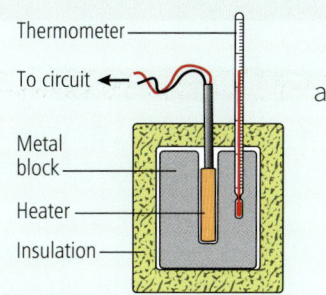

2. A joulemeter may be used as shown in Figure 6.3.1 to measure the energy supplied to the circuit. The joulemeter reading is set to zero and the power supply to the heater is then switched on for a certain time. After this time, the power supply is switched off and the joulemeter reading and the final temperature T_2 of the block are measured.

Figure 6.3.2 Measuring the specific heat capacity of a metal block

3. The specific heat capacity of the metal can then be calculated using the equation:

specific heat capacity: $c = \dfrac{E}{m(\Delta T)}$, where $\Delta T = T_2 - T_1$

Note An ammeter, a voltmeter and a stopwatch may be used instead of the joulemeter to measure the electrical energy supplied to the block. The circuit for this is shown in Figure 6.3.3. The ammeter is used to measure the heater current and the voltmeter is used to measure the heater voltage.

As explained in Topic 12.5, the electrical energy supplied $E =$ heater current $I \times$ heater p.d. $V \times$ heating time t

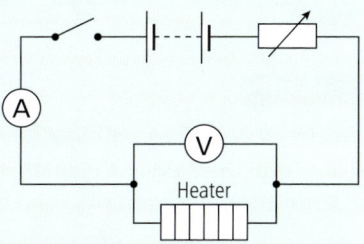

Figure 6.3.3 Circuit diagram

SUMMARY QUESTIONS

1. A bottle of water and a large bucket of water are left in strong sunlight.

   a. Which one warms up faster? Give a reason for your answer.

   b. The water in both containers heats up to the same temperature. Which container would cool faster if they are then placed in a cold room at the same time? Explain your answer.

2. Use the information below to answer this question.

substance	water	aluminium	lead	concrete
specific heat capacity / J/(kg °C)	4200	900	130	850

   a. Explain why a mass of lead heats up quicker than an equal mass of aluminium.

   b. Calculate the energy needed:

      i. to raise the temperature of 0.20 kg of aluminium from 15 °C to 40 °C

      ii. to raise the temperature of 0.40 kg of water from 15 °C to 40 °C

      iii. to raise the temperature of 0.40 kg of water in a 0.20 kg aluminium pan from 15 °C to 40 °C.

   c. A storage heater consists of a 20 kg concrete storage block in an aluminium case of mass 4.0 kg. Calculate the energy needed to heat a storage heater from 10 °C to 45 °C.

KEY POINTS

- When an object is heated, its temperature rise depends on how much energy is supplied to it, its mass and its specific heat capacity.

- The specific heat capacity, c, of a substance is defined as the energy needed to raise the temperature of 1 kg of the substance by 1 °C.

- To raise the temperature of mass m of a substance by ΔT, the energy needed, $E = mc\,\Delta T$.

6.4 Change of state

LEARNING OUTCOMES

- Describe what is meant by melting point and boiling point
- Recognise that energy is needed to melt a solid or boil a liquid
- **S** Explain the difference between boiling and evaporation

EXAM TIP

Students are often confused about the difference between evaporation and boiling.

Melting points and boiling points

When pure ice is heated and it melts, its temperature remains at 0 °C until all the ice has melted. When water is heated and it boils at atmospheric pressure, its temperature remains at 100 °C.

For any pure substance undergoing a change of state, its temperature stays the same. As shown in the table below, we refer to this temperature as the melting point or the boiling point of the substance, according to the type of change. The temperature at which a liquid changes to a solid is referred to as its freezing point and is the same temperature as the melting point of the solid.

change	initial and final state	temperature
melting	solid to liquid	melting point
freezing (also called solidification)	liquid to solid	
boiling	liquid to vapour	boiling point
condensation	vapour to liquid	

PRACTICAL

Measurement of the melting point of a substance

1 The substance in its solid state is placed in a test tube in a beaker of water, as shown in Figure 6.4.1a.

2 The water is heated and the temperature of the substance is measured when it melts.

3 If its temperature is measured every minute, the measurements can be plotted as shown in Figure 6.4.1b. The melting point is the temperature of the flat section of the graph, because this is when the temperature remains constant as the substance melts.

4 The same arrangement without the beaker of water may be used to find the boiling point of a liquid.

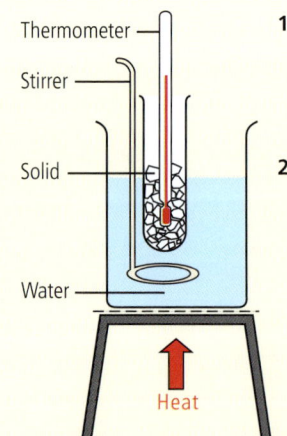

1 First, find the m.p. of the solid. Heat the water in the beaker. Stir. Watch the thermometer. When the solid melts, note the temperature. Stop heating.

2 Now find the f.p. of the liquid. Remove the beaker of water. Let the liquid in the tube cool. Watch the thermometer. When the liquid begins to freeze, the temperature stops falling. It stays the same until all the liquid has solidified.

Figure 6.4.1a Measuring the melting point of a substance

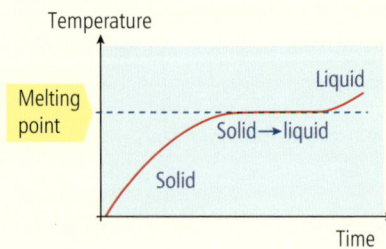

Figure 6.4.1b A graph of temperature v. time

Energy and change of state

Suppose a beaker of ice below 0 °C is heated steadily so that the ice melts and then the water formed from the melted ice boils. Figure 6.4.2 shows how the temperature changes with time.

The temperature:

1 increases until it reaches 0 °C when the ice starts to melt at 0 °C, then

2 remains constant at 0 °C until all the ice has melted, then

3 increases from 0 to 100 °C until the water in the beaker starts to boil at 100 °C, then

4 remains constant at 100 °C as the water turns to steam.

The energy supplied to a substance when it melts or boils enables the particles to overcome the force bonds between the particles.

The energy removed from a substance when it solidifies or condenses causes the particles to form force bonds between the particles.

Note The temperature of a pure substance does not change until all the substance has changed its state.

Most pure substances produce a graph with similar features to Figure 6.4.2. Note that:

1 **fusion** is often used to describe melting as different solids can be fused together when they melt

2 **evaporation** from a liquid occurs at its surface when the liquid is heated below its boiling point. At its boiling point, a liquid boils because bubbles of vapour form inside the liquid and rise to the surface to release the gas.

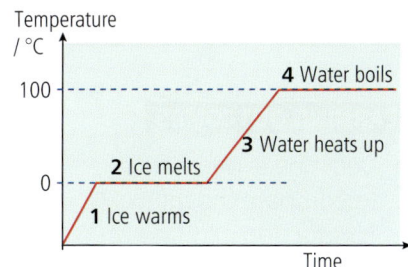

Figure 6.4.2 Melting and boiling of water

EXAM TIP

Boiling occurs only at the boiling point. Evaporation takes place at any temperature below this.

SUMMARY QUESTIONS

1 Copy and complete the sentences **a** to **d** below using words from the following list.

   decreases   increases   stays the same

   a When a liquid boils, its temperature _____.

   b When a solid is cooled below its melting point, its temperature _____.

   c When a liquid condenses, its temperature _____.

   d When a solid is heated at its melting point, its temperature _____.

2 A pure solid substance X was heated in a tube and its temperature was measured every 30 s. The measurements are given in the table below.

time / s	0	30	60	90	120	150	180	210	240	270	300
temp / °C	20	35	49	61	71	79	79	79	79	86	92

   a i Use the measurements in the table to plot a graph of temperature on the *y*-axis against time on the *x*-axis.

   ii Use your graph to find the melting point of X.

   b Describe the physical state of the substance as it was heated from 60 °C to 90 °C.

KEY POINTS

- For a pure substance:
  - its melting point is the temperature at which it melts or solidifies
  - its boiling point is the temperature at which it boils or condenses.
- Energy is needed to melt a solid or boil a liquid.
- Boiling occurs throughout a liquid at its boiling point. Evaporation occurs from the surface of a liquid when its temperature is below the boiling point.

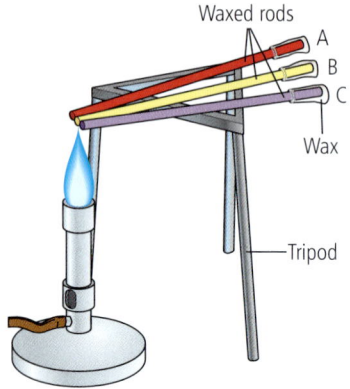

Figure 6.5.1 Comparing conductors

Waxed rods
A
B
C
Wax
Tripod

Insulating a loft

When you are cooking food, you need to know which materials are good **thermal conductors** and which are good insulators. If you can't remember, you are likely to burn your fingers.

Comparing rods of different materials as conductors

The rods need to be the same width and length for a fair comparison. Each rod is coated with a thin layer of wax near one end. The uncoated ends are then heated together, as shown in Figure 6.5.1. The wax melts fastest on the rod that conducts best.

Rods of different metals and non-metals (e.g. glass) could be used in the test. The results show that:

- metals conduct better than non-metals
- copper is a better conductor than steel.

PRACTICAL

Testing sheets of materials as insulators

1 Use different materials to insulate identical cans (or beakers) of hot water. Also, include a can without insulation. The volume of water and its temperature at the start should be the same.

2 Use a thermometer to measure the water temperature after the same time. The results should tell you which material is a good insulator and which insulator is best.

material	starting temperature	temperature after 300 s
no insulation	40 °C	22 °C
paper	40 °C	33 °C
felt	40 °C	38 °C

The results in the table show that:

- felt is a much better insulator than paper
- paper is better than no insulation.

Good insulators

Materials like wool and fibre glass are good insulators. This is because they contain air trapped between the fibres. Trapped air is a good insulator. We use insulators such as fibre glass to cut down energy transfer:

- from hot objects to colder surroundings (e.g. loft insulation in buildings, lagging hot water pipes)
- to cold objects from hotter surroundings (e.g. food in a freezer or cool box).

Supplement

Conductors in metals

Metals contain lots of **delocalised** (or 'free') **electrons**. These electrons have broken free from the atoms, leaving each atom as a positively charged ion. Free electrons move about at random inside the metal and bond the positive ions together. They collide with each other and with the positive ions.

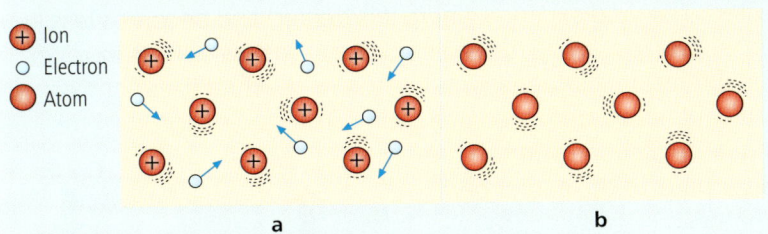

- ⊕ Ion
- ○ Electron
- 🔴 Atom

a   b

Figure 6.5.3 Energy transfer (a) in a metal, (b) in a non-metal

When a metal rod is heated at one end, the free electrons at the hot end gain kinetic energy and move faster.

- These electrons **diffuse** (i.e. spread out) and collide with other free electrons and the ions in the cooler parts of the metal.
- As a result, they transfer kinetic energy to these electrons and ions.

Energy is therefore transferred from the hot end of the rod to the colder end.

In a non-metallic solid, all the electrons are held in the atoms. There are no free electrons. Energy transfer takes place only because the atoms vibrate and shake each other. This is much less effective than energy transfer by free electrons. This is why metals are much better conductors than non-metals. Gases and most liquids are also poor conductors because they also have no free electrons.

PRACTICAL

How well does water conduct heat?

Heat a test tube of water near the top with a 'weighted' ice cube near the bottom. Even when the water at the top starts boiling, the ice cube does not melt. Water is a poor conductor of heat.

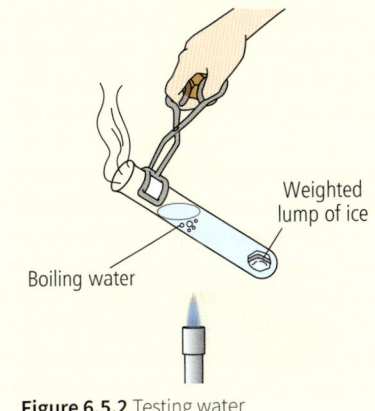

Weighted lump of ice

Boiling water

Figure 6.5.2 Testing water

EXAM TIP

You should be able to describe the process of thermal conduction in terms of conduction (or 'free') electrons.

SUMMARY QUESTIONS

1 Choose the best insulator or conductor from the list for a, b and c.

   fibre glass   plastic   steel   wood

   a _____ is used to insulate a house loft.

   b The handle of a frying pan is made of _____ or _____.

   c A radiator in a central heating system is made from _____.

2 a Why does a woolly hat keep your head warm in cold weather?

   b Choose a material you would use to line the inside of a cool box to be used to keep food cool in summer? Explain your choice of material.

   c Explain why an ice cube in water melts faster if the water is stirred.

KEY POINTS

- Materials such as fibre glass are good insulators because they contain pockets of trapped air.

- Conduction in a metal is due to free electrons transferring energy inside the metal. **S**

- Non-metals are poor conductors because they do not contain free electrons.

6.6 Convection

- Recall that convection happens in a gas or a liquid when it is heated
- Describe experiments that show convection
- Explain applications that use convection
- Relate convection to density changes

A natural glider

Make sure you are able to write a clear description of the process of convection.

A hot-air balloon carries its own burner

Observing convection

Glider pilots and birds know how to use **convection** to stay in the air. Convection currents of warm air can keep them high above the ground for hours.

Convection happens whenever a fluid (i.e. a gas or a liquid) is heated. The fluid expands where it is heated and becomes less dense. As a result, the heated fluid rises – just as objects in water float if they are less dense than water. As the fluid rises, it mixes with colder fluid and becomes cooler. The convection currents in a fluid transfer energy from the hotter parts to the cooler parts.

Figure 6.6.1 shows a simple demonstration of convection. The hot gases from the burning candle go straight up the chimney above the candle. The hot gases are replaced by cold air drawn down the other chimney to replace the air leaving the box.

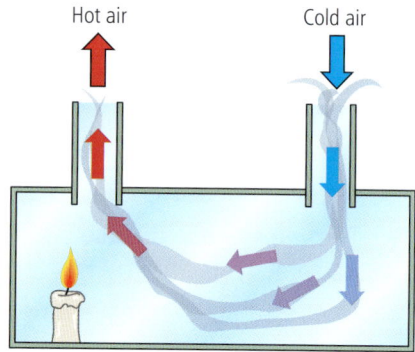

Figure 6.6.1 Convection

Using convection

● Hot water at home

Many homes have a hot water tank. Hot water from the boiler rises and flows into the tank where it rises to the top. Figure 6.6.2 shows the system. When you use a hot water tap at home, you draw hot water from the top of the tank.

● Hot air for fun

A hot air balloon carries its own burner to heat the air in the balloon. This makes air in the balloon less dense than the surrounding air. So the balloon rises and floats in the air. To keep the balloon floating, the air in the balloon needs to be heated every so often. If this isn't done, the air cools and the balloon descends.

Figure 6.6.2 Hot water at home

A **convector heater** has an electric heating element inside which heats the air in the heater. A thermostat in the heater prevents the room becoming too hot.

When the heater is on, the hot air in the heater rises and passes through a metal grille on the top of the heater.

- This hot air from the heater rises, causing the cold air in the room to sink.
- Some of the cold air is drawn in through the bottom of the heater where it is heated so the rising warm current of air continues.
- The air in the room continues to circulate in this way until all the air in the room is at the temperature set by the thermostat.

How convection works

Convection is an important method of energy transfer in liquids and gases. It takes place:

- only in liquids and gases (i.e. fluids)
- due to circulation (convection) currents within the substance.

The circulation currents are caused because fluids rise where they are heated (as heating makes them less dense) and fall where they cool (as cooling makes them more dense). Convection currents transfer thermal energy from the hotter parts to the cooler parts.

Why do liquids and gases rise when heated? Most liquids and gases expand when heated. This is because the particles move about more and take up more volume. Therefore the density decreases because the same mass of liquid or gas occupies a bigger volume. So heating a liquid or gas makes it less dense and it therefore rises.

Hot air

Figure 6.6.3 Convector heater

Sea breezes

Sea breezes keep you cool on a beach in hot weather. On a sunny day, the ground heats up faster than the sea. So the air above the ground warms up and rises. Cooler air from the sea flows in as a sea breeze to take the place of the warmer air.

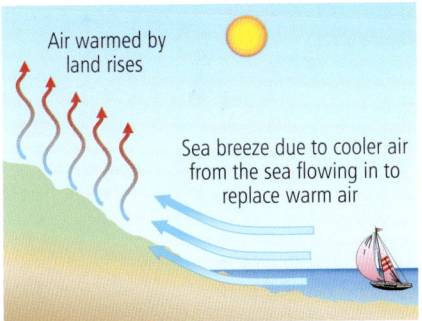

Air warmed by land rises

Sea breeze due to cooler air from the sea flowing in to replace warm air

Figure 6.6.4 Sea breezes

SUMMARY QUESTIONS

1 Use each word from the list once only to copy and complete the following sentences.

cools   falls   mixes   rises

When a fluid is heated, it _____ and _____ with the rest of the fluid. The fluid circulates and _____ then it _____.

2 Figure 6.6.3 shows a convector heater. It has an electric heating element inside and a metal grille on top.

  a What does the heater do to the air inside the heater?

  b Why is there a metal grille on top of the heater?

  c Where does air flow into the heater?

3 Some coloured crystals are placed on the bottom of a water-filled beaker. When the water is heated as shown in Figure 6.6.5, streams of coloured water are seen to rise from the crystals and travel across the surface and down the side of the beaker. Explain this observation.

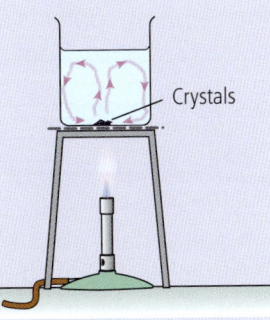

Crystals

Figure 6.6.5

KEY POINTS

- Convection takes place only in liquids and gases.
- Heating a liquid or a gas makes it less dense.
- Convection is due to a hot liquid or gas rising.

6.7 Infrared radiation

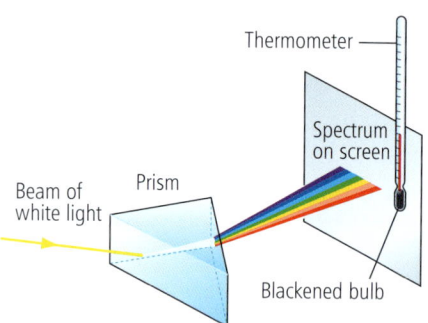

Figure 6.7.1 Detecting infrared radiation

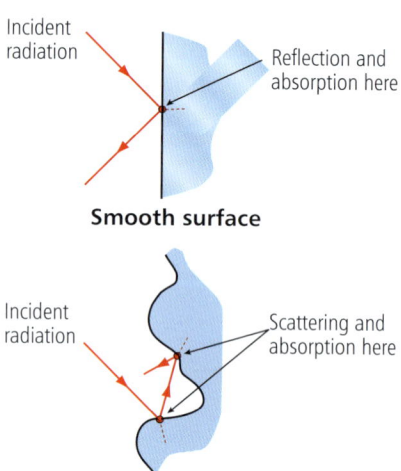

Figure 6.7.2 Absorbing infrared radiation

Seeing in the dark

We can use special TV cameras to 'see' animals and people in the dark. These cameras detect infrared radiation. Every object emits infrared radiation. The hotter an object is, the more infrared radiation it emits. If an object is very hot, it emits light as well. Light and infrared radiation are part of the spectrum of electromagnetic radiation. So too are radio waves, microwaves, ultraviolet rays and X-rays.

To detect infrared radiation, we can use a thermometer with a blackened bulb. Figure 6.7.1 shows how to do this.

- The glass prism splits a narrow beam of light from a lamp bulb into the colours of the spectrum.
- The thermometer reading rises when it is placed just beyond the red part of the spectrum. Some of the infrared radiation from the lamp bulb goes there. It is called infrared radiation because it is next to the red part of the visible spectrum.
- Our eyes cannot detect infrared radiation; the thermometer can.

The radiation emitted by an object due to its temperature is sometimes called **thermal radiation**. This is because it can include light as well as infrared radiation if the object is hot enough. Thermal radiation does not require a medium because all electromagnetic radiations can travel through vacuum.

Surfaces and radiation

Which surfaces are the best emitters of infrared radiation?

Rescue teams use thermal blankets to keep accident survivors warm. A thermal blanket has a shiny white outer surface. This emits much less radiation than a dull black surface. The colour and smoothness of a surface affects how much radiation it emits.

Dull black surfaces emit infrared radiation better than shiny white surfaces.

Which surfaces are the best absorbers of infrared radiation?

When a photocopier is used, the copies are warm. This is because infrared radiation from a lamp is used to dry the ink on the paper. Otherwise, the copies will be smudged. The black ink absorbs infrared radiation more easily than the white paper.

- A black surface absorbs infrared radiation better than a white surface.
- A dull surface absorbs infrared radiation better than a shiny surface because it has lots of cavities. Figure 6.7.2 shows why these cavities trap and absorb the infrared radiation.

Dull black surfaces absorb infrared radiation better than shiny white surfaces.

Supplement

Testing different surfaces

1 To compare the radiation from two different surfaces, measure how fast two beakers (or cans) of hot water cool. One beaker needs to be shiny and silvery and the other dull black. Figure 6.7.3 shows the idea. At the start, the volume and temperature of the water in each beaker need to be the same. The dull black beaker cools faster than the other one.

2 To compare absorption by different surfaces, use two beakers as shown with cold water in. Place the beakers in a sunlit room in the sunlight.

You should find that a beaker with a shiny, silvery surface warms up slower than a beaker painted dull black.

3 In these tests, the two cans must have equal surface areas and be at the same initial temperatures, as well as containing equal volumes of water. This is because the amount of radiation emitted from an object depends on the **area** and **temperature** of its surface, as well as on the surface colour and how shiny or dull it is.

Figure 6.7.3 Testing different surfaces

Keeping watch in the darkness

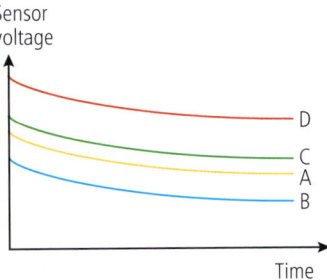

Figure 6.7.5 Graph for Summary Question 2

SUMMARY QUESTIONS

1 Explain why:

 a penguins huddle together to keep warm

 b houses in hot countries are usually painted white

 c solar heating panels are painted black.

2 A metal cube filled with hot water was used to compare the thermal energy radiated from its four vertical faces, A, B, C and D.

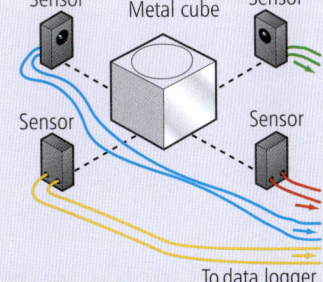

Figure 6.7.4

An infrared sensor was placed opposite each face at the same distance, as shown in Figure 6.7.4. The more intense the infrared radiation detected by a sensor, the greater its output voltage was. The sensors were connected to a computer. The results of the test are shown in the graph in Figure 6.7.5.

 a Which face radiated **i** most? **ii** least?

 b Why was it important for the distance from each sensor to the face to be the same?

 c One face was white and shiny, one was white and dull, one was black and shiny and one was dull black.

  Which face was **i** white and shiny? **ii** dull and black?

KEY POINTS

- Infrared radiation is part of the spectrum of electromagnetic radiation.

- All bodies emit infrared radiation.

- Dull black surfaces are better emitters and absorbers of infrared radiation than shiny white surfaces.

- The hotter a body is, the more infrared radiation it emits. The greater the surface temperature or the surface area, the more infrared radiation it emits.

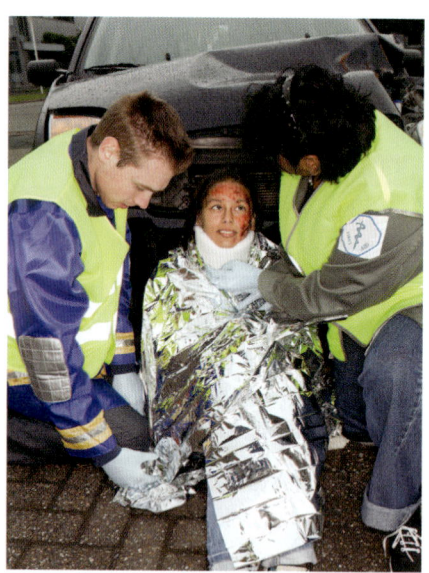

Figure 6.8.1 An emergency blanket in use

Absorption and emission of infrared radiation

Every object absorbs and emits infrared radiation, whatever its temperature is. If an object has a constant temperature, the object emits infrared radiation at the same rate as it absorbs it. When an object absorbs radiation faster than it emits radiation, its temperature increases.

Rescue teams use light-coloured, shiny blankets to keep accident survivors warm (Figure 6.8.1). A light, shiny outer surface emits a lot less radiation than a dark, matt (non-glossy) surface. This keeps the patient warm, as less infrared radiation is emitted than if an ordinary blanket had been used.

Radiation and the Earth's temperature

The temperature of the Earth depends on many factors, such as the rate that light and infrared radiation from the Sun are:

- reflected back into space or absorbed by the Earth's atmosphere or by the Earth's surface
- emitted from the Earth's surface and from the Earth's atmosphere into space.

If the Earth had no atmosphere, the temperature on the surface would plunge to about −180 °C at night, the same as the Moon's surface at night. This would happen because the surface would not be receiving any radiation from the Sun – it would be emitting radiation into space.

Some gases in the atmosphere, such as water vapour, methane, and carbon dioxide (greenhouse gases), absorb longer wavelength infrared radiation from the Earth and prevent it escaping into space. These gases absorb the radiation and then emit it back to the surface (Figure 6.8.2). This process makes the Earth warmer than it would be if these gases were not in its atmosphere.

More about global warming

The temperature in thousands of places around the world has been measured for many years. Scientists have used these measurements to calculate the average global temperature for each year since about 1880. Their calculations show that the average global temperature has increased by about 0.8°C since 1880 with most of the increase since about 1960. This increase is reckoned to be caused by increasing use of fossil fuels to meet the world's increasing demand for energy. The burning of fossil fuels creates more greenhouse gases in the atmosphere. Such gases in the atmosphere, particularly carbon

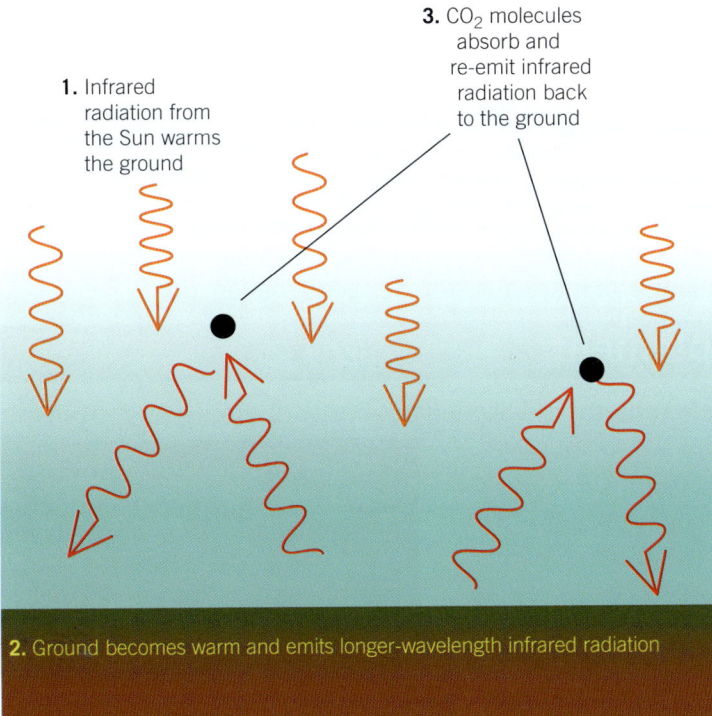

3. CO_2 molecules absorb and re-emit infrared radiation back to the ground

1. Infrared radiation from the Sun warms the ground

2. Ground becomes warm and emits longer-wavelength infrared radiation

Figure 6.8.2 The absorption and emission of infrared radiation

dioxide, make the atmosphere warmer by preventing infrared radiation from the surface from escaping into space. The graph in Figure 6.8.3 shows how the amount of carbon dioxide in the atmosphere has increased since about 1960.

The increase in global warming is causing the polar ice caps to shrink and sea levels to rise as well as changing weather patterns. See Topic 4.7 *Energy and the environment* for more about climate change.

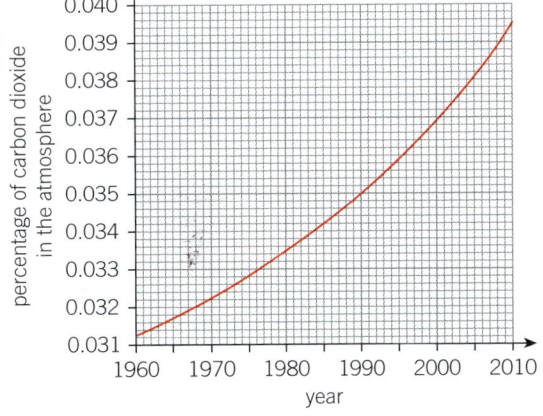

Figure 6.8.3 Carbon dioxide levels in the atmosphere

SUMMARY QUESTIONS

1 The surface of the Earth absorbs radiation from the Sun and it emits radiation.

 a Give one similarity and one difference in the radiation the Earth emits and the radiation it absorbs from the Sun.

 b Describe what happens to the radiation that the surface of the Earth emits on a very clear night.

2 On a hot day in summer, the interior of a parked car in sunlight with its windows closed becomes unbearably hot.

 a Explain why the temperature inside the car becomes much higher than the outside temperature.

 b Explain why the inside of the car would not become as hot if it had been parked in a shaded area.

3 Explain why the presence of greenhouse gases in the atmosphere makes the Earth warmer than it would be if there were no greenhouse gases in the atmosphere.

KEY POINTS

- The temperature of an object increases if it absorbs more radiation than it emits.

- The Earth's temperature depends on a lot of factors, including the absorption of infrared radiation from the Sun, and the emission of radiation from the Earth's surface and atmosphere.

6.9 Thermal energy at work

LEARNING OUTCOMES

- Recognise that we need to increase energy transfer to stop devices like engines from over-heating
- Recall the main methods of increasing thermal energy transfer
- Recognise that we need to reduce thermal energy transfer to keep objects warm
- Explain the main methods of reducing thermal energy transfer

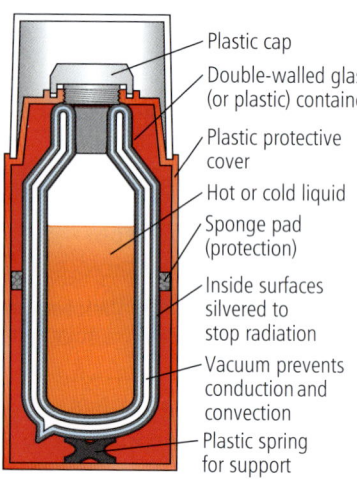

- Plastic cap
- Double-walled glass (or plastic) container
- Plastic protective cover
- Hot or cold liquid
- Sponge pad (protection)
- Inside surfaces silvered to stop radiation
- Vacuum prevents conduction and convection
- Plastic spring for support

Figure 6.9.1 A vacuum flask

Cooling by design

Lots of things can go wrong if thermal energy transfer isn't controlled. A car engine that overheats can go up in flames. An electronic component in an electrical device that overheats will stop working and may cause the device to malfunction.

Supplement

A vehicle radiator is designed to transfer thermal energy from the engine to the surroundings.

Most vehicles have a water-cooled engine. Water pumped through channels in the engine case is heated by the engine before flowing through the radiator. As the hot water flows through the radiator, it loses thermal energy as a result of thermal conduction through the metal case of the radiator. The radiator is flat, so it has a large surface area. This increases the amount of air heated by the radiator, so increasing the energy loss through convection in the air and through radiation. Most cars also have a cooling fan that switches on when the engine is too hot. This increases the flow of air over the radiator surface.

A car radiator

Keeping warm

A fire burning wood or coal can also keep you warm in cold weather. When the fuel burns, energy is transferred by radiation and convection from the store of chemical energy to the surroundings. Some of the energy released from the burning fuel heats and ignites the unburned fuel.

The vacuum flask

If you are outdoors in cold weather, a hot drink from a vacuum flask keeps you warm. The same flask can also be used in hot weather to keep a cold drink cold.

The liquid is in the double-walled glass or plastic container.

- The vacuum between the two walls of the container cuts out energy transfer by conduction and convection between the walls. However infrared radiation travels through a vacuum as it is electromagnetic radiation.
- Glass is a poor conductor so there is little conduction through the glass.
- The glass surfaces are silvery to reduce radiation from the outer wall.

Reducing thermal energy transfer at home

In cold weather, home heating bills can be expensive. Figure 6.9.2 shows how people in cold climates can reduce energy losses in cold weather in their homes and cut home heating bills.

Loft insulation such as fibre glass reduces energy loss through the roof. Air between the fibres also helps to reduce energy loss by conduction.

Cavity wall insulation reduces energy loss through the walls. We place insulation between the two layers of brick that make up the walls of a house.

Aluminium foil between a radiator panel and the wall reflects thermal radiation away from the wall.

A double-glazed window has two glass panes with dry air or a vacuum between the panes. Dry air is a good insulator so it cuts down energy transfer by conduction. A vacuum cuts out energy transfer by convection as well.

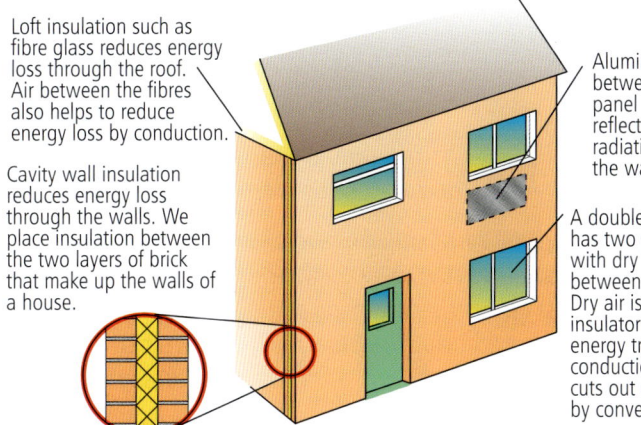

Figure 6.9.2

On a hot day, energy transfer into buildings from outside needs to be reduced. Air-conditioning units keep the interior of buildings cool but they are expensive to run. Measures to reduce energy transfer into a building would reduce the need for air-conditioning units. Such measures include:

- painting outside walls and roofs white so they absorb less energy from the Sun because they would reflect sunlight better than dark buildings
- fitting white blinds or window shutters to reflect sunlight and thus stop rooms becoming too hot
- building thick walls (when the building is constructed) to cut down energy transfer due to conduction through the walls.

KEY POINTS

- Thermal energy transfer to or from an object can be increased by increasing the surface area of the object or making its surface dull black.

- Thermal energy transfer from a building in cold weather can be reduced using cavity wall insulation, double glazing, loft insulation and aluminium foil behind radiators.

- Thermal energy transfer into a building in summer can be reduced by painting outdoor surfaces white, using white window shutters or blinds, and having thick walls.

SUMMARY QUESTIONS

1   Hot water is pumped through a central heating radiator like the one in Figure 6.9.3.

    Copy and complete the sentences below about the radiator.

    a   Energy transfer through the walls of the radiator is due to _____.

    b   Hot air in contact with the radiator causes energy transfer to the room by _____.

2   a   A double-glazed window has a plastic frame and a vacuum between the panes.

        i   Why is a plastic frame better than a metal frame?

        ii Why is a vacuum between the panes better than air?

    b   Give two reasons why a building with thick external walls heats up less in hot weather than an identical building with thinner walls.

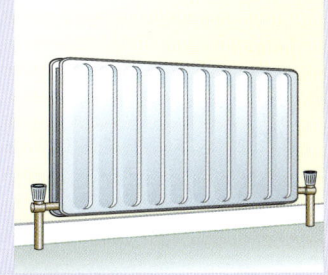

Figure 6.9.3

1 The photograph shows an expansion gap in a motorway.

(a) Describe what could happen if expansion gaps were not included in the design of the road.

(b) Explain why the expansion gap works.

2 (a) State the temperatures of the following:

   (i) boiling point of water

   (ii) freezing point of water.

(b) Suggest a typical temperature for each of the following:

   (i) room temperature

   (ii) air temperature on a very hot day

   (iii) normal body temperature.

3 The graph shows the results obtained by a student who was heating a solid until it melted.

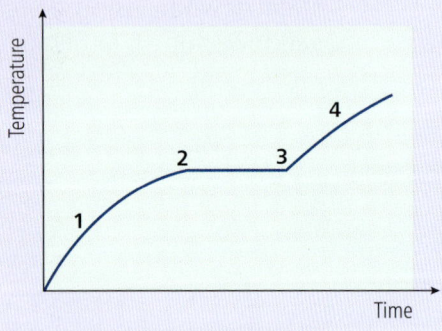

(a) Describe what is happening to the substance that is being heated:

   (i) at point 1 on the graph

   (ii) between 2 and 3 on the graph

   (iii) at point 4 on the graph.

(b) Copy the sketch graph and mark on the temperature axis the position of the melting point of the solid.

Practice Questions

1 Which of the following states of matter has the greatest expansion when heated by 1°C?

  A Solids

  B Liquids

  C Gases

  D All states expand by the same amount

2 Which of the following changes of state requires a decrease in the energy of the particles?

  A Melting   C Freezing

  B Boiling   D Evaporation

3 Which of these surfaces will be the worst emitter of infrared radiation?

  A Matt black surface

  B Shiny black surface

  C Matt white surface

  D Shiny white surface

4 Ice melts at 0°C. What is this temperature in kelvins?

  A −273 K   C 100 K

  B 0 K   D 273 K

5 In an experiment, a student uses two identical kettles to heat water from a starting temperature of 30°C to boiling point. Kettle A contains 0.5 kg of water and kettle B contains 0.75 kg of water.

(c) Explain in terms of the movement of molecules what is happening:

   (i) between 1 and 2 on the graph

   (ii) between 2 and 3 on the graph

   (iii) between 3 and 4 on the graph.

4 An object of mass 2.2 kg is supplied with 6 000 J of thermal energy and its temperature increases from 15°C to 18°C. Calculate the specific heat capacity of the object.

5 (a) Define the term specific heat capacity.

(b) A block of metal of mass 800 g is heated from 20°C to a temperature of 35°C. The amount of thermal energy supplied is 9600 J. Calculate the specific heat capacity of the metal.

(a) What is the temperature change required for the water to reach boiling point? [1]

(b) Which kettle will boil the water first? [1]

(c) Explain your answer to part (b). [3]

(d) To measure the starting temperature of the water the student uses a liquid-in-glass thermometer. Describe how this thermometer works. [2]

6 A student carries out an experiment to measure the melting point of a substance. The melting point is below 100°C. The results of the experiment are shown in the figure below.

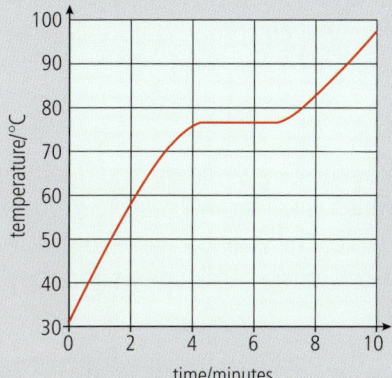

(a) Describe how the student could obtain the data to make this graph. You should include a sketch of the equipment needed. [4]

(b) What is the melting point of the substance tested? [1]

(c) Explain why the temperature of the substance does not change between 4 and 7 minutes. [3]

7 A large container of water is heated by placing an electrical heating element in the centre. The temperature of the water at the top of the beaker increases quickly while the temperature of the water at the bottom increases slowly.

(a) Name a process which causes the temperature of the water at the bottom of the water to increase. [1]

(b) Describe the process which causes the temperature of the water at the top of the container to increase. [4]

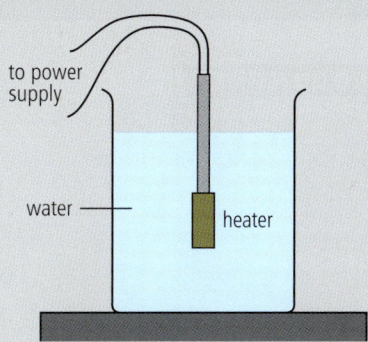

The heating element is switched off and the beaker of water gradually cools.

(c) Describe the two processes which allow the water to cool. [5]

S **8** A student attempts to measure the specific heat capacity of water by heating 0.20 kg of water electrically in an insulated cup and monitoring the temperature change.

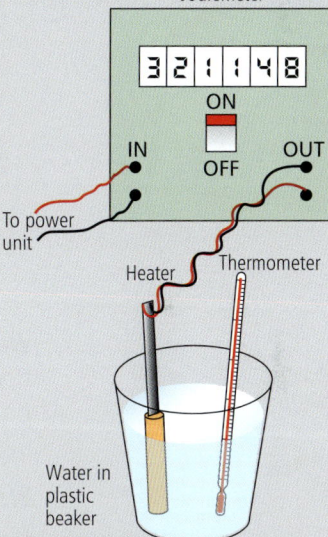

The student notes that the temperature of the water increases from 32°C to 42°C when provided with 8.6 kJ of energy.

(a) Use the data to calculate the specific heat capacity of the water. [3]

(b) The accepted value for the specific heat capacity of water is 4200 J/ kg °C. Suggest why the value the student found is higher than this accepted value. [2]

(c) Suggest an improvement which would allow the student to gain a more accurate value for their result. [1]

(d) A 50 W heater was used to heat the water. Estimate the time taken to heat the water from 32°C to 42°C. [2]

7.1 Wave motion

LEARNING OUTCOMES

- Describe and observe the wave motion of different types of waves
- Explain the meaning of speed, frequency, wavelength and amplitude
- Use the equation 'speed = frequency × wavelength'

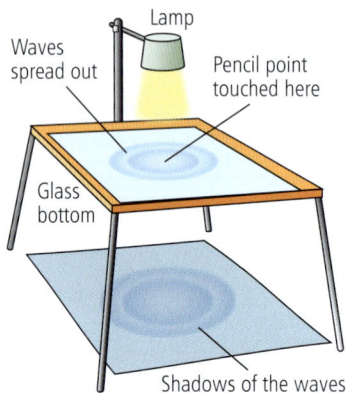

Figure 7.1.1 The ripple tank

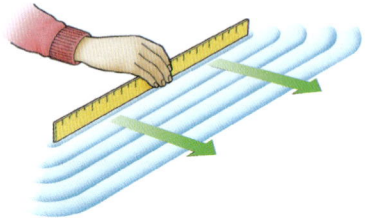

Figure 7.1.2 Making water waves

EXAM TIP

A common error is to think that the amplitude of a wave is the distance from the top of a crest to the bottom of the next trough.

To understand wave motion, we can study waves on water and waves on ropes and on springs. Then we can apply what we learn to any wave motion. All the examples below show waves as repeated disturbances that travel along a rope or spring or across water. The examples show that waves transfer energy without transferring matter.

Water waves can be produced in many ways. If you drop a pebble in a pond, the waves spread out in expanding circles. The expanding circles are called **wavefronts.** Anything floating on the water will bob up and down as successive wave fronts pass by. Figure 7.1.1 shows how we can study water waves in controlled conditions using a ripple tank. Touch the centre of the water surface and you will see the waves spread out in all directions.

PRACTICAL

Making straight waves

1 Make straight waves by moving a rule up and down on the water surface in a ripple tank (Figure 7.1.2). The wavefronts travel away from the rule. The straight wavefronts are called plane waves. They move at the same speed and they keep the same distance apart.

2 Observe the effect on the waves of moving the rule up and down faster then slower. More wavefronts are produced every minute and they become closer together. Use a stopwatch to find out if their speed has changed.

Waves on a rope can be produced by moving one end of the rope from side to side repeatedly. This action sends waves along it. Each part of the rope vibrates from side to side. The strings of musical instruments vibrate in this way to create sound waves in air.

Waves on a long stretched spring, such as a slinky coil, can be produced by fixing one end of the spring and moving the other end repeatedly to and fro in any direction. The motion of this part of the spring creates waves that travel along the spring. Each part of the spring vibrates and makes the next part vibrate, which makes the next part vibrate, and so on. So the vibrations travel along the spring.

Measuring waves

The speed of a wave, v, is the distance travelled by a wave crest (also called a wave peak) or a wave trough every second. For example, sound waves in air travel at a speed of 340 m/s. In 5 s, sound waves travel a distance of 1700 m (= 340 m/s × 5 s).

Figure 7.1.3 shows a snapshot of waves travelling along a rope.

Figure 7.1.3 Making waves on a rope

Each point on the rope vibrates up and down. The words we use to describe waves are shown in Figure 7.1.4 below.

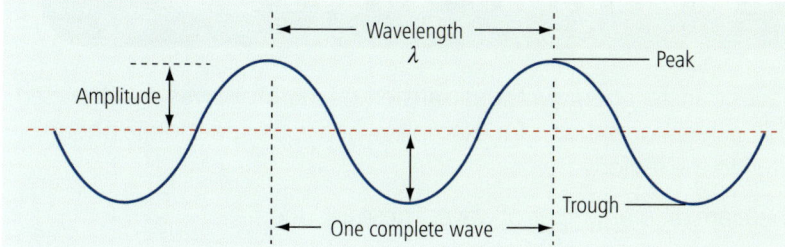

Figure 7.1.4 Features of a wave

- **One complete wave** is from one wave peak to the next.
- **The frequency, *f*,** of the waves is the number of complete waves (or wave peaks) passing a point in one second. The unit of frequency is the hertz (Hz) named after Heinrich Hertz, the scientist who discovered how to produce and detect radio waves.
- **The wavelength, λ** (Greek, pronounced 'lambda'), of the waves is the distance from one wave peak to the next.
- **The amplitude** of the waves is the height of the wave peak or the depth of the wave trough from the middle. The bigger the amplitude of the waves, the more energy the waves carry.

The speed of the waves depends on the frequency and the wavelength according to the following equation:

speed = frequency × wavelength

$$v = f\,\lambda$$

The proof of this equation is not required in your examination. However, awareness of the proof gives a deeper understanding of the link between frequency, speed and wavelength.

Figure 7.1.5 shows a surfer riding on the crest of some unusually fast waves. Suppose the frequency of the waves is 3 Hz and the wavelength of the waves is 4 m. At this frequency, three wave crests pass a fixed point once every second.

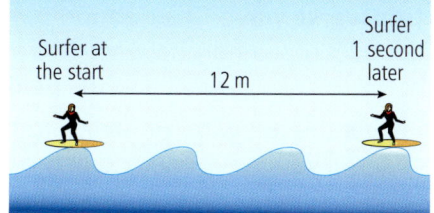

Figure 7.1.5 Surfing

The surfer on a wave crest therefore moves forward a distance of 3 wavelengths every second, and therefore in 1 s travels a distance of 12 m (= 3 wavelengths × 4 m for each wavelength). So the speed of the surfer is 12 m/s, which is equal to the frequency × the wavelength of the waves.

7.2 Transverse and longitudinal waves

LEARNING OUTCOMES

- Describe what waves can be used for
- Explain what transverse waves are
- Explain what longitudinal waves are
- Recognise which types of ave are transverse and which are longitudinal

Figure 7.2.1 Waves in water are examples of mechanical waves

MAKING MORE WAVES

If you pluck a guitar string, it oscillates because you send transverse waves along the string. The oscillating string sends sound waves into the surrounding air. The sound waves are longitudinal.

For more information on electromagnetic waves, see 8.7 'The electromagnetic spectrum'.

EXAM TIP

- Make sure that you understand the difference between transverse waves and longitudinal waves.
- Remember that electromagnetic waves are transverse and sound waves are longitudinal.

Waves transfer energy without transferring matter. Waves are also used to transfer information – for example, when you use a mobile phone or listen to the radio.

There are different types of wave. These include:

- sound waves, water waves, waves on springs and ropes, and seismic waves produced by earthquakes. These are examples of **mechanical waves**, which are vibrations that travel through a **medium** (a substance).
- light waves, radio waves, and microwaves. These are examples of **electromagnetic waves**, which can all travel through a vacuum at the same speed of 300 000 kilometres per second. No medium is needed.

PRACTICAL

Observing mechanical waves

Figure 7.2.2 shows how you can make waves on a rope by moving one end up and down.

Figure 7.2.2 Transverse waves

Tie a ribbon to the middle of the rope. Move one end of the rope up and down. You will see that the waves move along the rope but the ribbon doesn't move along the rope – it just moves up and down. This type of wave is known as a **transverse wave**. The ribbon vibrates or oscillates. This means it moves repeatedly between two positions. When the ribbon is at the top of a wave, it is at the **peak** (or crest) of the wave.

Repeat the test with the slinky. You should observe the same effects if you move one end of the slinky up and down.

However, if you push and pull the end of the slinky as shown in Figure 7.2.3, you will see a different type of wave, known as a **longitudinal wave**. Notice that there are areas of **compression** (coils squashed together) and areas of **rarefaction** (coils spread further apart) moving along the slinky.

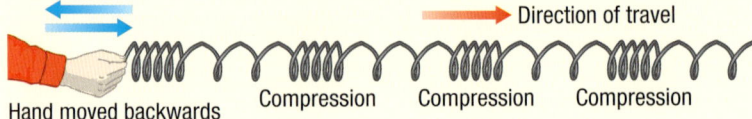

Hand moved backwards and forwards along the

Figure 7.2.3 Making longitudinal waves on a slinky

- How does the ribbon move when you send longitudinal waves along the slinky?

Safety: Handle the slinky spring carefully.

Transverse waves

The waves on a rope are called **transverse waves** because any point on the rope oscillates along a line **perpendicular** to the direction in which the waves transfer energy. All electromagnetic waves are transverse waves.

For a transverse wave the oscillations are perpendicular to the direction of energy transfer.

Longitudinal waves

The slinky spring in Figure 7.2.3 is useful to demonstrate how sound waves travel. When one end of the slinky is pushed in and out repeatedly, vibrations travel along the spring. These oscillations are parallel to the direction in which the waves transfer energy, along the spring. Waves that travel in this way are called **longitudinal waves**.

Sound waves are longitudinal waves. When an object vibrates in air, it makes the air around it vibrate as it pushes and pulls on the air. The oscillations (**compressions** and **rarefactions**) that travel through the air are sound waves. The oscillations are along the direction in which the wave travels.

For a longitudinal wave the oscillations are parallel to the direction of energy transfer.

Seismic waves are the 'shock' waves caused by earthquakes. The waves travel through the Earth from the 'focus' or location in the Earth where the earthquake originated. Seismic waves are recorded by an instrument called a seismometer. Figure 7.2.4 shows a typical seismometer recording of the tremors from an earthquake:

- **Primary (P) waves** arrive first. These are **longitudinal** waves that push and pull on the material they pass through.

- **Secondary (S) waves** arrive next. They are **transverse** waves that shake the material they pass through from side to side. They are stronger than P waves. The P waves arrive first because longitudinal seismic waves travel faster than transverse seismic waves.

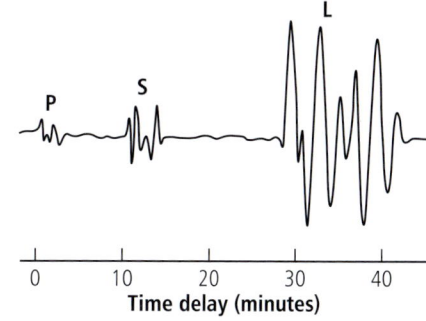

Figure 7.2.4 A seismogram
More waves called long (L) waves arrive after the S-waves. They travel round the Earth not through it and they shake the surface violently.

SUMMARY QUESTIONS

1 **a** What is the difference between a longitudinal wave and a transverse wave?
   b State one example of each type of wave.
   c When a sound wave passes through air, what happens to the air particles at a compression?

2 A long rope with a knot tied in the middle lies straight along a smooth floor. A student picks up one end of the rope. This sends waves along the rope.
   a Are the waves on the rope transverse or longitudinal waves?
   b What can you say about:
      i the direction of energy transfer along the rope?
      ii the movement of the knot?

7.3 Wave properties – reflection and refraction

- Describe the reflection and refraction of plane waves in a ripple tank
- Recall that refraction is due to a change in the speed of waves when they cross a boundary

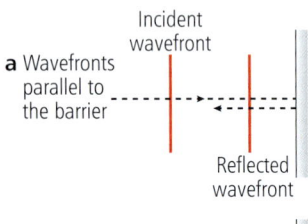

a Wavefronts parallel to the barrier

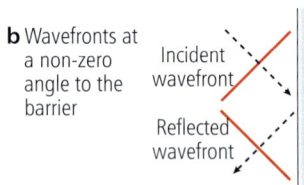

b Wavefronts at a non-zero angle to the barrier

Figure 7.3.1 Reflection of plane waves

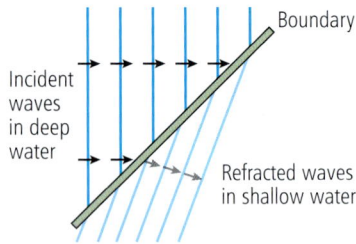

Figure 7.3.2 Observing refraction

Investigating waves using a ripple tank

Reflection of plane waves can be investigated using the ripple tank shown in Topic 7.1. Plane (i.e. straight) waves produced by dipping a rule in water are directed at a metal barrier in the water. These waves are referred to as the incident waves to distinguish them from the reflected waves. The incident waves are reflected by the barrier. Figures 7.3.1a and b each show a wavefront before and after hitting the barrier.

- In Figure 7.3.1a, the incident wavefront is parallel to the barrier as it approaches the barrier. It is still parallel to the barrier after reflection as it travels away from the barrier.
- In Figure 7.3.1b, the incident wavefront is not parallel to the barrier before or after reflection. The reflected wavefront moves away from the barrier at the same angle to the barrier as the incident wavefront.

PRACTICAL

A reflection test using a ripple tank

Use a rule to create and direct plane waves at a straight barrier, as shown in Figure 7.3.1. Find out if the reflected waves are always at the same angle to the barrier as the incident waves. You could align a second rule with the reflected waves and measure the angle of each rule to the barrier. Repeat the test for different angles.

Refraction of waves is the change of the direction in which they are travelling when they cross a boundary. This can be observed in a ripple tank when water waves cross a boundary between 'deep' and 'shallow' water. An area of shallow water can be created by placing a glass plate flat in the water. The water above the plate is shallower than the water outside the plate area. Plane waves are directed at a non-zero angle to a boundary. The wavefronts change direction as they cross the boundary, as shown in Figure 7.3.2.

PRACTICAL

Refraction tests

1 Use a vibrating beam to create plane waves continuously in a ripple tank containing a transparent plastic plate. Arrange the plate so the waves cross a boundary between the deep and shallow water.

2 **At a non-zero angle to a boundary**, as shown in Figure 7.3.2, the waves bend when they cross the boundary. The water over the plate needs to be very, very shallow. Find out if plane waves bend towards or away from the boundary when they cross from deep to shallow water.

3 Parallel to a boundary. The waves cross the boundary without bending or changing direction. However, their speed changes. Find out if the waves travel slower or faster when they cross the boundary. As we shall see later, this change of speed explains why the waves are refracted when they cross a boundary at a non-zero angle.

Describing reflection and refraction

To explain how a wavefront moves forward, imagine each tiny section creates a wavelet which travels forwards. The wavelets move forward together to recreate the wavefront that created them.

Refraction: when plane waves cross a boundary at a non-zero angle to the boundary where they slow down, each wavefront changes its direction.

In Figure 7.3.3, the wavefronts move more slowly after they have crossed the boundary. They are not as far from the boundary as they would have been if their speed had not changed. So the refracted wavefronts are at a smaller angle to the boundary than the incident wavefronts.

Reflection: when plane waves reflect from a flat barrier, the reflected waves are at the same angle to the barrier as the incident waves. When each point on the wavefront reaches the barrier, it creates a wavelet moving away from the barrier. This wavelet lines up with the previous 'reflected' wavelets to form a reflected wavefront moving away from the barrier. All parts of a wavefront move at the same speed. This means that the reflected wavefront is at the same angle to the barrier as the incident wavefront.

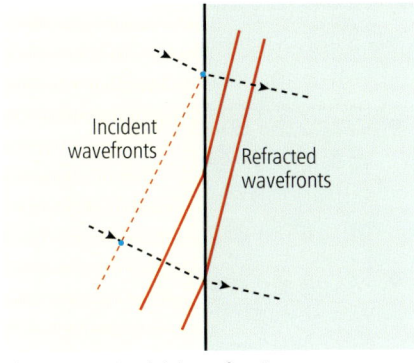

Figure 7.3.3 Explaining refraction

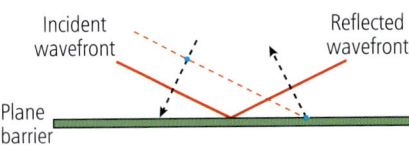

Figure 7.3.4 Explaining reflection

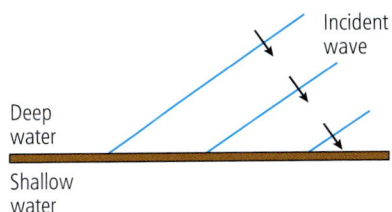

Figure 7.3.5 Diagram for Summary Question 2

SUMMARY QUESTIONS

1 Copy and complete the following sentences using words from the list below.

more than   less than   the same as

a When plane waves reflect from a straight barrier, the angle of each reflected wave to the barrier is _____ the angle of each incident wave to the barrier.

b When waves are refracted, the angle of each refracted wavefront is not _____ the angle of each incident wave front to the barrier.

c When waves speed up on crossing a boundary, the angle of each refracted wavefront is _____ the angle of each incident wave front to the barrier.

2 Copy Figure 7.3.5 which shows plane waves passing from deep to shallow water, where they move more slowly than in the deep water. Draw in some refracted wavefronts, indicating their direction.

3 Sea waves rolling up a sandy beach are not reflected. Why are the sides of a ripple tank sloped? What would happen if the sides were vertical rather than sloped?

KEY POINTS

- Plane waves reflect from a straight barrier at the same angle to the barrier as the incident waves.

- Refraction is the change of direction of waves when they cross a boundary due to a change in the speed of waves when they cross a boundary.

7.4 Wave properties – diffraction

LEARNING OUTCOMES

- Describe diffraction of plane waves in a ripple tank

S - Recall the effect on gap width and wavelength on the diffraction of waves passing through a gap

- Recall the effect of wavelength on the diffraction of waves passing the edge of an obstacle

EXAM TIP

The definition of diffraction is simple, but you must learn it.

A Hubble Space Telescope image of two colliding galaxies

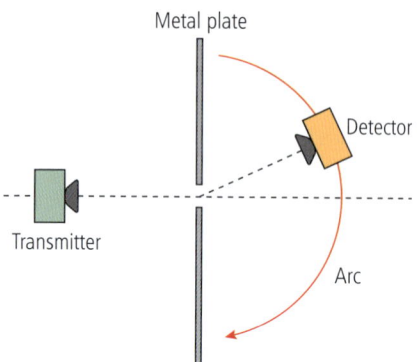

Figure 7.4.2 Diffraction of microwaves (top view)

Investigating diffraction using a ripple tank

Diffraction is the spreading of waves when they pass through a gap or move past an obstacle. The waves that pass the edges of the gap or of the obstacle spread out. Figure 7.4.1 shows waves in a ripple tank spreading out after they pass through a gap. Notice that:

- the narrower the gap, the more the waves spread out
- the wider the gap, the less the waves spread out.

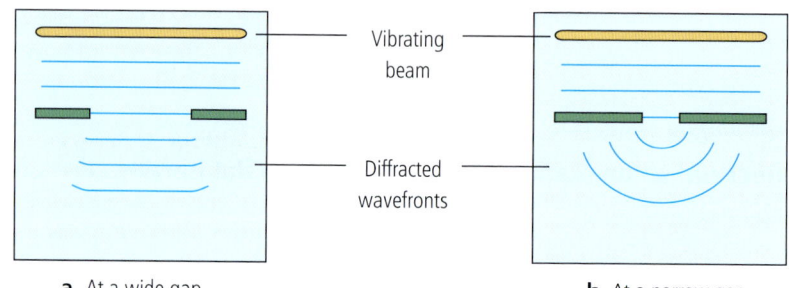

a At a wide gap   **b** At a narrow gap

Figure 7.4.1 Diffraction

Diffraction is important in any optical instrument. The Hubble Space Telescope in its orbit above the Earth has provided amazing images of objects far away in space. Its focusing mirror is 2.4 m in diameter. When it is used, astronomers can see separate images of objects which are far too close to be seen separately using a narrower telescope. Little diffraction occurs when light passes through the Hubble Space Telescope because it is so wide. So its images are very clear and very detailed.

PRACTICAL

Investigating diffraction

Use a ripple tank as in Figure 7.4.1 to direct plane waves continuously at a gap between two metal barriers.

1 Change the gap spacing and observe the effect on the diffraction of the waves that pass through the gap. You should find that the diffraction of the waves increases as the gap is made narrower, as shown in Figure 7.4.1.

2 Keep the gap spacing constant and change the wavelength of the waves by altering the frequency of the vibrating beam. Observe the effect on the diffraction of the waves. You should find that the smaller the wavelength of the waves, the more they are diffracted.

3 Remove one of the barriers and observe the waves diffract round the edge. Observe how a change of wavelength affects the diffracted waves.

Tests using microwaves

A microwave transmitter and a detector can be used to demonstrate diffraction of microwaves. The transmitter produces microwaves of wavelength 3.0 cm. The detector includes a meter which gives a non-zero reading when it detects microwaves.

1 The detector is placed in the path of the microwave beam from the transmitter. When a metal plate is placed between the transmitter and the detector, the meter reading drops to zero. This shows that microwaves cannot pass through metal.

2 Two metal plates with a narrow gap between them are placed in the path of the beam from the transmitter. When the detector is moved along an arc at a constant distance beyond the gap, as in Figure 7.4.2, it detects microwaves that have spread out from the gap. When the gap is made wider, the microwaves passing through the gap spread out less. The detector needs to be nearer the centre of the arc to detect the microwaves.

Supplement

Diffraction factors

When waves pass through a gap, they spread out. For diffraction at a gap to be noticeable, the gap must be similar in width to the wavelength of the waves. The waves spread out more if:

● the gap is made narrow, as shown in Figure 7.4.1

● the wavelength of the waves is increased, as shown in Figure 7.4.3.

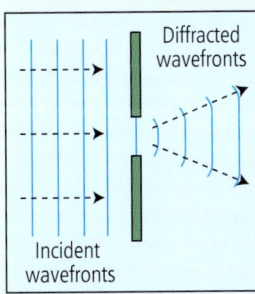

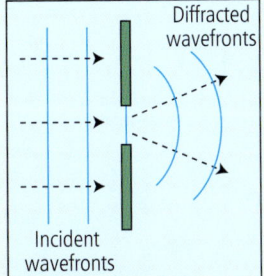

a Smaller wavelength than b with the same gap width

b Longer wavelength than a with the same gap width

Figure 7.4.3 The effect of wavelength on diffraction and wavefronts

When waves pass near an edge of an obstacle, they diffract and spread out behind the obstacle, as shown in Figure 7.4.4. The waves spread out more if the wavelength of the waves is increased.

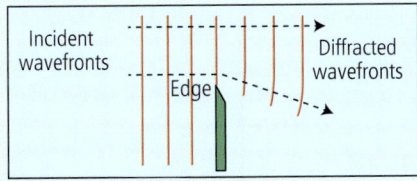

Figure 7.4.4 Diffraction at an edge

Notice that the wavelength does not change when diffraction occurs.

1 a A small portable radio in a room can be heard all along a corridor outside the room when the room door is open. Explain why it can be heard by someone in the corridor who is not near the door.

   b State what is meant by diffraction.

2 Copy and complete the following sentences about waves passing through a gap using words from the list below.

   more than   less than
   the same as

   a Diffracted waves spread out _____ they would if the gap was made wider.

   b When waves pass through a gap, their wavelength is _____ it was before it passed through the gap.

KEY POINTS

● Diffraction is the spreading of waves when they pass through a gap or around an obstacle. For noticeable diffraction the gap must be similar in size to the wavelength.

● The wavelength does not change on diffraction.

● The narrower the gap, the more the waves spread out.

● When waves pass through a gap, or an edge, the larger the wavelength, the more they diffract.

1 When a wave is (a) refracted or (b) diffracted, which of the following, if any, will change?

- the speed of the waves
- the frequency of the waves
- the wavelength of the waves.

2 During a thunderstorm a student sees a flash of lightning and hears the thunder 4 seconds later. If the speed of sound in air is 330 m/s, calculate how far the storm is from the student.

3 Draw diagrams to illustrate a transverse wave and a longitudinal wave.

On each diagram mark a distance of one wavelength.

4 Write down three differences between sound waves and radio waves.

5 A microwave oven has a glass door with a metal grid. The grid has a large number of small holes. Explain why, when the door is closed and the oven switched on, you can see into the oven but microwaves cannot get out.

6 A ball is floating in a pond. The ball is out of reach. A person tries to make the ball move to one side of the pond by hitting the water with a stick and making waves.

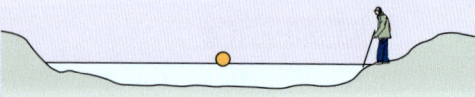

Explain whether or not this is an effective way to move the ball across the pond.

7 The person in question 6 hits the water at regular intervals 10 times in 20 s. The waves produced travel across the water at 0.5 m/s. Calculate the wavelength of the waves. Show your working.

Practice Questions

1 A transverse wave is shown in the diagram.

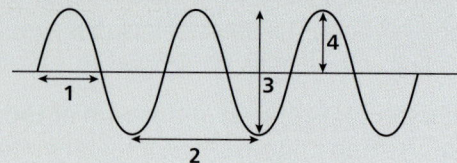

Which of the labels represents the wavelength of the wave?

A 1

B 2

C 3

D 4

2 Which of these is an example of a longitudinal wave?

A microwaves

B radio waves

C water waves

D sound waves

3 Which of the following statements about waves is correct?

A Waves transfer both energy and matter.

B Waves do not transfer energy or matter.

C Waves transfer matter without transferring energy.

D Waves transfer energy without transferring matter.

4 Ripples on the surface of a pond travel at a velocity of 0.30 m/s and have a frequency of 4.5 Hz. What is the wavelength of these ripples?

A 6.6 cm

B 6.6 m

C 1.4 m

D 15 m

5 Which of the following behaviours apply to longitudinal waves but not transverse waves?

A reflection

B refraction

C compression and rarefaction

D diffraction

6 The diagram below shows a longitudinal wave travelling in a spring. The waves are caused by a student pushing the spring in and out.

(a) Copy the diagram and label the following **(i)** a wavelength, **(ii)** a compression **(iii)** the direction of travel for the wave. [3]

(b) Describe how the student could increase the amplitude of the longitudinal wave. [1]

(c) Give another example of a longitudinal wave. [1]

(d) Describe the relationship between the vibrations and the direction the wave travels for a transverse wave. [1]

(e) Name two applications of transverse waves. [2]

(f) The wavelength of the wave is 0.20 m and the frequency is 1.5 Hz. Calculate the speed of the wave in the spring. [2]

7 The behaviour of waves can be represented by wavefront diagrams.

The figure below shows the wavefront of a water ripple approaching a solid barrier.

(a) Copy and complete the diagram showing the position of the wavefront after it has been reflected by the barrier. [2]

Water waves can change speed and/or direction when they pass a boundary between deeper and shallower water.

(b) What is the name of the process which causes a change in speed or direction? [1]

(c) In what circumstances will there be a change in speed but not direction? [1]

(d) Describe how you can demonstrate in a school laboratory the change in water waves as they pass a boundary. [4]

8 The diagram below shows a microwave detector moving through an arc near a gap between two metal plates. The gap size is similar to the wavelength of the wave.

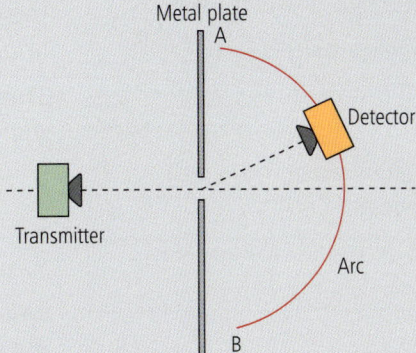

(a) What is the name of the process that causes the waves to spread as they pass through the gap? [1]

(b) Describe how the strength of the signal on the detector varies as it moves around the arc from point A to point B. [3]

(c) Describe the effect of increasing the gap size so that it is significantly larger than the wavelength of the microwaves. [2]

9 (a) The behaviour of waves is often demonstrated by teachers using microwaves with a wavelength of 3 cm. Microwaves have a wave speed of 3×10^8 m/s. Calculate the frequency of these microwaves. [4]

(b) A teacher demonstrated the reflection of microwaves using the apparatus below.

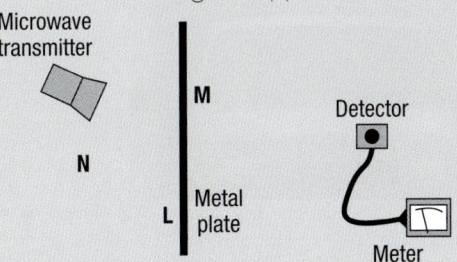

The meter read zero when the detector was at M but had a high reading when the detector was at N.

i Explain why the meter read zero when the detector was at M. [3]

ii Explain why the meter had a high reading when the detector was at N. [2]

iii What would the meter read if the detector was placed at L? Explain your answer. [3]

LEARNING OUTCOMES

- Describe the formation of an image by a plane mirror
- Describe the properties of the image formed by a plane mirror
- Recall and use the law of reflection
- **S** Explain the formation of the image by a plane mirror

Mirror images

If you have visited a 'Hall of Mirrors' at a funfair, you will know that the shape of a mirror affects what you see. To see an undistorted image of yourself, you need to look in a plane (i.e. flat) mirror.

The photograph shows the images of a toy in front of two plane mirrors. Whichever mirror you look at, you can see:

- the image in the mirror is the same size as the object
- the image is behind the mirror at the same distance from the mirror as the object.

Which way would the images move if the toy was moved away from the mirror?

Images in a plane mirror

MIRROR IMAGES

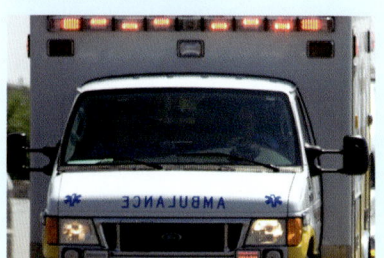

Ambulances and police cars often display a 'mirror image' sign at the front. This is so that the driver of a vehicle in front looking in the rear-view mirror can read the sign.

Investigating reflection of light by a plane mirror

To understand the nature of a mirror image, we need to understand how light behaves when it reflects from a mirror. Figure 8.1.1a shows how we can use a ray box and a mirror to investigate reflection. In the diagram, a narrow beam (or ray of light) reflects from the mirror. Alternatively, optical pins may be used as in Figure 8.1.1b.

- The straight line at right angles to the mirror is called the **normal**.
- The angle between the incident ray and the normal is called the **angle of incidence**, labelled i in the diagram.
- The angle between the reflected ray and the normal is called the **angle of reflection**, labelled r in the diagram.

If you make measurements of angle r for different measured angles i as described on the next page, you should find the two angles are always equal. Check this for yourself.

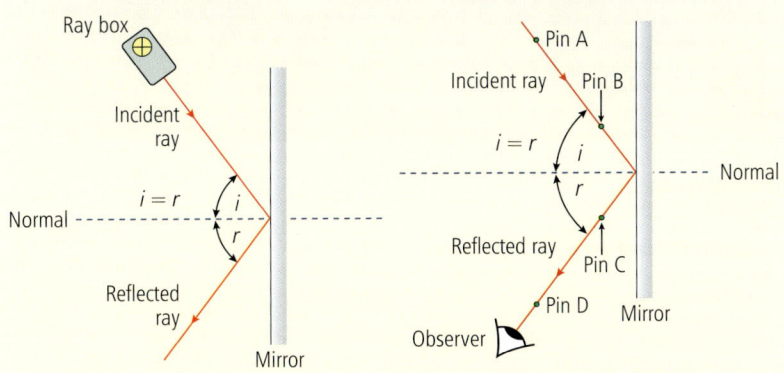

Figure 8.1.1a Using a ray box   Figure 8.1.1b Using optical pins

1 Mark a pencil line along the mirror and draw a line at right angles for the normal.

2 Mark the path of the incident ray at a measured angle of incidence to the normal.

3 Use the ray box to direct a ray (i.e. the incident ray) at the mirror or place optical pins A and B along the path of the incident ray. Measure and record the angle of incidence *i*.

4 Mark the path of the reflected ray either directly, if you are using a ray box, or observe the images of pins A and B such that A is directly behind B; then place two further pins C and D in line with the images of A and B.

5 Repeat the procedure for different angles of incidence.

The law of reflection

Measurements show that for any ray of light reflected by a mirror:

the angle of incidence, *i* = the angle of reflection, *r*

The light ray in Figure 8.1.1 gives the direction of the wavefronts in the light beam. To explain the law of reflection, remember that plane waves reflect from a straight barrier at the same angle as they move towards it. So the angle of the reflected waves and the incident waves to the reflecting barrier is the same. Therefore the reflected and incident rays are at the same angle to the normal.

Image formation by a plane mirror

Figure 8.1.2 shows how an image of a point object is formed by a plane mirror. The diagram shows the path of two rays of light from the object that both reflect off the mirror.

The angle of reflection of each ray is equal to its angle of incidence. The two reflected rays diverge as if they have come from the same point behind the mirror. This point is the image of the object. An observer looking along the two rays into the mirror sees an image of the object at this point.

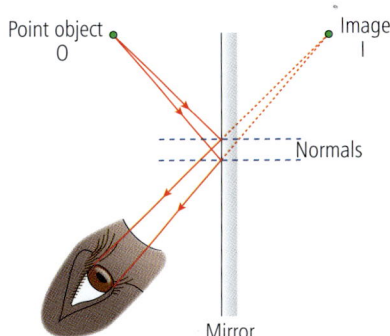

Figure 8.1.2 Image formation by a plane mirror

CONTINUED

LEARNING OUTCOMES

- Describe the formation of an image by a plane mirror
- Describe the properties of the image formed by a plane mirror
- Recall and use the law of reflection
- **S** Explain the formation of the image by a plane mirror

PRACTICAL

Constructing a ray diagram. **S**

The diagram in Figure 8.1.2 on the previous page is called a ray diagram. Construct your own ray diagram on a white sheet of paper to show the image is at the same distance behind the mirror as the object in front. Use a rule and a protractor. Make sure the angle of incidence is equal to the angle of reflection for each incident light ray.

Label the mirror, the normal, the object O and the image I on your diagram.

Measure the perpendicular distance from O and from I on the diagram to the mirror. The two distances should be equal.

Supplement

Real and virtual images

The image seen in a mirror is a **virtual image**. When you look at a mirror image, the rays that reflect off the mirror diverge and appear to come from the image behind the mirror. A virtual image cannot be projected onto a screen by a lens. An image projected onto a screen is described as a **real** image because it is formed by focusing rays onto a screen (see Topic 8.5). Large screens like giant TV screens and 'virtual reality' headsets are electronic and don't use lenses.

EXAM TIP

The difference between a real image and a virtual image is an important distinction to understand.

An image on a screen

SUMMARY QUESTIONS

1 A point object O is placed in front of a plane mirror, as shown in Figure 8.1.3.

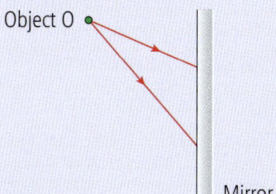

Figure 8.1.3

a Copy the diagram and complete the path of the two rays shown from O after they have reflected off the mirror.

b i Use the reflected rays to locate the image of O.

   ii Show that the image and the object are equidistant (i.e. the same distance) from the mirror.

2 a Two plane mirrors are placed perpendicular to each other.

   i Draw a ray diagram to show the path of a light ray at an angle of incidence of 60° that reflects off both mirrors.

   ii Measure the angle between the final reflected light ray and the initial incident ray.

3 A woman stands 0.5 m in front of a plane mirror on a wall. How far is her image **i** behind the mirror, **ii** away from her?

4 In **b**, the woman is unable to see below her knees in the mirror. Use Figure 8.1.4 to explain why she cannot see her feet in the mirror.

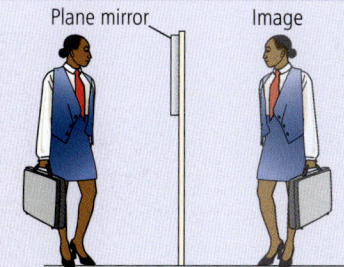

Figure 8.1.4

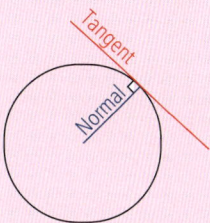

Refraction tests

When you have your eyes tested, the optician might test different lenses in front of each of your eyes. Each lens changes the direction of light passing through it. The change of direction is due to refraction which happens when light enters or leaves the lens. People with good eyesight don't need extra lenses. Each eye already has a perfectly good lens in it.

PRACTICAL

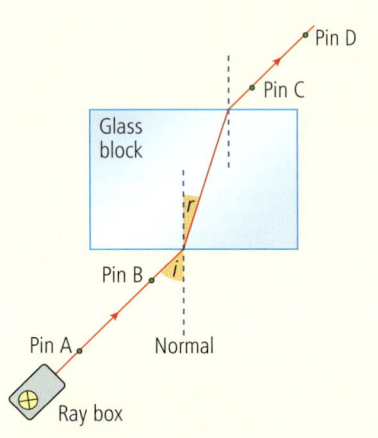

Figure 8.2.1 Refraction of light

Investigating refraction of light

1 Use a ray box and a rectangular glass block or optical pins as in Figure 8.2.1 to investigate refraction. If you use optical pins, pins C and D should be placed so they are in line with the images of pins A and B seen in line through the block.

2 Pins C and D should then be used to locate the point where the ray from A and B leaves the glass block. You should find that a light ray changes its direction at the boundary between air and glass unless it is along the normal.

Your investigation should show that a light ray directed along the normal passes straight through without being refracted. If the light ray is at a non-zero angle to the normal, it:

- bends towards the normal when it travels from air into glass. The angle of refraction, r, is smaller than the angle of incidence, i.

- bends away from the normal when it travels from glass into air. The angle of refraction, r, is greater than the angle of incidence, i.

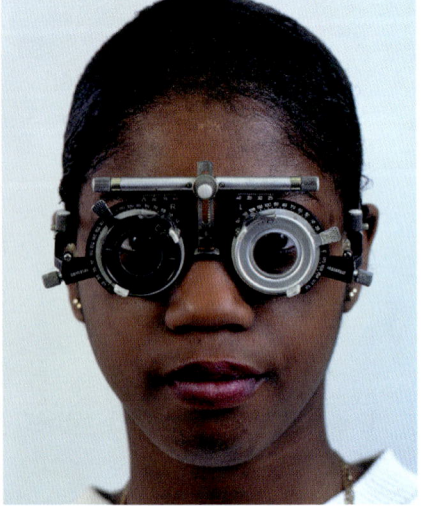

An eye test

Explaining refraction

Refraction is a property of all forms of waves including light and sound. Figure 7.3.2 in Topic 7.3 shows how we can demonstrate refraction using water waves in a ripple tank. The waves are slower in shallow water than in deep water. Because they change speed when they cross the boundary, they change direction:

- towards the normal when they cross from deep to shallow water and slow down
- away from the normal when they cross from shallow to deep water and speed up.

Light travels more slowly in glass than in air.

- When a light ray travels from air to glass, it refracts towards the normal because it slows down on entering the glass.
- When a light ray travels from glass to air, it refracts away from the normal because it speeds up on leaving the glass.

Real and apparent depth

Don't jump into water unless you know how deep it is. Otherwise, you might find the water is much deeper than you thought. Light from the bottom refracts at the surface. This makes the water appear shallower than it really is. Figure 8.2.2 shows the refraction of a light ray from an object on the bottom of a swimming pool. The light ray bends away from the normal when it reaches the surface. The swimmer standing at the side sees a virtual image of the object above the object. To the swimmer, the apparent depth of the pool is less than the real depth.

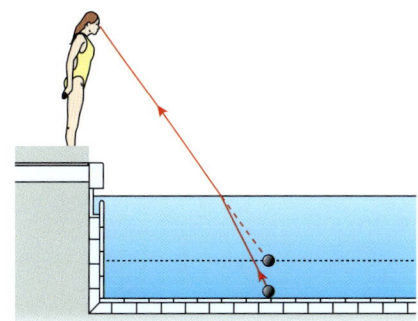

Figure 8.2.2 At a swimming pool

PRACTICAL

Investigating refraction by a rectangular glass block

Test 1 Outline a rectangular glass block on a sheet of white paper. Use the arrangement in Figure 8.2.1 to observe a refracted light ray when it passes through opposite sides of the glass block. The light ray refracts towards the normal when it enters the block and it refracts away from the normal when it leaves the block.

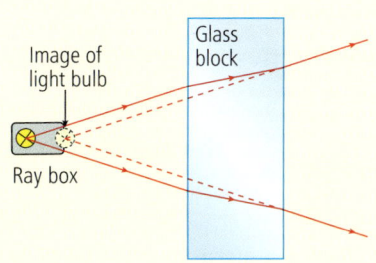

Figure 8.2.3 Refraction by a glass block

Make measurements to find out if the direction of the light ray after leaving the block is the same as it was before it entered the block. You should find the direction is unchanged. This is because the opposite sides are parallel.

Test 2 Replace the slit plate of the ray box with a plate with two slits so the ray box produces two diverging rays. Direct the two rays into the side of the glass block as shown in Figure 8.2.3, so they emerge on the opposite side of the block. Look into the block along the outgoing rays to see an image of the ray box nearer to you. Trace the rays and locate the image.

KEY POINTS

- Refraction of light is the change in direction of a ray of light when it crosses a boundary between two transparent substances including air.

- Refraction is towards the normal when light travels from air to glass (or any other transparent substance).

- Refraction is away from the normal when light travels from glass (or any other transparent substance) to air.

EXAM TIP

You must know which way the light will be refracted at the boundary between different materials.

SUMMARY QUESTIONS

1 Copy and complete the sentences using words from the following list.

   away from   **decreases**
   increases   **towards**

   a When a light ray travels from air into glass, its speed _____ and it bends _____ the normal.

   b When a light ray travels from glass into air, its speed _____ and it bends _____ the normal.

   c When a light ray travels from water into glass, it bends towards the normal because its speed _____.

2 In Figure 8.2.3:

   a Is the image real or virtual? Give a reason for your answer.

   b Explain why the image of the light bulb is nearer the glass block than the actual light bulb.

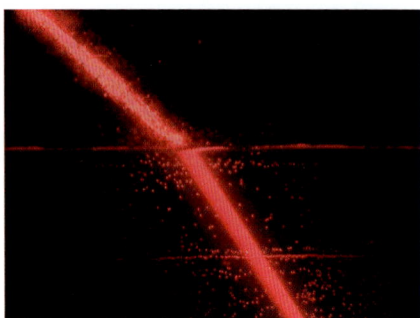
A laser beam entering water

Refractive index

When a ray of light travels from air into a transparent substance, its speed decreases. If the incident light ray is at a non-zero angle to the air-substance boundary, the light ray changes direction at the boundary. The change of direction is due to the change of speed at the boundary.

The speed of light in a transparent substance depends on the substance. For example, the speed of light in glass is about 12% faster than the speed in water. Therefore, the change of direction of a ray when it travels from air into a transparent substance depends on the substance.

The refractive index of a substance, n, is defined as:

$$\frac{\text{the speed of light in air}}{\text{the speed of light in the substance}}$$

For example,

- In air, light travels at a speed of 300 000 km/s, which means it travels a distance of 300 km in a thousandth of a second.

- In glass, light travels at a speed of 200 000 km/s, which means it would travel a distance of 200 km in glass in a thousandth of a second.

Therefore, the ratio $\dfrac{\text{speed of light in air}}{\text{speed of light in glass}} = \dfrac{300\,000\ \text{km/s}}{200\,000\ \text{km/s}} = 1.50$

So the refractive index of glass is 1.50.

The refractive index of a substance is a measure of the change of direction of a ray at non-normal incidence when it passes from air into the substance. For example, the refractive index of water is 1.33 and the refractive index of glass is 1.50. So glass refracts light more than water does.

Investigating how the angle of refraction varies with the angle of incidence

We can use a rectangular transparent block as shown in Figure 8.2.1.

The angle of refraction, r, is measured for different angles of incidence, i.

To make the measurements as accurately as possible:

1 Outline the block on a white sheet of paper and draw the normal at one of the longer flat sides of the rectangular block (whichever is used).

2 Use a protractor to draw straight lines at angles of 10°, 20°, 30° etc. to the normal. Label on the diagram as point P the point where the normal intersects the outline of the block.

3 Direct a ray along each straight line at P in turn. Label these lines 1, 2, 3, etc. corresponding to angles of incidence 10°, 20°, 30°, etc.

4 Mark on the outline of the block where the refracted ray in each case leaves the block for each angle of incidence. Label these points 1, 2, 3, etc. corresponding to angles of incidence 10°, 20°, 30°, etc.

5 Remove the block and draw straight lines from P to each marked point where a ray left the block. In each case, measure the angle of refraction.

6 Record all your measurements in a table. Some typical results are shown below. The results depend on the substance in the block.

$i\,/°$	$r\,/°$	sin i	sin r	$\dfrac{\sin i}{\sin r}$
10.0	6.5	0.174	0.113	1.54
20.0	13.0	0.342	0.225	1.52
30.0	19.0			

The measurements can be used to show that $\dfrac{\sin i}{\sin r}$ always has the same value, regardless of the angle of incidence. This relationship was first discovered in 1618 and is known as Snell's law, after its discoverer. Calculate the mean value of sin i/sin r for your own measurements. As explained below, this is the refractive index of the block you tested.

The law of refraction

In Topic 7.3, we saw in Figure 7.3.3 how the refraction of waves is explained. Applying the explanation to rays of light, it can be shown that:

$$\frac{\sin i}{\sin r} = \frac{\text{the speed of the incident light waves}}{\text{the speed of the refracted light waves}},$$

where angles i and r are, respectively, the angles of incidence and of refraction.

Therefore, for light travelling from air into a transparent substance:

$$\frac{\sin i}{\sin r} = \frac{\text{the speed of light in air}}{\text{the speed of light in the substance}}$$

Since the refractive index of the substance, n, is defined as:

$$\frac{\text{the speed of light in air}}{\text{the speed of light in the substance}}$$

$$\frac{\sin i}{\sin r} = n$$

Note Given values of i and n, we can calculate r by rearranging the above equation to give:

$$\sin r = \frac{\sin i}{n}$$

and substituting the known values in the rearranged equation to find r.

SUMMARY QUESTIONS

1 a In the table above, when $i = 30°$ and $r = 19°$, calculate the value of the refractive index.

b Determine the mean value of refractive index given by the data in the table above.

c The speed of light in air is 300 000 km/s. Use the value of refractive index calculated in **b** to work out the speed of light in the glass block that gave the results in the table.

2 The refractive index of water is 1.33.

a A ray of light enters a flat water surface at an angle of incidence of 35°. Calculate the angle of refraction of the light ray.

b The speed of light in air is 300 000 km/s. Calculate the speed of light in water.

8.4 Total internal reflection

LEARNING OUTCOMES

LEARNING OUTCOMES

- State the meaning of critical angle
- Describe total internal reflection
- **S** Describe the use of optical fibres

The critical angle

When a ray of light is refracted as it passes from glass to air, as shown at point P in Figure 8.4.1, it bends away from the normal.

If the angle of incidence in the glass is gradually increased, the angle of refraction increases until the refracted ray emerges along the boundary, as shown in Figure 8.4.2. The angle of incidence at this position is referred to as the **critical angle,** labelled c in Figure 8.4.2.

If the angle of incidence is increased beyond the critical angle, the light ray is **totally internally reflected** at P, as shown in Figure 8.4.3. The angle of reflection at P, r, is equal to the angle of incidence, i, when total internal reflection occurs.

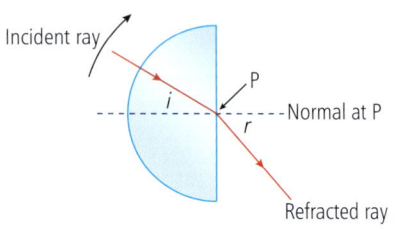

Figure 8.4.1 From glass to air

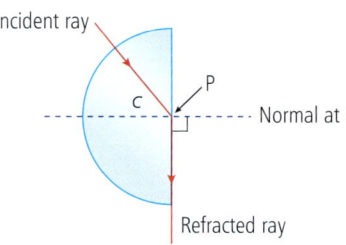

Figure 8.4.2 At the critical angle

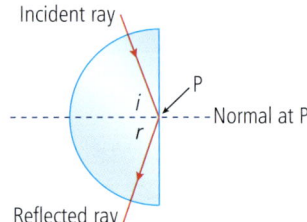

Figure 8.4.3 Total internal reflection

In Figures 8.4.1 and 8.4.2, because the rays travel from glass to air at point P,

$$\frac{\sin i}{\sin r} = \frac{\text{speed of light in transparent substance}}{\text{speed of light in air}} = \frac{1}{n}$$

as n is defined as the speed of light in air ÷ speed of light in the substance.

In Figure 8.4.2, the angle of incidence, i, = the critical angle, c, and the angle of refraction = 90°.

Applying these values to the above equation gives: $\frac{\sin c}{\sin 90°} = \frac{1}{n}$. Since sin 90° = 1, then sin c = $\frac{1}{n}$ or $n = \frac{1}{\sin c}$.

PRACTICAL

Investigating total internal reflection

Outline a semi-circular glass block on a sheet of white paper.

1 Use a ray box to direct a light ray at the centre of the flat side of the block (i.e. point P in Figure 8.4.1) through the curved side. Note that the light ray passes straight through the curved surface of the block without refraction. This is because, at the point where it enters the block, the light ray is directly along the normal at that point.

2 Adjust the angle of incidence of the light ray so the ray refracts into the air, as in Figure 8.4.1. Note that some light does reflect **partially**

at P back into the block but most of the light refracts. This partial internal reflection becomes total internal reflection when the angle of incidence exceeds the critical angle.

3 Increase the angle of incidence until the ray refracts along the boundary, as in Figure 8.4.2. Trace this ray into the block and measure the critical angle.

4 Increase the angle of incidence beyond the critical angle and observe that total internal reflection occurs, as in Figure 8.4.3. Make measurements to show that the angle of incidence is equal to the angle of reflection.

Supplement

Optical fibres

Optical fibres are very thin glass fibres. We use them to transmit light or infrared radiation. The rays can't escape from the fibre. Each ray entering a fibre at one end leaves the fibre at the other end even if the fibre bends around. This is because a ray in the fibre is totally internally reflected each time it reaches its boundary. Optical fibres are used in medicine to see inside the body without cutting the body open and in telecommunications to send signals securely.

- The medical endoscope is used by a surgeon to see inside a body cavity such as the stomach. The endoscope is inserted into the stomach via the patient's throat (see the photo). The endoscope contains two bundles of fibres, one to shine light into the cavity and the other to see the internal surfaces in the cavity. A tiny lens over the second bundle is used to form an image on the ends of the fibres in the bundle. The image can then be seen at the other end of the fibre bundle and viewed through a TV camera.

- A telecommunications optical fibre is used to carry digital signals in the form of light pulses. Such signals may be produced from an electrical signal by a 'light emitting diode' (or LED) in an electrical circuit. The pulses of light stay inside the fibre because any light reaching the fibre boundary is totally internally reflected. When the pulses reach the far end of the fibre, they are detected by a light sensor and converted into an electrical signal. The signals carried by an optical fibre can be detected only by the optical fibre receiver, unlike radio wave signals which can be detected by any receiver in the path of the radio waves. In addition, an optical fibre can carry much more information than a radio wave signal.

EXAM TIP

Make sure you understand the terms critical angle and total internal reflection.

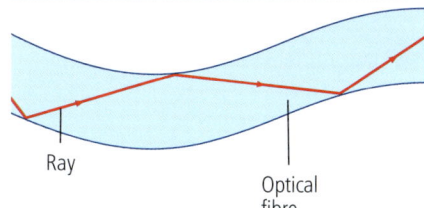

Ray

Optical fibre

Figure 8.4.4 A ray of light in an optical fibre

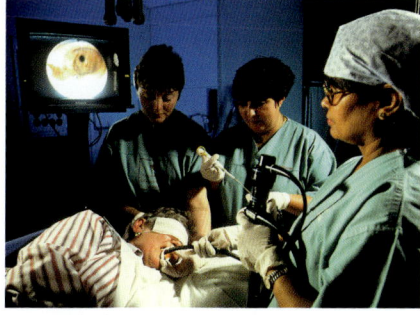

A stomach ulcer viewed through an endoscope

KEY POINTS

When a light ray inside a transparent substance reaches the surface at an angle of incidence which is:

- less than the critical angle, it mostly refracts away from the normal (and some partial reflection occurs)

- equal to the critical angle, it refracts along the surface

- greater than the critical angle, it undergoes total internal reflection at the surface.

SUMMARY QUESTIONS

1 Copy and complete the following sentences using words from the list below.

refraction   reflection   partial reflection
total internal reflection

a When a ray of light travels from air into glass and changes direction, it undergoes _____.

b When a ray inside glass reaches the surface with air and stays in the glass, it undergoes _____.

c When a ray passes from glass to air, some of the light undergoes _____.

d When a ray inside glass reaches the surface with air, it undergoes _____ if the angle of incidence is less than the critical angle.

S 2 a Copy Figure 8.4.4 and show the path of an additional ray along the fibre.

b State two advantages of using optical fibres instead of radio waves to carry signals.

- Describe how a thin converging lens works
- Recall the meaning of focal length
- Draw ray diagrams to show how an image is formed by a thin converging lens

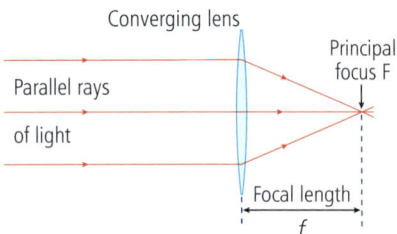

Figure 8.5.1 The converging lens

A diverging lens causes a beam of light to spread out. Figure 8.5.2 shows its effect on parallel rays of light. The refracted rays diverge and appear to originate from a single point. This point is the principal focus of a diverging lens. You will learn in Topic 8.7 that diverging lenses and converging lenses are used to correct sight defects.

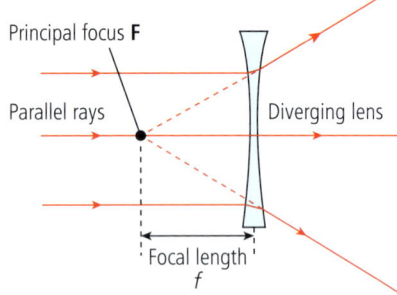

Fig 8.5.2 The diverging lens

Converging lenses are used in optical devices such as the camera. Although a digital camera is very different from the very first cameras made over 160 years ago, they both contain a **converging lens** that is used to form an image.

A converging lens works by changing the direction of light passing through it. Each surface of the lens is a convex surface. The curved shape of the lens surface refracts the rays so they meet at a point. Figure 8.5.1 shows the effect of a converging lens on parallel rays.

- The point to where **parallel** rays directed straight at the lens are focused is the **principal focus F (or focal point)** of the lens.

- The distance from the lens to the principal focus is the **focal length, *f*,** of the lens.

Investigating the converging lens

1 Use the arrangement in Figure 8.5.3 to investigate the image formed by a converging lens.

2 With the object at different distances from the lens, adjust the position of the screen until you see a clear image of the object on it. The image is real because it is formed on the screen where the rays meet.

- When the object is a long distance away, the image is formed at the principal focus on the other side of the lens. This is because the rays from any point of the object are effectively parallel to each other when they reach the lens.

- If the object is moved nearer the lens, the screen must be moved further from the lens to see a clear image. The nearer the object is to the lens, the larger the image is. However, if the object is moved too near the lens, an image cannot be formed on the screen because the rays from the lens do not converge.

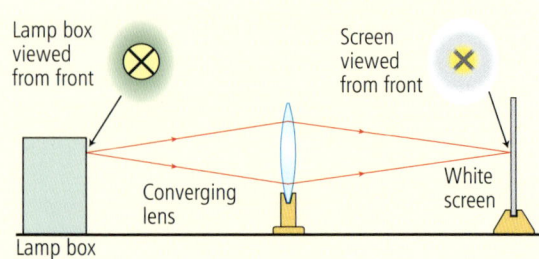

Figure 8.5.3 Investigating images

Ray diagrams

The position and nature of the image formed by a lens depends on:

- the focal length of the lens, and
- the distance from the object to the lens.

If we know the focal length and the object distance, we can draw a ray diagram to find the position and nature of the image. The ray diagram in Figure 8.5.4 shows how a converging lens forms a **real image** of an object. The object must be beyond the principal focus F of the lens. The image is formed beyond the principal focus on the other side of the lens.

The diagram shows that:

- three key 'construction' rays from a single point of the object are used to locate the image,
- the image is real, inverted and smaller than the object.

Notice that:

1 Ray 1 is refracted through F, the focal point of the lens, because it is parallel to the lens axis before the lens.

2 Ray 2 passes through the lens at its centre without change of direction. This is because the lens surfaces are parallel to each other at the axis.

3 Ray 3 passes through F, the focal point of the lens, before the lens so it is refracted by the lens parallel to the axis.

The image is diminished compared with the object because the object distance is greater than 2F. This is how a **camera** is used.

EXAM TIP

When you draw a ray diagram, draw the lens as a flat line with arrows at each end, outward arrows for a converging lens and inward arrows for a diverging lens. Then draw **the principal axis**, which is the straight line perpendicular to the lens through its centre.

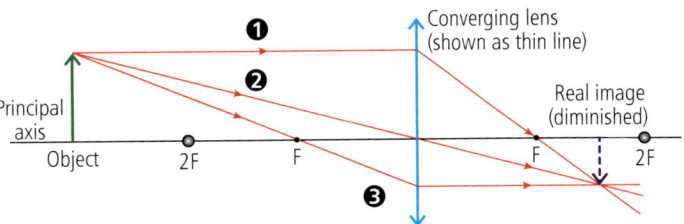

Ray ❶ is parallel to the axis and is refracted through F
Ray ❷ passes straight through the centre of the lens
Ray ❸ passes through F and is refracted parallel to the axis

Figure 8.5.4 Formation of a real image by a converging lens

KEY POINTS

- The principal focus of a converging lens is the point where parallel rays directed straight at the lens converge.

- The focal length of a converging lens is the distance from the lens to the principal focus of the lens.

- A real image is formed by a converging lens if the object is further away than the principal focus of the lens.

SUMMARY QUESTIONS

1 Copy and complete the following sentences using words from the list below.

at   near   far   from

a A converging lens forms a real image of an object on a screen.
    i If the object is far from the lens, the image is formed on the other side of the lens _____ its principal focus.
    ii If the object is near the lens, the image is formed on the other side of the lens _____ its principal focus.

b A camera is used to photograph an object.
    i If the object is not far from the camera, the distance from the lens to the film needs to be adjusted so that the film is _____ the principal focus of the lens.
    ii If the object is far from the camera, the distance from the lens to the film needs to be adjusted so that the film is _____ the principal focus of the lens.

2 a Draw a ray diagram to show how a converging lens forms a real image of an object.

b State whether the image is **i** upright or inverted, **ii** magnified or diminished.

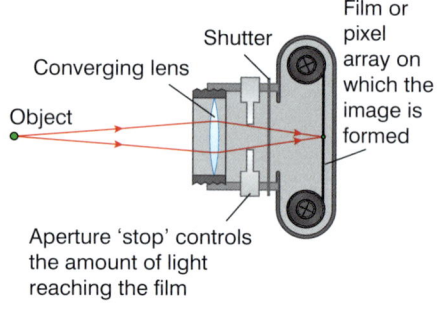

Figure 8.6.1 The camera

Investigating magnification

1 Use the arrangement in Figure 8.5.3 in Topic 8.5 to investigate how the size of the image depends on the distance from the lens to the screen. For different distances from the object to the lens:

- measure the image distance (i.e. the distance from the lens to the screen when the image is in focus on the screen)

- measure the diameter of the image.

2 What conclusion can you draw from your readings?

The image diameter and the object diameter are equal when the object and the image are equidistant from the lens. This is when the object is at 2F.

Magnification

The image formed by a converging lens may be smaller or larger than the object according to the distance from the object to the lens and the focal length of the lens. In general, we say that the image is:

- diminished if it is smaller than the object
- magnified if it is larger than the object.

If the object is further from the lens than the principal focus F of the lens, its image will always be real and inverted and on the other side of the lens. The image will be:

- diminished if the object is beyond 2F (i.e. beyond twice the distance from the lens to F)
- the same size as the object if the object is exactly at 2F
- magnified if the object is between F and 2F.

The camera

In a camera, a converging lens is used to produce a diminished, inverted, real image of an object on a film (or on an array of 'pixels' in the case of a digital camera). The position of the lens is adjusted to focus the image on the film, according to how far away the object is.

- For a distant object, the distance from the lens to the film must be equal to the focal length of the lens.

- The nearer an object is to the lens, the greater the distance from the lens to the film.

The projector

A projector lens produces a magnified real image of an object (such as a transparent slide) on a screen. Figure 8.6.2 shows the ray diagram for an object which gives a magnified, inverted, real image. The image can be seen in sharp focus on a screen placed where the image is located. If the screen is moved away from this position, the image becomes blurred and out-of-focus.

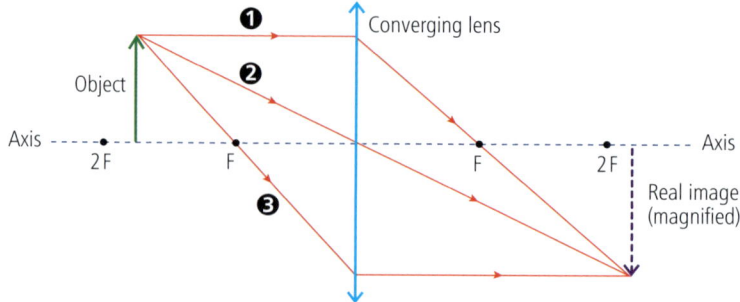

Figure 8.6.2 A real inverted magnified image

Supplement

The magnifying glass

A magnifying glass is useful if you want to see fine detail on a small object such as a coin. A magnifying glass is a converging lens held close to the object. When the object is viewed through the magnifying glass, a magnified image of the object is seen. The photograph shows the image of a frond (divided leaf) seen through a magnifying glass. Notice that the image is upright (i.e. the same way up as the object).

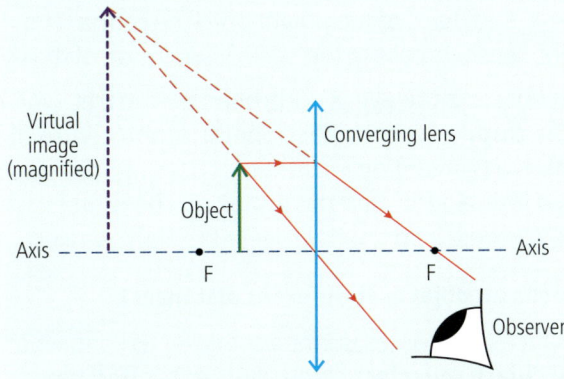

Figure 8.6.3 Ray diagram for a magnifying glass

The image is a **virtual image** (like an image seen in a mirror) because it cannot be formed on screen like a real image can. The ray diagram in Figure 8.6.3 shows how the virtual image is formed. The lens refracts the rays from the object so they appear to come from an image beyond the object. Provided the object is between the lens and its principal focus F, the image formed by the lens is always magnified, upright and virtual.

Investigating the magnifying glass

Hold a converging lens over a printed page to observe a magnified virtual image of the print. Observe how the image is changed as the lens is moved away from the page. You should find the magnification increases until the image becomes too large and too blurred to see.

A converging lens as a magnifying glass

SUMMARY QUESTIONS

1 a Draw a ray diagram to show how a converging lens forms a magnified real image of an object.

b State whether the image is upright or inverted.

c Describe an application of the lens used in this way.

2 a Draw a ray diagram to show how a converging lens forms a magnified virtual image of an object.

b State whether the image is upright or inverted.

c Describe an application of the lens used in this way.

KEY POINTS

object–lens distance	image–lens distance	image description	application
beyond 2F	between F and 2F	real, inverted, diminished	camera
at 2F	at 2F	real, inverted, same size	
between F and 2F	beyond 2F	real, inverted, magnified	projector
at F	at infinity		
between F and the lens	same side of the lens as the object	virtual, upright, magnified	magnifying glass

8.7 Correction of sight defects

Supplement

LEARNING OUTCOMES

- Describe the eye as an optical instrument
- Recognise how the eye forms an image
- Explain what short sight is and how it is corrected
- Explain what long sight is and how it is corrected

Inside the eye

Figure 8.7.1 shows a simplified view of the inside of a human eye. Light enters the eye through a tough transparent layer called the **cornea**. The cornea protects the eye and helps to focus light onto the **retina**. The retina is a layer of light-sensitive cells at the back of the inside of the eye.

The amount of light entering the eye is controlled by the **iris**, which adjusts the size of the **pupil** – the circular opening at the centre of the iris. The **eye lens** is a converging lens that focuses light to give a sharp image on the retina. Although the image on the retina is inverted, the brain interprets it so you can see it the right way up.

How does the eye focus on objects at different distances?

Your eye lens automatically becomes thinner or thicker to keep what you see in focus. The **ciliary muscle** is a ring of muscle round the perimeter of the eye lens. It alters the thickness of the eye lens. When the muscle contracts, the fibres shorten and squeeze the eye lens, making the lens thicker; when it relaxes, the eye lens becomes thinner.

- The normal human eye has a **range of vision** from 25 cm to infinity. This means it can see clearly any object that is 25 cm or more from the eye. In other words, the normal eye has a **near point** of 25 cm and a **far point** at infinity.

- To see a nearby object clearly, the eye lens has to be thicker than if the object is far away. Figure 8.7.2a and b show this.

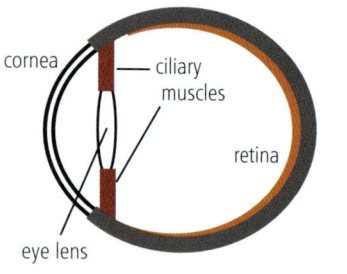

Figure 8.7.1 The eye as an optical instrument

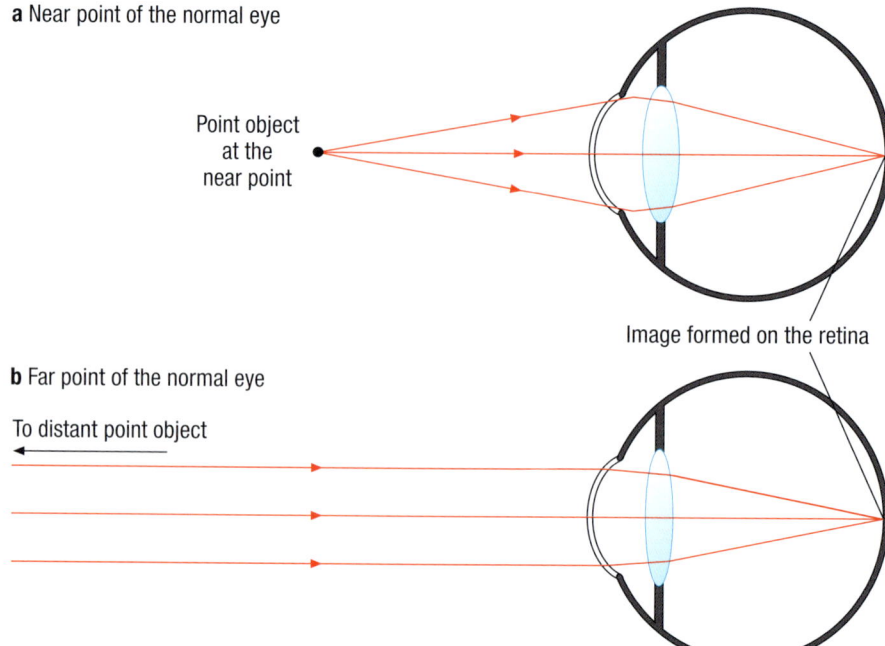

a Near point of the normal eye

Point object at the near point

Image formed on the retina

b Far point of the normal eye

To distant point object

Figure 8.7.2 The normal eye

Sight defects

Short sight occurs when an eye cannot focus on **distant** objects. The uncorrected image is formed in front of the retina, as shown in Figure 8.7.3. This is because the eyeball is too long or the eye lens is too strong. The eye muscles cannot make the eye lens thin enough to focus the image of a far-away object on the retina of the eye. The eye can focus nearby objects so the defect is referred to as 'short sight'.

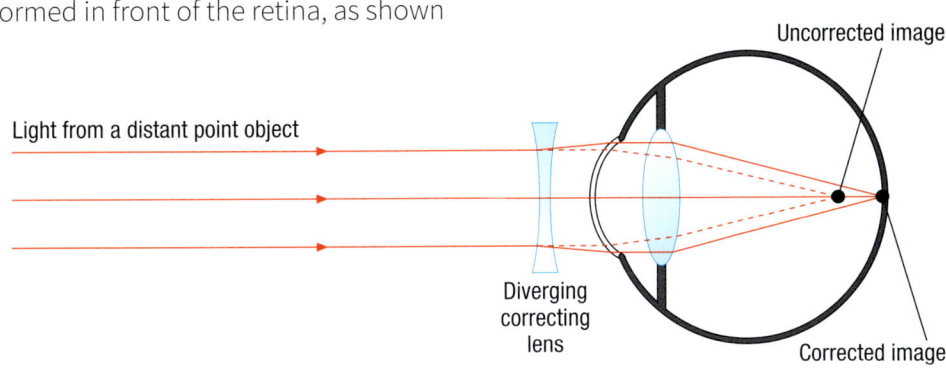

Figure 8.7.3 Short sight and its correction

Short sight is corrected by placing a diverging lens of a suitable focal length in front of the eye as shown in Figure 8.7.3. The diverging lens counteracts some of the 'excess' focusing power of the eye lens.

Long sight occurs when an eye cannot focus on nearby objects. The uncorrected image is 'formed' behind the retina, as shown in Figure 8.7.4. The eye lens cannot be made thick enough to focus an image on the retina. The eye can focus distant objects hence the defect is referred to as 'long sight'.

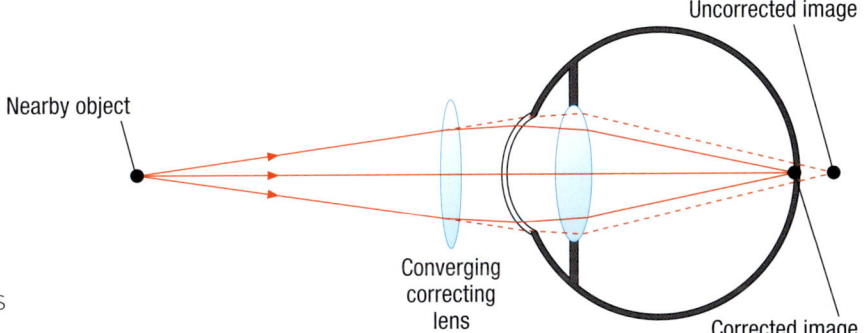

Figure 8.7.4 Long sight and its correction

Long sight is corrected by placing a converging lens of a suitable focal length in front of the eye, as shown in Figure 8.7.4. The correcting lens makes the rays from the object diverge less. The eye lens can then focus the rays onto the retina. The correcting lens adds to the focusing power of the eye lens.

SUMMARY QUESTIONS

1 Copy and complete sentences a to c using the words below.

   cornea iris lens retina

   a The front of the eye is protected by the ....
   b The pupil of the eye is at the centre of the ....
   c The .... of the eye focuses light onto the ....

2 Using his left eye, a student can only see the writing on a board at the front of the class if he sits near the board.

   a What sight defect is he suffering from in this eye?
   b What type of lens should be used to correct this defect?

KEY POINTS

- Light is focused on to the retina by the cornea and the eye lens which is a variable focus lens.

- The normal human eye has a range of vision from 25 cm to infinity.

- A short-sighted eye is an eye that can only see near objects clearly. We use a diverging lens to correct it.

- A long-sighted eye is an eye that can only see distant objects clearly. We use a converging lens to correct it.

8.8 Electromagnetic waves

LEARNING OUTCOMES

- Describe the spectrum of white light and how to produce it using a prism
- Recall the parts of the electromagnetic spectrum
- Recognise that all electromagnetic waves travel at the same speed through space
- **s** State the approximate value of the speed of electromagnetic waves in air

Dispersion of white light by a prism.

The white light spectrum

Light from ordinary lamps and from the Sun is called **white light**. This is because it has all the colours of the visible spectrum in it. You see the colours of the spectrum when you look at a rainbow. You can also see them if you use a glass prism to split a beam of white light, as shown in the photograph, into separate colours. Light of a single colour is called **monochromatic** light. **s**

Each colour of light is refracted slightly differently. This is because the speed of light in glass depends on its colour. As a result, the refractive index of glass (= speed of light in air/speed of light in glass) depends on its colour. The result is that the beam of white light is split into separate colours by the prism. This occurs due to refraction where it enters and where it leaves the prism. This effect is called **dispersion**.

PRACTICAL

Investigating the white light spectrum

1. Use a prism to split a narrow beam of white light from a ray box into the colours of the spectrum. Display the spectrum on a white screen. You should be able to see all the colours in the following order, as shown in Figure 8.8.1:

   Red **O**range **Y**ellow **G**reen **B**lue **I**ndigo **V**iolet

   (or Roy G Biv – the presence of indigo is sometimes disputed!)

2. Which colour is refracted most? The colour that is refracted most travels slowest in glass. You should find (as shown in the photograph) that violet is refracted most. So it travels slowest in glass. The speed of light in glass increases from violet to red across the spectrum.

3. Place a blackened thermometer on the screen just beyond each end of the spectrum. You should find a slight warming effect just beyond the red end of the spectrum. This is because the lamp produces **infrared radiation** as well as light.

   Infrared   Ultraviolet

   Figure 8.8.1 The visible spectrum.

4. If a sensitive ultraviolet detector is placed just beyond the violet part of the spectrum, it should be possible to detect **ultraviolet radiation** there. This radiation does not have a heating effect like infrared radiation, but it is harmful to the eyes.

The electromagnetic spectrum

Light, infrared radiation and ultraviolet radiation are all part of the spectrum of electromagnetic waves. Figure 8.8.2 shows the electromagnetic spectrum. All electromagnetic waves travel through space at the same speed.

Notice the spectrum is continuous. The frequencies and the wavelengths at the boundaries between different parts are approximate as the different parts of the spectrum are not precisely defined.

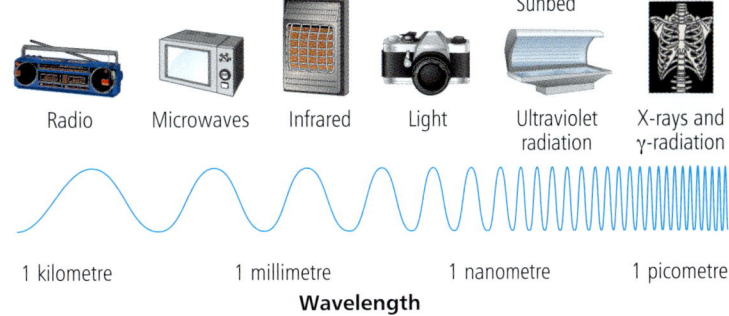

Radio	Microwaves	Infrared	Light	Ultraviolet radiation	X-rays and γ-radiation

1 kilometre   1 millimetre   1 nanometre   1 picometre

Wavelength

(1 nanometre = 0.000001 millimetres, 1 picometre = 0.001 nanometres)

Figure 8.8.2 Using electromagnetic waves.

Electromagnetic waves are electric and magnetic disturbances that transfer energy from one place to another.

S • All electromagnetic waves travel through space at a speed of 300 000 km/s. Their speed through air is almost the same as their speed through space.

• Electromagnetic waves do not transfer matter. The energy they transfer depends on the **wavelength** of the waves. This is why waves of different wavelengths have different effects.

As explained in Topic 7.1, we can calculate the frequency from the wavelength (or the other way around) using the equation:

speed = frequency × wavelength

1 We can work out the wavelength if we know the frequency and the wave speed. To do this, we rearrange the equation into:

$$\text{wavelength (in metres)} = \frac{\text{wave speed (in m/s)}}{\text{frequency (in Hz)}}$$

2 We can work out the frequency if we know the wavelength and the wave speed. To do this, we rearrange the equation into:

$$\text{frequency (in Hz)} = \frac{\text{wave speed (in m/s)}}{\text{wavelength (in metres)}}$$

Note that 1 MHz = 1000 kHz = 1 000 000 Hz

EXAM TIP

LONG wavelength waves have a LOW frequency. SHORT wavelength waves have a HIGH frequency. Use the correct words!

KEY POINTS

• Dispersion is the splitting of white light into the colours of the spectrum using a prism.

• The main parts of the electromagnetic spectrum are, in order of decreasing wavelength:
  – radio waves
  – microwaves
  – infrared radiation
  – light
  – ultraviolet radiation
  – X-rays
  – gamma rays

• **S** All electromagnetic waves travel at a speed of 300 000 km/s through space.

SUMMARY QUESTIONS

1 Copy and complete the sentences below using these words.

   greater than   smaller than   the same as.

   a The wavelength of light waves is _____ the wavelength of radio waves.

   b The speed of radio waves in a vacuum is _____ the speed of gamma rays.

   c The frequency of X-rays is _____ the frequency of infrared radiation.

2 a Copy and complete the electromagnetic spectrum below.
   radio; _____; infrared; visible; _____; X-rays; _____.

   b Work out **i** the wavelength of radio waves of frequency 600 MHz, **ii** the frequency of microwaves of wavelength 0.30 m. The speed of electromagnetic waves in a vacuum = 300 000 km/s.

8.9 Applications of electromagnetic waves

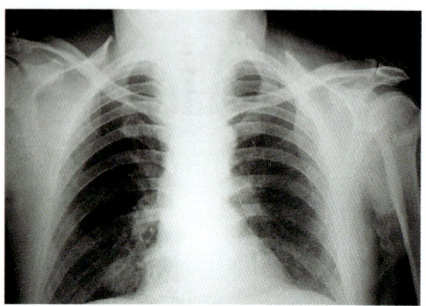

A chest X-ray.

Light is just a small part of the electromagnetic spectrum. Our eyes cannot detect the other parts. The table below summarises the main sources, detectors and uses of the different parts of the electromagnetic spectrum. Notice that light covers a very narrow range of wavelengths from about 350 nm for blue light to about 650 nm for red light, where 1 nanometre (nm) = 1 millionth of 1 mm.

More about X-rays

X-rays pass through body tissue but they are absorbed by bones and thick metal plates. To make a **radiograph** or X-ray picture, X-rays from an X-ray tube are directed at the patient. A light-proof cassette containing a photographic film is placed on the other side of the patient.

- When the X-ray tube is switched on, X-rays from the tube pass through the patient's body and leave a 'shadow' image of the bones on the film.

- When the film is developed, the parts exposed to X-rays are darker than the other parts. So the bones appear lighter than the surrounding tissue which appears dark. The developed film shows a 'negative image' of the bones.

The electromagnetic spectrum

type of radiation	wavelength range	sources	detectors	main properties	uses
radio waves	> about 0.1 m	radio, TV transmitters	receivers with aerials	reflected by metal sheets	radio and TV broadcasting, astronomy, radio frequency identification
microwaves	about 0.1 m–1 mm	microwave transmitters and ovens		reflected by metal sheets, absorbed by water	mobile phones, satellite TV and communications, microwave ovens for heating food
infrared radiation	about 1mm–650 nm	hot objects	blackened thermometer bulb, infrared camera	reflected by metal sheets, absorbed at or near the surface	electric grills and heaters, thermal imaging, intruder alarms, optical fibre communications, TV remote control
light	about 650 nm–350 nm	glowing objects	eye, camera	reflected by mirrors, refracted by transparent substances	optical instruments, photography, lighting, vision
ultraviolet radiation	about 350 nm–1 nm	UV lamps, the Sun	fluorescent chemicals, UV film	makes fluorescent chemicals glow, absorbed by ozone layer	UV sun tan lamps, UV ink driers, security markers, sterilising water, detecting fake bank notes
X-rays	< about 1 nm	X-ray tubes	film, Geiger tube (see Topic 15.1)	ionises substances, penetrates substances such as body tissue	X-radiography, security scans
gamma radiation		radioactive isotopes			detection and treatment of cancer, radioactive tracers, sterilising equipment, crack detection in metals

Safety matters

X-radiation or gamma radiation is dangerous and causes cancer. This is because X-rays and gamma rays create ions (charged atoms) in substances they pass through. High doses kill living cells and low doses can damage cells, cause cell mutation and cancerous growth. There is no evidence of a lower limit below which living cells would not be damaged. You will learn more about ionising radiation in Topic 15.1.

Anyone using equipment or substances that produce any form of ionising radiation must wear a film badge (see page 229).

Ultraviolet radiation is harmful to human eyes and can cause eye conditions and blindness. UV wavelengths are smaller than light wavelengths. UV rays carry more energy than rays of light. Too much UV radiation causes sunburn and can damage skin cells and cause skin cancer.

- If you stay outdoors in hot weather, use skin creams to block UV radiation and prevent it reaching the skin.

- If you use a sunbed to get a suntan, don't exceed the recommended time. In addition, wear special 'goggles' to protect your eyes.

Radio waves and microwaves penetrate into the body and have a heating effect on body tissues (and any other substance that absorbs them). **Infrared radiation** can cause skin burns as it has a heating effect on skin tissue (as well as on the surface of any other substance that absorbs it).

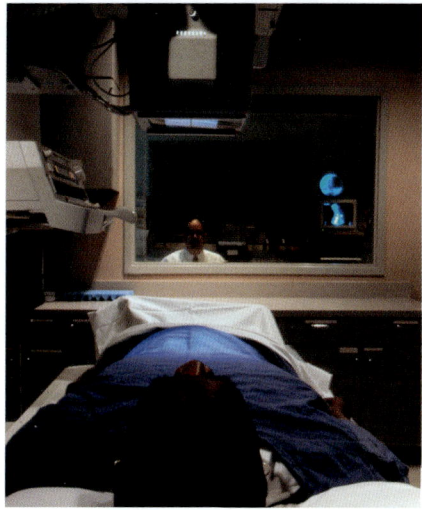
Taking a chest X-ray.

EXAM TIP

Make sure you understand the process of ionisation. It is described in Topic 15.2.

PRACTICAL

An infrared radiation test

Can infrared radiation pass through paper? Use a TV remote handset to find out.

KEY POINTS

- X-rays and gamma radiation damage living tissue when they pass through it.
- X-rays are used in hospitals to take radiographs.
- Ultraviolet radiation harms the skin and the eyes.
- Light and infrared radiation is used to carry signals in optical fibres.
- Microwaves and radio waves are used for communications.

SUMMARY QUESTIONS

1 Copy and complete the table below showing the type of electromagnetic radiation produced by each device **a** to **d**.

Device	infrared	microwave	radio
a electric toaster			
b microwave oven			
c TV broadcast transmitter			
d remote handset			

2 Use words from the list to copy and complete the sentences below.

**infrared radiation   gamma rays   light   microwaves
radio waves   ultraviolet radiation   white light   X-rays**

a _____ from the Sun is absorbed by the ozone layer.
b _____ includes all the colours of the spectrum.
c In a TV set, the aerial detects _____ and the screen emits _____.
d In a microwave oven, food absorbs _____, heats up and emits _____.
e _____ and _____ ionise living tissue.

8.10 More about electromagnetic waves

Supplement

LEARNING OUTCOMES

- Recognise that we use radio waves of different frequencies for different purposes
- Describe how the range of a radio signal depends on its frequency
- Describe how optical fibres are used in communications.

Figure 8.10.1 Sending microwave signals to a satellite

PRACTICAL

Demonstrating an optical fibre

Observe light shone into an optical fibre. You should see the reflection of light inside the optical fibre.

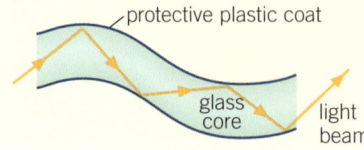

Figure 8.10.2 Optical fibres

Radio communications

Microwaves and radio waves of different wavelengths are used for different communications purposes. Examples are given below:

Microwaves are used by mobile phones (cell phones), as well as by wireless internet devices that require short aerials and electromagnetic waves that can pass through house walls. Microwaves are also used for satellite communications and broadcasting because they can travel between satellites in space and the ground.

Also, they spread out less than radio waves do, so the signal doesn't weaken as much.

Radio waves of wavelengths less than about 1 metre are used for TV broadcasting from TV masts because they can carry more information than longer radio waves. In contrast, Bluetooth devices send out shorter-wavelength radio waves that are weakened when they pass through walls.

Radio waves of wavelengths from about 1 metre up to about 100 m are used by local radio stations (and for the emergency services) because their range is limited to the area round the transmitter.

Radio waves of wavelengths greater than 100 m are used by national and international radio stations because they have a much longer range than shorter-wavelength radio waves.

Mobile phone radiation

The radio waves to and from a mobile phone have a wavelength of about 30 cm. Radio waves at this wavelength are not quite in the microwave range, but they do have a similar heating effect to microwaves so they are usually referred to as microwaves.

Optical fibre communications

Optical fibres are very thin glass fibres. We use them to transmit signals carried by light or infrared radiation. The light rays can't escape from the fibre. When they reach the surface of the fibre, they are reflected back into the fibre, as shown in Figure 8.10.2.

In comparison with radio waves and microwaves:

- optical fibres can carry much more information. This is because light and infrared radiation have a much smaller wavelength than radio waves so can carry more pulses of waves.

- optical fibres are more secure because the signals stay in the fibre.

Analogue and digital signals

Computers, calculators, mobile phones, TVs and many other electronic devices are all 'digital' although many radio stations still transmit analogue signals.

Figure 8.10.3 shows an analogue signal and a digital signal. The waves that are used to carry a signal are called **carrier waves**. A digital signal is sent by switching the carrier wave on and off repeatedly. An analogue signal is sent by varying (modulating) the amplitude or the frequency of the carrier wave.

Analogue signals (for example the electrical waves from a microphone signal) can be converted to digital signals which can then be transmitted. All signals pick up unwanted variations called '**noise**'. The key advantages of digital signals compared with analogue signals are:

- digital signals can be regenerated accurately with noise removed, increasing their range as a result.

- digital signals can transmit data at an increased rate because the duration of each pulse can be made shorter so more data can be transmitted each second.

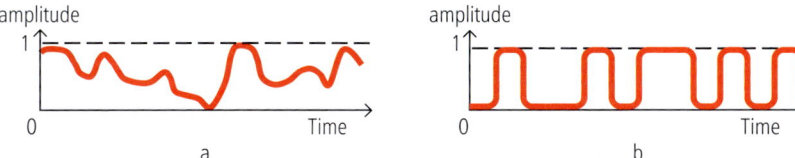

Figure 8.10.3 a) An analogue signal b) A digital signal

SUMMARY QUESTIONS

1 a What types of electromagnetic wave are used to carry:
   i mobile phone signals?
   ii signals along a thin transparent fibre?

b i Why should signals to and from a mobile phone be at different frequencies?
   ii Why are signals in an optical fibre more secure than radio signals?

2 a Explain why microwaves are used for satellite TV and radio waves for terrestrial TV.

b Why can microwave signals be sent a long distance between two transmitter dishes provided that the two transmitter masts are within sight of each other?

1 (a) Write down the law of reflection.

(b) Draw a labelled diagram showing a plane mirror, an incident ray, a normal where the incident ray hits the mirror, the reflected ray, the angle of incidence and the angle of reflection.

2 Which way is light bent (if at all) at the boundary between air and glass in the following circumstances:

(a) The light travels along a normal.

(b) The light hits the surface at an angle of incidence of 30° travelling from air to glass.

(c) The light hits the surface at an angle of incidence of 30° travelling from glass to air.

3 (a) Name the colours of the spectrum of white light in the correct order beginning with red.

(b) The diagram shows a ray of white light hitting a glass prism.

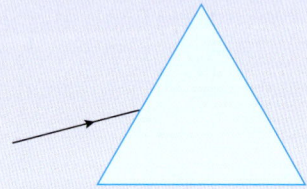

Copy and complete the diagram to show the dispersion of white light into the colours of the spectrum. Label the diagram so that it is clear which colour is refracted least and which colour is refracted most.

4 Suggest a source and a use for each of the following types of electromagnetic wave:

(a) microwaves

(b) infrared

(c) ultraviolet

(d) X-rays.

5 Light enters a transparent rectangular block at an angle of incidence I = 20°. The angle of refraction is 13°.

(a) Calculate the refractive index of the material of the block.

(b) Draw a diagram to show the effect described.

6 Draw a ray diagram to show how a converging lens can be used as a magnifying glass.

Practice Questions

1 The figure below shows a ray of light being reflected from a plane mirror. The angle of incidence is shown. What is the angle of reflection?

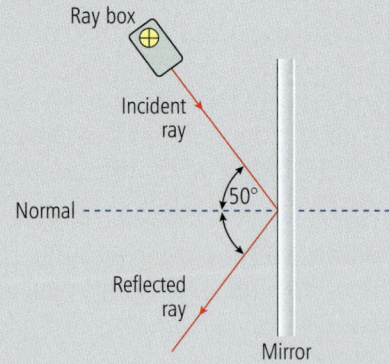

A   30°   C   45°

B   50°   D   0°

2 Which of these rows describe the properties of the image formed in a plane mirror?

Answer	Size of image	Type of image
A	Same as object	Real
B	Smaller than object	Virtual
C	Same as object	Virtual
D	Larger than object	Real

3 A beam of light enters and leaves a glass block as shown in the diagram. Which row in the table correctly describes the behaviour of the beam of light as it enters and leaves the block?

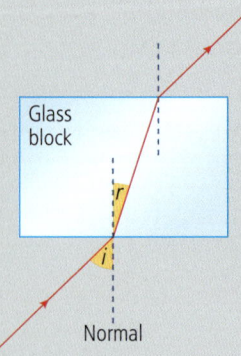

Answer	Entering the glass block	Leaving the glass block
A	Slows and changes direction towards the normal	Speeds up and changes direction towards the normal
B	Slows and changes direction towards the normal	Speeds up and changes direction away from the normal
C	Speeds up and changes direction towards the normal	Slows down and changes direction towards the normal
D	Slows down and changes direction away from the normal	Speeds up and changes direction away from the normal

4 Which of the following parts of the electromagnetic spectrum has the longest wavelength?

A Microwaves   C X-rays

B Visible light   D Radio waves

5 (a) A person with normal eyesight who is reading a book looks up to observe a distant object. Describe and explain how the shape of each eye lens alters in this change. [3]

(b) (i) Explain is meant by short sight. [2]

(ii) State the type of lens used to correct short sight. [1]

6 A converging lens is used in a camera to form a real image of an object placed far outside its focal length.

(a) Carefully draw a ray diagram showing how the real image is formed by the lens. The image should include:

(i) the principal focus of the lens [1]

(ii) an object placed beyond two focal lengths of the lens [1]

(iii) three construction lines [3]

(iv) the position of the image formed. [1]

(b) The image formed is real. State two other properties of the image. [2]

7 A student decides to measure the critical angle of a transparent plastic material. They shine a ray of red light into a semicircular block of the material and adjust the angle of incidence until the refracted ray just escapes the block as shown in the diagram.

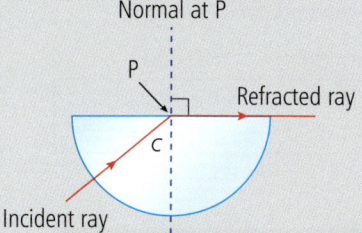

(a) Why doesn't the incident ray change direction as it enters the semicircular block? [1]

(b) Describe how the student could measure the critical angle c accurately. [4]

The student finds that the critical angle is 36°.

(c) Use the critical angle to find the refractive index of the plastic. [2]

The student repeats the process but uses a beam of white light. They notice that the light splits into a range of colours as it leaves the block.

(d) What is the name of this colour splitting process? [1]

8 All electromagnetic radiation travels at 3.0×10^8 m/s in a vacuum and very close to this speed in air.

A radio wave transmitter produces waves with a wavelength of 1.25 m. The radio waves are used to communicate with a robotic probe placed on the surface of the moon.

(a) What is the frequency of the radio waves? [1]

(b) Name the item on the probe that is used to detect the signal on the moon. [1]

The probe is exposed to high levels of ultraviolet radiation because the moon has no atmosphere.

(c) What is the source of the ultraviolet radiation? [1]

(d) Why is ultraviolet radiation harmful to living organisms? [2]

9.1 Sound waves

LEARNING OUTCOMES

- Describe how sound waves are produced by vibrating sources
- Describe the longitudinal nature of sound waves
- **S** • Describe what compressions and rarefactions are

Making sound waves

EXAM TIP

You should know in what ways sound is similar to light and in what ways it is different.

Making sound waves

Sound waves are easy to produce. Your vocal cords vibrate and produce sound waves every time you speak. Any object vibrating in air creates sound waves. You can use a loudspeaker to produce sound waves by passing alternating current through it. Figure 9.1.1 shows how to do this using a signal generator. This is an alternating current supply unit with a variable frequency dial. The frequency of the alternating current and hence the frequency of the sound waves can be changed by adjusting the frequency dial.

If you observe the loudspeaker closely, you can see it vibrating. It pushes the surrounding air backwards and forwards.

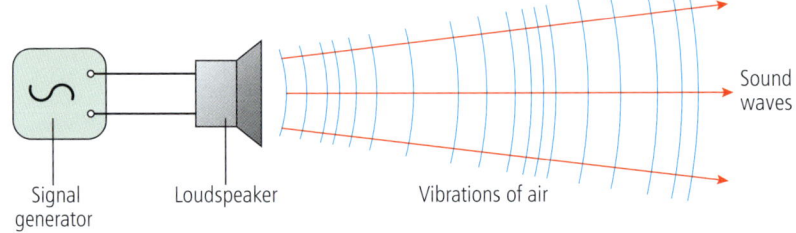

Signal generator | Loudspeaker | Vibrations of air | Sound waves

Figure 9.1.1 Using a loudspeaker

- If the frequency of the alternating current is decreased, the loudspeaker vibrates at a slower rate so it pushes the surrounding air backwards and forwards at a slower rate. The sound waves are therefore created with a lower frequency.
- If the frequency of the alternating current is increased, the loudspeaker vibrates faster so it pushes the surrounding air backwards and forwards at a faster rate. The sound waves are therefore created with a higher frequency.

Sound waves are longitudinal waves. When sound waves pass through air, the air molecules in the path of the sound waves move forwards and backwards along the direction in which the waves are moving. In other words, the molecules vibrate along the direction of motion of the waves. Sound waves are therefore longitudinal waves because the vibrations are along the direction they travel in.

Note This vibrating motion is much more significant than the individual random motion of the molecules just as the up-and-down motion of water molecules in sea waves is more significant than the random motion of the molecules.

PRACTICAL

Making longitudinal waves

1 Use a slinky to demonstrate how sound waves travel. If you move one end backwards and forwards, longitudinal waves travel along the slinky.

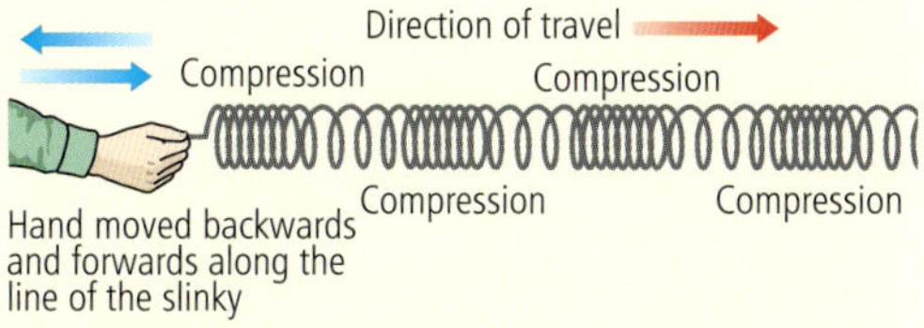

Direction of travel

Compression Compression

Compression Compression

Hand moved backwards and forwards along the line of the slinky

Figure 9.1.2 Making longitudinal waves in a slinky spring

2 Blow over the top of an empty bottle and see if you can make the bottle resonate with sound. Find out how the sound is affected by altering the water level in the bottle.

Supplement

The nature of sound waves

A loudspeaker sends sound waves through the air because its surface pushes and pulls repeatedly on the air. When the waves reach your ears, your eardrums vibrate so you hear sound.

As the surface vibrates, it moves forwards then backwards repeatedly. Each time it moves forwards into the surrounding air, it sends a **compression** wave through the surrounding air as a result of:

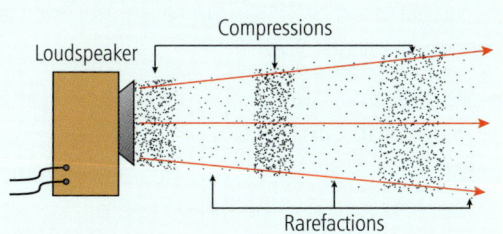

Loudspeaker

Compressions

Rarefactions

Figure 9.1.3 Compressions and rarefactions

- pushing the nearby air molecules forwards, so
- they bump into air molecules further away pushing them forwards, so they bump into air molecules further away pushing them forward … and so on.

Each time a vibrating surface moves backwards away from the surrounding air, it sends a **rarefaction** wave, the opposite of a compression wave, through the surrounding air as a result of:

- leaving a space which is then filled by nearby air molecules moving in, so
- they leave a space which is then filled by air molecules further away moving backwards,
- these air molecules leave space which is then filled by air molecules even further away moving in … and so on.

The result is that as the surface vibrates, it repeatedly sends a compression wave followed by a rarefaction wave into the surrounding air.

KEY POINTS

- Sound waves are produced when a vibrating surface pushes and pulls on the surrounding substance.
- Sound waves are longitudinal waves which means they vibrate along the direction in which the wave travels.
- Sound waves in air consist of alternate compressions and rarefactions passing through the air. **S**

SUMMARY QUESTIONS

1 Copy and complete the following sentences using words from the list below.

**compressions
longitudinal waves
rarefactions   vibrations**

a Sound waves are caused by the _____ of an object.

b _____ are caused when a vibrating surface moves into the surrounding air.

c Sound waves are _____ because the _____ of the air molecules are along the direction in which the waves move.

d Sound waves consist of alternate _____ and _____. **S**

2 a Describe the motion of a point on a slinky coil when longitudinal waves pass along the slinky.

b Describe how the motion of air molecules is affected when sound passes through the air. **S**

9.2 Properties of sound

- Recognise that a medium is required to transmit sound waves

- Recall that sound waves cannot pass through a vacuum

- State the approximate range of frequencies that can be heard by the human ear

- Relate the loudness and pitch of a sound to the amplitude and frequency of the sound waves

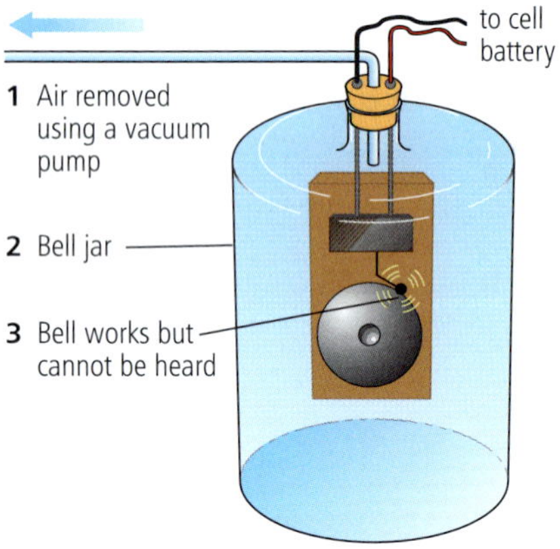

If you listen to a recording of your own voice, you may be surprised by the difference between your recorded voice and what you hear when you speak. The reason is that the sound of your own voice is carried to your ears by two routes:

- sound waves that pass from your larynx via your mouth into the air and then through the air to your ears, and

- sound waves that pass from your larynx to your ears directly through your head.

In comparison, your recorded voice is carried only by sound waves that pass through the air from the recorder to your ears. No sound waves pass directly through your head. Your recorded voice is the sound others hear when you speak.

Sound waves travel through solids, liquids and gases. You can test the passage of sound through a solid by placing a ticking watch in contact on a table and listening to it by placing your ear in contact with the table. The ticking of the watch can be clearly heard in this way even when it cannot be heard directly.

Sound waves cannot travel through a vacuum. This can be demonstrated by listening to an electric bell ringing in a 'bell jar', as shown in Figure 9.2.1.

- As the air is pumped out of the bell jar, the ringing sound fades away and can no longer be heard.

- When the air is let back into the bell jar, the ringing sound returns and can be heard again.

Without air in the bell jar, no sound waves can be created inside the jar.

Figure 9.2.1 A sound test
Does this experiment show that light can travel through a vacuum? Yes, it does. You can still see the bell when the air is removed!

1 Air removed using a vacuum pump

2 Bell jar

3 Bell works but cannot be heard

4 Wires to cell battery

PRACTICAL

Hearing tests

1 Use a loudspeaker connected to a signal generator as in Topic 9.1 to find out the lowest and the highest frequency you can hear.

2 Work in a small group with each person in turn as the 'subject' at a fixed distance from the loudspeaker. For each subject, adjust the signal generator so the sound can be heard comfortably, then increase the frequency of the signal generator gradually until the subject can no longer hear the sound. This is the upper frequency limit of the subject's hearing. Repeat the test to find the lower frequency limit for the subject.

Young people can usually hear sound frequencies from about 20 Hz to about 20 000 Hz. Older people in general cannot hear frequencies at the higher end of this range. Sound waves above the upper limit of human hearing are referred to as **ultrasonic** waves or **ultrasound**.

Investigating different sounds

Use a microphone connected to an oscilloscope to display the waveforms of different sounds.

Figure 9.2.2 Investigating different sound waves

1 Test a tuning fork to see the waveform of a sound of constant frequency.

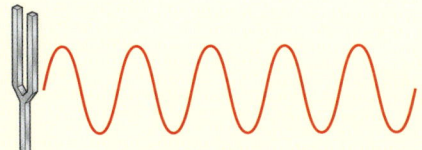

Figure 9.2.3 Tuning fork waves

2 Compare the 'pure' waveform of a tuning fork with the sound you produce when you talk or sing or when you whistle. You may be able to produce a pure waveform when you whistle or sing but not when you talk.

3 Use a signal generator connected to a loudspeaker to produce sound waves. The waveform on the oscilloscope screen should be a pure waveform.

Your investigations should show you that

- increasing the **loudness** of a sound makes the waves on the screen taller. This is because the **amplitude** of the sound waves (the maximum disturbance) is bigger the louder the sound is

- increasing **the frequency** of a sound (the number of waves per second) increases its **pitch**. This makes more waves appear on the screen.

SUMMARY QUESTIONS

1 Copy and complete the sentences below using words from the list.

decreases   **increases**   **stays the same**

a If the sound from a loudspeaker is made louder without changing its pitch, the amplitude of the sound waves _____ and the frequency of the waves _____.

b If the pitch of the sound from a loudspeaker is raised and the loudness is reduced, the amplitude of the sound waves _____ and the frequency of the waves _____.

c If a vibrating guitar string is shortened, the pitch of the sound it produces _____ and the frequency of the sound waves _____.

2 a What is the highest frequency the human ear can hear?

b A sound meter is used to measure the loudness of sound. Describe how you would use the meter and the arrangement shown in Figure 9.2.4 to find out if high frequency sounds travel further than low frequency sounds. Assume the signal generator has a loudness control knob and a frequency dial that can be used to change the frequency of the sound from the loudspeaker.

KEY POINTS

- Sound waves can travel through liquids and gases and in solids.

- Sound waves cannot travel in a vacuum.

- The loudness of a sound increases if the amplitude of the sound wave increases.

- The pitch of a sound increases if the frequency of the sound increases.

- Ultrasound waves are sound waves of frequency above 20 kHz.

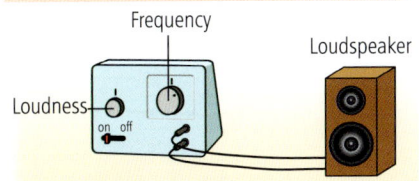

Figure 9.2.4

9.3 The speed of sound

LEARNING OUTCOMES

- Describe how to measure the speed of sound in air
- Explain how echo sounding can be used to measure distances
- **S** State 'order of magnitude' values for the speed of sound in air, in liquids and in solids

When you see a flash of lightning from a lightning strike some distance away, you will hear the clap of thunder from the lightning strike a short time later. The light waves from the lightning strike reach you almost instantly, because light takes only about 3 millionths of a second to travel each kilometre. In comparison, sound waves in air take about 3 seconds to travel each kilometre.

A lightning strike

PRACTICAL

Measuring the speed of sound in air

1   You need two people for this. You and a friend should stand on opposite sides of a field at a measured distance apart. You should be as far away from each other as possible but within sight of each other.

Figure 9.3.1 Measuring the speed of sound in air

2   If your friend bangs two cymbals together, you will see them crash together straightway but you will not hear them straightaway. This is because the 'bang' will be delayed as sound travels much slower than light. Use a stopwatch to time the interval between seeing the impact and hearing the sound.

Work out the speed of sound in air using the formula:

$$\text{speed} = \frac{\text{distance}}{\text{time taken}}$$

Note If possible, swap positions so your friend creates the sound where you were at first and you detect the sound where your friend was at first. This is in case the wind affects the speed of sound in air. Use the two timing measurements to calculate the average time taken by the sound to travel the measured distance and then calculate the speed of sound from speed = distance/average time taken.

EXAM TIP

You can estimate how far away a lightning strike is by counting the seconds between the lightning flash and the thunder clap then dividing by 3. The result gives you the approximate distance in kilometres.

In an experiment to measure the speed of sound in air, a student fired a starting pistol. Another student 600 m away measured the time between the flash and the bang at 1.8 s. Calculate the speed of sound in air.

Solution

$$\text{speed} = \frac{\text{distance}}{\text{time}} = \frac{600\,\text{m}}{1.8\,\text{s}} = 330\,\text{m/s}$$

Echoes

A sound **echo** is an example of reflection of sound. Echoes can be heard in a large hall or gallery which has bare smooth walls or in a cave with smooth walls. For example, if you clap your hands in a large hall, you should be able to hear an echo of the clap.

The echo is due to sound waves created when you clap your hands reflecting off the wall and returning to you. The further away you are from the wall, the longer the sound waves take to travel to the wall and back to you, so the longer the delay between the clap and the echo.

Unwanted echoes in buildings and halls can be eliminated by:

- covering the walls in soft fabric, which absorbs sound waves instead of reflecting them so no echoes will be heard
- making the wall surface uneven, not smooth, so reflected sound waves are scattered (i.e. 'broken up') so they cannot cause echoes.

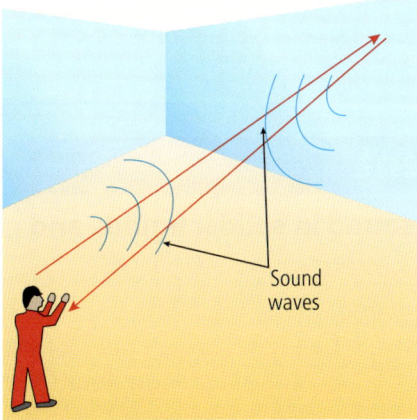

Sound waves

Figure 9.3.2 An echo

Echo sounding

The depth of water beneath a boat or a ship can be measured by directing pulses of sound vertically downwards from a transmitter beneath the ship. The system is called **sonar**. Each pulse reflects from the sea bed and is detected by a detector such as a microphone at the same depth as the transmitter. Each detected pulse is an echo due to reflection from the sea bed (see Figure 9.3.3 on the next page).

CONTINUED

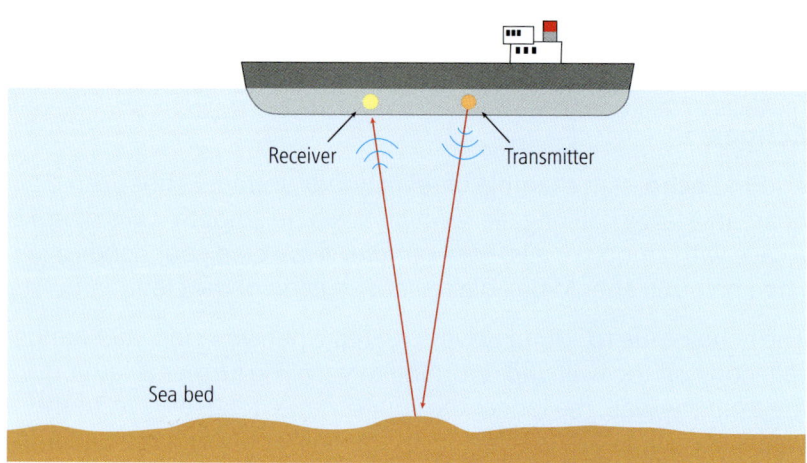

Figure 9.3.3 Echo sounding

To find the depth, d, of the sea bed below the transmitter, the time taken, t, by each pulse to travel from the transmitter to the sea bed and back to the microphone is measured.

- In time t the distance travelled by the pulse $= vt$, where v is the speed of sound in water.
- The depth of the sea bed below the transmitter, $d = \frac{1}{2} \times$ the distance travelled by the pulse in time t.

Therefore, $d = \frac{1}{2}vt$

EXAM TIP

Remember that the sound has to travel down to the sea bed and back up again to reach the receiver!

Supplement

The speed of sound in solids, liquids and gases

The speed of sound depends on the substance the sound is travelling through and on the temperature of the substance. The table on the opposite page shows the speed of sound in some different substances, all at the same temperature of 20 °C.

The table shows that in general, for the same temperature:

- sound travels much faster in solids and liquids than in gases
- sound travels faster in solids than in liquids.

The table shows there are exceptions to these two general statements, for example sound travels faster in hydrogen gas than in liquid alcohol at the same temperature. However, the speed of sound in most gases is less than in most liquids at the same temperature.

substance	physical state	speed of sound / m/s
air	gas	340
carbon dioxide	gas	270
hydrogen	gas	1320
alcohol	liquid	1210
water	liquid	1460
copper	solid	3560
iron	solid	5130
lead	solid	1230

EXAM TIP

You do not need to learn the numbers in the table but you do need to have an awareness of the relative speeds of sound in air, water and in solids.

KEY POINTS

- An echo is sound reflected from a smooth hard surface.
- The speed of sound in air is about 340 m/s.
- The speed of sound in water is about 1400 m/s.
- The speed of sound in many solids is more than 3000 m/s.

SUMMARY QUESTIONS

Use speed of sound in air = 340 m/s.

1 a A clap of thunder from a lightning strike was heard 6.0 s after the lightning flash was observed. Calculate the distance from the observer to where the lightning strike occurred.

  b In an experiment to measure the speed of sound in air, a student blew a short blast on a whistle and raised an arm at the same time. An observer 500 m away heard the whistle blast 1.5 s after seeing the arm raised.

    i Explain why the whistle blast was heard by the observer after the arm was raised.

    ii Use the data to calculate the speed of sound in air.

2 In a test to measure the depth of the sea bed, sound pulses took 0.40 s to travel from the surface to the sea bed and back. Given that the speed of sound in sea water is 1350 m/s, calculate the depth of the sea bed.

3 A ship in fog near a cliff sounded its horn, as in Figure 9.3.4. An echo was heard 5.0 s later.

  a Calculate the distance from the ship to the cliff at the time when the horn was sounded.

  b The ship's captain repeated the test a minute later and found the echo was heard 5.6 s later. Explain without calculation whether or not this and the previous measurement mean that the ship is moving away from the cliff.

Cliff

Figure 9.3.4

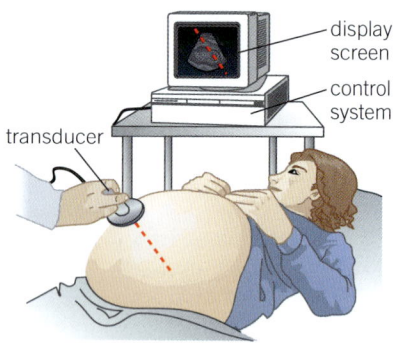

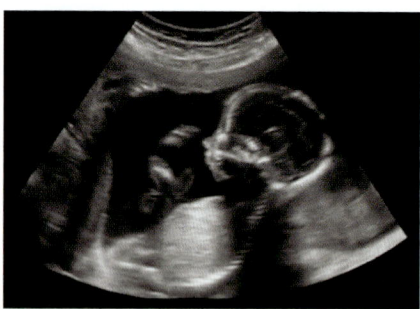

Figure 9.4.1 a An ultrasound scanner system b An ultrasound image of a baby in the womb

EXAM TIP

Ultrasound is often reflected – it goes there and back – so in calculations be careful about distance.

The human ear can detect sound waves in the frequency range of about 20 Hz to about 20 kHz. Sound waves above the highest frequency that humans can detect are called **ultrasound waves**.

Ultrasound scanners

Ultrasound waves are used for prenatal scans of a baby in the womb. They are also used to get an image of organs in the body such as a kidney, or damaged ligaments and muscles. An ultrasound scanner is made up of an electronic device called a transducer placed on the body surface, a control system, and a display screen. The transducer produces and detects sets (or pulses) of ultrasound waves.

Each ultrasound wave pulse from the transducer:

- is partially reflected from the different tissue boundaries in its path
- returns to the transducer as a sequence of ultrasound waves reflected by the tissue boundaries, arriving back at different times.

The transducer is moved across the surface of part of the body. The ultrasound waves are then detected by the transducer. They are used to build up an image on a screen of the internal tissue boundaries in the body.

The advantages of using ultrasound waves for certain types of medical scanning are that ultrasound waves, unlike X-rays, are:

- reflected at boundaries between different types of tissue (different media), so they can be used to scan organs and other soft tissues in the body
- non-ionising. Non-ionising radiation is radiation that does not have enough energy to remove an electron to ionise an atom or molecule. So it is harmless when used for scanning.

Measuring ultrasound waves

Ultrasound waves can be very useful in industrial imaging. Flaws in metal castings can be detected using ultrasound waves. A flaw might be an internal crack, which creates a boundary inside the metal. The ultrasound waves are partly reflected from this boundary. A transducer at the surface of the block sends ultrasound waves into the block. The reflected waves are detected by the transducer and displayed on an oscilloscope screen or on a computer monitor, as shown in Figure 9.4.2.

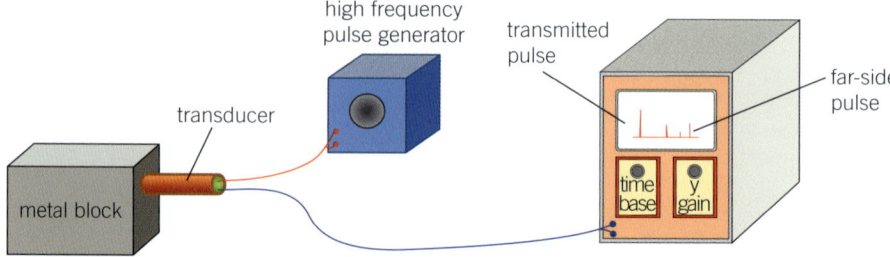

Figure 9.4.2 Detecting flaws in a metal

The oscilloscope screen in Figure 9.4.2 shows:

- a strong ultrasound wave pulse due to partial reflection of the transmitted pulse at the surface, then

- some further ultrasound wave pulses – in this case, two due to partial reflection at internal boundaries, and the last wave pulse due to partial reflection at the far side of the metal object.

The further away a boundary is from the transmitter, the longer a reflected wave takes to return. You can use the oscilloscope to measure the time taken by the wave to travel from the transmitter at the surface to and from the boundary that reflected it. To calculate distance travelled, use the equation:

$$\begin{array}{c}\text{distance travelled}\\\text{by the wave, }s\\\text{(metres, m)}\end{array} = \begin{array}{c}\text{speed of ultrasound waves}\\\text{in body tissue, }v\\\text{(metres per second, m/s)}\end{array} \times \begin{array}{c}\text{time taken, }t\\\text{(seconds, s)}\end{array}$$

Because the ultrasound waves travel from the surface to the boundary then back to the surface, the depth of the boundary below the surface is half the distance travelled by each wave to and

$$\begin{array}{c}\text{the depth of the}\\\text{boundary below}\\\text{the surface, }s\\\text{(metres, m)}\end{array} = \frac{1}{2} \times \begin{array}{c}\text{speed of the}\\\text{ultrasound waves, }v\\\text{(metres per}\\\text{second, m/s)}\end{array} \times \begin{array}{c}\text{time taken, }t\\\text{(seconds, s)}\end{array}$$

from the boundary. So:

SUMMARY QUESTIONS

1 a Explain why ultrasound waves are partly reflected by body organs.

b Explain why an ultrasound scanner is better than an X-ray scanner for scanning a body organ.

c Give one reason why the amplitude of the ultrasonic waves becomes smaller as the waves travel across the body.

2 Look at the screen in Figure 9.4.2. The oscilloscope shows the reflected wave pulses that are detected for each transmitted wave pulse.

a How many internal boundaries are present?

b I The oscilloscope beam takes 2.0×10^{-4} s to travel across the screen. The distance across the block is 90 mm. Calculate the speed of ultrasound in the block.

II Calculate the distance from the flaw nearest to the transducer to the surface of the block where the transducer is placed.

WORKED EXAMPLE

In Figure 9.4.2 the distance across the block is 90 mm. Estimate the distance from the transducer at the surface to the nearest internal boundary.

Solution

On the screen, the pulse due to the nearest internal boundary is halfway between the transmitted pulse and the far-side pulse.

Therefore, the distance d from the transducer to the nearest internal boundary is about half the distance across the block.

Distance $d = 0.5 \times 90$ mm = **45 mm**

KEY POINTS

- Ultrasound waves are sound waves of frequency above 20 kHz.

- Ultrasound waves are partly reflected at a boundary between two different types of body tissue.

- Ultrasound waves reflected at boundaries are timed, and the timings are used to calculate distances.

- An ultrasound scan is non-ionising, so it is safer than an X-ray.

1 (a) Copy and complete the sentence.

Sound waves are produced when objects
_____ .

(b) List four musical instruments and describe in each case what is vibrating to cause the sound.

2 Describe and explain the difference between transverse and longitudinal waves.

Give one example of each type of wave.

3 (a) Describe an experiment to demonstrate that sound does not travel through a vacuum. Include a diagram in your answer.

(b) How can you show with the same apparatus that light does travel through a vacuum?

4 (a) Explain what is meant by the term 'echo'.

(b) A person is 30 m away from a reflecting surface. He claps his hands and hears the echo. If the speed of sound in air is 340 m/s, how long does it take from the time the person claps his hands to hearing the echo?

5 Describe an experiment to measure the speed of sound in air. Include a diagram in your answer and an explanation of how you would calculate the result from the readings.

6 Sound travels at different speeds in different substances.

Which of the following gives the correct order for listing the speed of sound in different substances, from greatest to lowest speed?

(a) air, iron, water

(b) air, water, iron

(c) iron, water, air

Practice Questions

2 What would you hear as the frequency of a sound wave is increased?

A The loudness of the sound would increase as the frequency increases.

B The pitch of the sound decreases as the frequency increases.

C The pitch of the sound increases as the frequency increases.

D The loudness of the sound would decrease as the frequency increases.

3 Which row in the table shows the approximate range of human hearing?

Answer	Lowest frequency in Hz	Highest frequency in Hz
A	2	20 000
B	20	2 000
C	20	20 000
D	200	20 000

4 An electric bell is placed inside a jar as shown in the diagram. The air is removed from the jar with a vacuum pump. Which of these is the best explanation of why the bell cannot be heard even though it can be seen to be ringing?

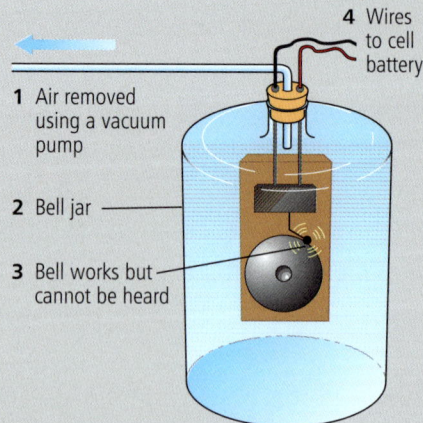

4 Wires to cell battery

1 Air removed using a vacuum pump

2 Bell jar

3 Bell works but cannot be heard

A The metal of the bell cannot vibrate in a vacuum.

B Sound waves need a medium to travel through.

C The vacuum is slowing down the hammer, so it makes less sound.

D The vacuum absorbs the sound wave energy.

1 What type of wave is a sound wave?

A Longitudinal wave

B Transverse wave

C Electromagnetic wave

D Vibrating wave

5 Which row of the table shows the correct values for the speed of sound in the different materials?

Answer	Air (gas)	Water (liquid)	Iron (solid)
A	5130 m/s	1460 m/s	340 m/s
B	513 m/s	146 m/s	34 m/s
C	34 m/s	146 m/s	513 m/s
D	340 m/s	1460 m/s	5130 m/s

6 A student decides to measure the speed of sound in air by producing echoes as shown in the figure. They measure the distance between themselves and the wall, then clap and time how long it takes for the echo to reach them.

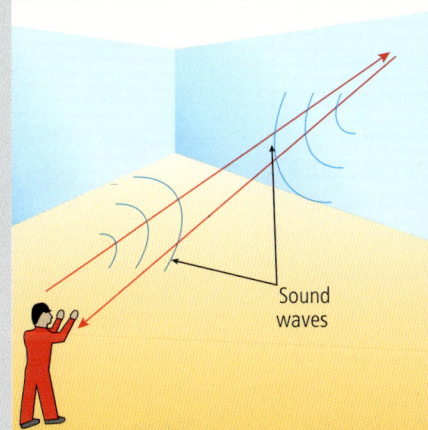

Sound waves

(a) Explain how the student should use the distance and the time data to find the speed of sound. [2]

The student finds that their reaction time is not quick enough to start and stop the stopwatch.

(b) How can the student increase the time it takes for them to hear the echo? [1]

(c) How might this change make it more difficult for the student to hear the sound? [1]

(d) Describe a more accurate method to measure the speed of sound in air. [2]

7 A student is testing a range of instruments to examine the sound waves they produce. They start by examining the 'pure' waveforms produced by tapping tuning forks.

(a) Name the device which can show the waveforms produced by sound or electrical signals on a screen. [1]

The student taps a tuning fork near a microphone and this produces the waveform shown below.

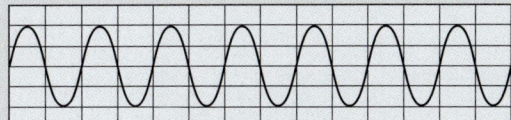

(b) Copy the diagram and mark the period and amplitude of the wave. [2]

(c) The student taps a tuning fork which produces a lower pitched and quieter sound. Sketch a waveform which shows what the student would see on the screen. [3]

8 An echo sounding system is used to measure the depth of a lake as shown in the figure. A signal is produced by the transmitter and received back at the receiver 0.50 s later.

The speed of sound in water is 1500 m/s.

(a) Calculate the depth of the water. [3]

(b) Explain why the positioning of the transmitter and receiver on this boat will result in inaccuracies in the depth measurement. [2]

(c) Suggest one other reason why the depth measurement may be inaccurate. [1]

Sound waves travel as longitudinal waves.

(d) Describe how a sound wave travels through the water to reach the bottom of the lake. [5]

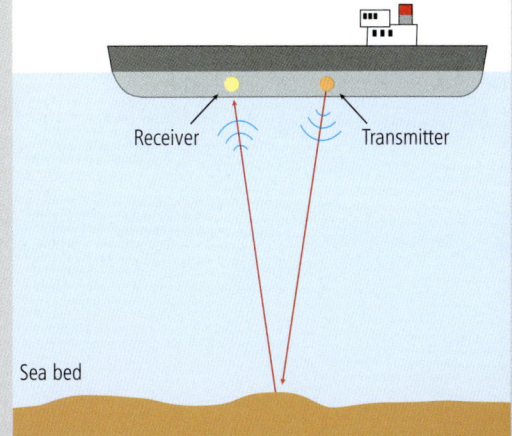

Receiver   Transmitter

Sea bed

10.1 Magnets

Magnetic poles

A magnetic compass is a very useful device if you lose your way outdoors. The compass needle is a tiny 'bar' magnet that is pivoted at its centre of gravity so it is free to turn horizontally.

When outdoors, the needle turns until one end points north and the other end points south. The effect is caused by the Earth's magnetic field. The end of the plotting compass that points north is called the 'north-seeking' pole and the other end the 'south-seeking' pole (usually referred to as the magnet's **north pole** (N) and **south pole** (S), respectively).

Using a compass

PRACTICAL

Investigating bar magnets

1. Suspend a bar magnet as shown in Figure 10.1.1 to check which end is its N-pole and which is its S-pole. Make sure the bar magnet is suspended horizontally and there are no other magnets nearby. If the ends are unmarked, label them N and S respectively.

2. Hold the N-pole of a second bar magnet near the suspended bar magnet as shown in Figure 10.1.2. You should find it attracts the S-pole of the suspended bar magnet and repels the N-pole.

3. Repeat the above test by holding the S-pole of the second bar magnet near the suspended bar magnet. This time, you should find it attracts the N-pole of the suspended bar magnet and repels its S-pole.

The tests show the general rule that:

like poles repel; unlike poles attract

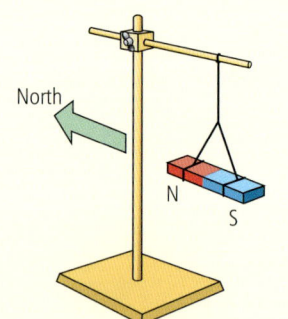

Figure 10.1.1 Checking the poles of a bar magnet

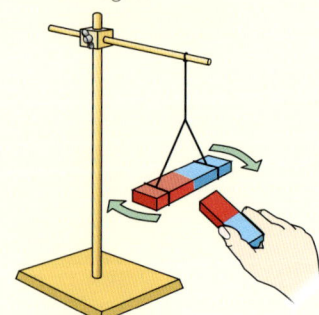

Figure 10.1.2 Like poles repel; unlike poles attract

Magnetic induction

An unmagnetised iron bar placed near a bar magnet is always attracted to the bar magnet. For example, Figure 10.1.3 shows an unmagnetised bar XY placed end-on with a bar magnet a few centimetres away. Whichever way around the magnet is placed, there is always a force of attraction between XY and the bar magnet.

- In Figure 10.1.3a, the magnet's N-pole induces (i.e. brings about) an S-pole at the nearest end of XY (i.e. end Y). So end Y is attracted to the bar magnet.
- In Figure 10.1.3b, the bar magnet has been reversed so the magnet's S-pole is nearest to end Y. This induces an N-pole at end Y. So end Y is attracted to the S-pole of the bar magnet.

In other words, the magnet induces magnetism in XY such that the end of XY nearest to a pole of the bar magnet has the opposite polarity to that pole and is therefore attracted by the bar magnet.

When the bar magnet is removed, the magnetism induced in XY is lost.

Magnetic materials

Any iron or steel object can be magnetised (or demagnetised if it is already magnetised). Steel is referred to as a **ferrous** material because it contains iron. Any ferrous material can be magnetised or demagnetised. Only a few non-ferrous materials can be magnetised and demagnetised. Cobalt and nickel are two such examples. Oxides of these metals are used to make ceramic magnets and to coat magnetic tapes and discs.

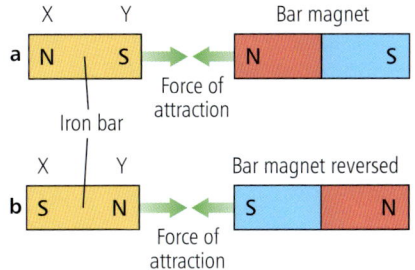

Figure 10.1.3 Magnetic induction

SUMMARY QUESTIONS

1 A bar magnet XY is freely suspended in a horizontal position so end X points north and end Y points south.

 a State the magnetic polarity of **i** end X, and **ii** end Y.

 b End P of a second bar magnet PQ placed near end X of bar magnet XY repels end X and attracts end Y. Explain this observation and state the magnetic polarity of end P.

2 A bar magnet is suspended horizontally as shown in Figure 10.1.1. When one end of an unmagnetised iron bar is placed near the N-pole of the bar magnet, the bar magnet is attracted to the iron bar.

 a Explain this observation.

 b State and explain what would be observed if the bar magnet was reversed.

KEY POINTS

- Like poles repel and unlike poles attract.

- A ferrous material, such as steel, contains iron. Any ferrous material can be magnetised and demagnetised.

- Magnetism is induced in a bar of ferrous material when a magnet is placed near it.

10.2 Magnetic fields

LEARNING OUTCOMES

- Recall the pattern of magnetic field lines around a bar magnet
- Describe an experiment to plot the field lines of a magnetic field
- Describe how to magnetise and demagnetise a magnetic material

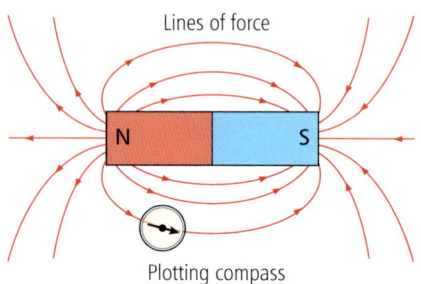

Figure 10.2.1 The magnetic field of a bar magnet

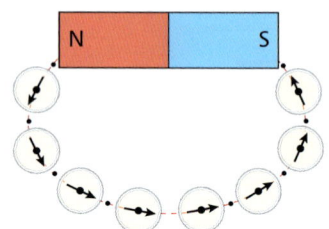

Figure 10.2.2 Plotting magnetic field lines

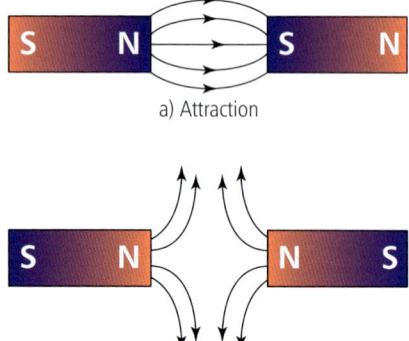

a) Attraction

b) Repulsion

Figure 10.2.3 Magnets in line

Magnetic field patterns

If a sheet of paper is placed over a bar magnet and iron filings are sprinkled onto the paper, the filings form a pattern of lines. The space around the magnet is called a **magnetic field.** Any other magnet placed in this space experiences a force due to the first magnet.

In Figure 10.2.1:

- iron filings form lines that end at or near the poles of the magnet. These lines are called **magnetic lines of force** (or sometimes 'magnetic field lines').

- **S** the stronger the magnet, the more lines of force there are and the closer together they are.

- a plotting compass placed in the field would align itself along a line of force, pointing in a direction away from the N-pole of the magnet and towards the magnet's S-pole. For this reason, the direction of a line of force at any point is in the direction of the force on the north pole of a compass at that point.

- **S** magnetic forces are due to the interaction between magnetic fields.

PRACTICAL

Plotting a magnetic field

1 Draw an outline of a bar magnet on a sheet of white paper.

2 Mark a dot near the north pole of the bar magnet. Place the tail of the compass needle above the dot and mark a second dot at the tip of the compass needle. Move the compass so the tail of the needle is above the second dot and mark a new dot at the tip of the needle. Figure 10.2.2 shows the idea. Repeat the procedure until the compass reaches the S-pole of the magnet. Draw a line through the dots and mark the direction of the line from the N-pole to the S-pole of the bar magnet.

3 Repeat the procedure for further lines.

The magnetic field near two magnets

The strength of the magnetic field is represented by the spacing between the magnetic field lines. In Figure 10.2.1, the lines are closest near the poles where the field is strongest. Figure 10.2.3 shows the field pattern between two bar magnets placed end-on.

- In **a** the field between the two magnets is very strong because the field of each magnet there is from left to right.

- In **b** the field between the two magnets is very weak because the field of one magnet is in opposite directions to that of the other magnet. If both magnets are equally strong, the field midway between them would be zero.

Making a magnet

An iron nail can be magnetised by moving one end of a bar magnet along the nail repeatedly in the same direction, as shown in Figure 10.2.4.

In Figure 10.2.4, the N-pole of a bar magnet is moved repeatedly along the nail from its head to its tip. When the nail is magnetised in this way:

- the head of the nail becomes an N-pole (as if it tries to repel the magnet as it approaches)
- the tip of the nail becomes an S-pole (as if it tries to attract the magnet it as it moves away).

The magnetic polarity of each end of the magnetised nail can be checked by holding each end of the nail in turn near a compass needle. If one end of the magnetised nail attracts the N-pole of the plotting compass and repels the S-pole, that end of the nail must be an S-pole (because unlike poles always attract).

To demagnetise the nail, the S-pole of the bar magnet (instead of the N-pole) can be moved repeatedly along the nail from head to tip. However, if the process is repeated too many times, the nail will be demagnetised then magnetised in the opposite direction.

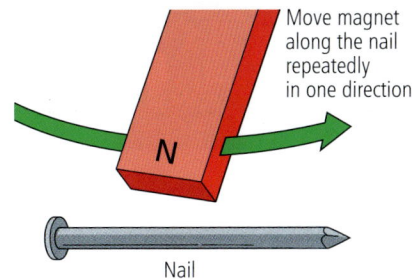

Move magnet along the nail repeatedly in one direction

N

Nail

Figure 10.2.4 Making a magnet

Extension

A magnet also loses its magnetism if it is repeatedly hammered or heated strongly!

SUMMARY QUESTIONS

1 **a** Sketch the pattern of the magnetic field lines around a bar magnet. On your diagram, label the north and south poles of the bar magnet and indicate the direction of the field lines.

  b Figure 10.2.5 shows a bar magnet XY and a plotting compass near end Y of the bar magnet. The needle of the plotting compass points towards end Y.

   i State the magnetic polarity of each end of the bar magnet.

   ii If the bar magnet was rotated gradually about its centre through 180°, describe and explain the effect on the direction of the plotting compass needle.

2 **a** Describe how you would use a bar magnet to magnetise an iron nail so its tip is an S-pole.

  b The tip of an iron nail is held in turn near each end of a plotting compass needle. State whether the tip of the nail is an N-pole, an S-pole or is unmagnetised in each of the following possible observations.

   i The N-pole of the plotting compass is repelled by the tip of the nail and the S-pole is attracted by the tip.

   ii The N-pole of the plotting compass is attracted by the tip of the nail and the S-pole is repelled by the tip.

   iii The N-pole of the plotting compass is attracted by the tip of the nail and the S-pole is also attracted by the tip.

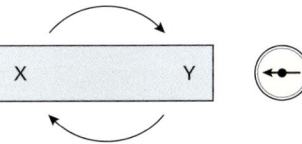

X   Y

Figure 10.2.5 Magnetic field lines

KEY POINTS

- A line of force (also called a field line) of a magnetic field is a line along which a plotting compass points.
- The magnetic field lines of a bar magnet curve from the north pole of the bar magnet around to the south pole.
- An iron or steel bar can be magnetised by:
  - repeatedly moving a pole of a bar magnet along the bar in the same direction
  - withdrawing the bar from a long coil carrying a direct current.

10.3 More about magnetic materials

LEARNING OUTCOMES

- Recall the meaning of 'hard' and 'soft' magnetic materials

- Distinguish between iron and steel in terms of their magnetic properties

- Explain why iron is used in an electromagnet and steel is used to make a permanent magnet

Magnetic atoms

Each atom of a magnetic material has its own magnetic field like that of a tiny bar magnet because of the arrangement and motion of the electrons in the atom. The atoms exert forces on each other due to their magnetism. In a magnet, the atoms are all lined up with each other, so the whole bar is magnetised. If the magnet is demagnetised, the overall pattern of alignment breaks up and there is no overall magnetisation.

Comparison of the magnetic properties of iron and steel

Steel is harder to magnetise and to demagnetise than iron, so it is referred to as a **hard magnetic material**. In comparison, iron is said to be a **softer magnetic material** because it is easy to magnetise and demagnetise. Iron that loses its magnetism very easily is called **soft iron**.

- Steel rather than iron is used to make permanent magnets because iron loses its magnetism more easily; for example when it becomes hot or when it is struck repeatedly with a hammer. Hitting a magnetised iron bar repeatedly makes its atoms vibrate more, with the result that the atoms lose their overall alignment and so the bar loses its overall magnetism. Steel is much harder to demagnetise than iron because much more energy is needed to make the atoms in steel move out of their overall alignment.

- Soft iron rather than steel is used as the core of an electromagnet because iron is easier to magnetise and demagnetise than steel. An iron bar can be magnetised using a bar magnet or a current-carrying coil. In both cases, an external magnetic field is applied to the iron bar. This field aligns all the atoms in the bar in the direction of the magnetic field. As a result, the whole bar becomes magnetised in the direction of the external magnetic field. An **electromagnet** consists of a soft iron 'core' with insulated wire coiled around it. When an electric current is passed through the wire, the core becomes magnetised. When the current is switched off, the core loses its magnetism. An electromagnet attached to a crane is used in scrap yards to lift car bodies or any other objects made of a ferrous material.

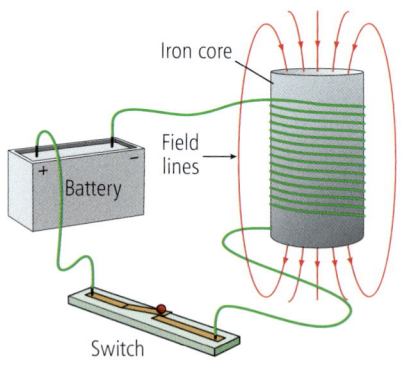

Figure 10.3.1 (a) An electromagnet

(b) Using an electromagnet

More about permanent magnets

Permanent magnets are made in different shapes including bars, flat rectangles, discs, cylinders, flat rings and U-shapes.

U-shaped magnets (sometimes referred to as horseshoe magnets) are bars bent into a U shape and then magnetised with opposite polarity at each end. The magnetic field in the gap between the two ends is very strong, as shown in Figure 10.3.2.

A loudspeaker magnet is a flat ring with a central cylinder of opposite polarity, as in Figure 10.3.3. The loudspeaker coil moves in the space between the flat ring and the central cylinder where the field is strongest. When an alternating current passes through the loudspeaker coil, the coil is forced to move in and out of the magnetic field as the current alternates. As a result, the diaphragm attached to the coil is made to vibrate, creating sound waves. See Topic 14.2 *The motor effect*.

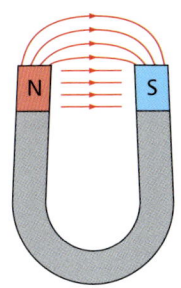

Figure 10.3.2 A horseshoe (U-shaped) magnet

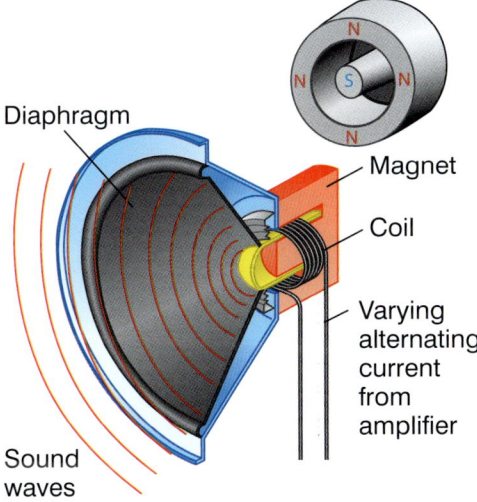

Diaphragm

Magnet

Coil

Varying alternating current from amplifier

Sound waves

Figure 10.3.3 Inside a loudspeaker

SUMMARY QUESTIONS

1 Copy and complete the following sentences using words from the list below.

hard   iron   keeps   loses   soft   steel

a A permanent magnet is made from _____ which is magnetically _____ and therefore _____ its magnetism.

b A _____ magnetic material such as _____ is easier to magnetise and demagnetise than a _____ magnetic material such as _____.

2 a Describe the essential features of an electromagnet.

b Explain why the core of an electromagnet is made of iron rather than steel.

KEY POINTS

- A 'hard' magnetic material is harder to magnetise and demagnetise than a softer magnetic material.
- In magnetic terms, steel is hard and iron is soft.
- Soft iron is used in an electromagnet and steel is used to make a permanent magnet.

1 Describe an experiment to plot the shape and direction of the magnetic field lines around a bar magnet.

2 Copy and complete the diagrams to show the shape and direction of the magnetic field between the magnetic poles.

(a)  N   S

(b) S   N

(c) 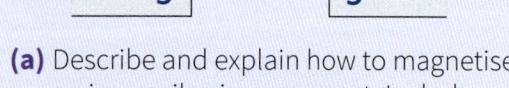 N   N

(d) S   S

3 (a) Describe and explain how to magnetise an iron nail using a magnet. Include a diagram in your answer.

(b) Describe how to demagnetise a magnetised iron bar.

4 Describe and explain the difference between a 'hard' magnetic material and a 'soft' magnetic material. Give an example of each.

5 (a) Describe what is meant by the term 'electromagnet'. Include a diagram in your answer.

(b) Give three examples of the uses of electromagnets.

Practice Questions

1 Which of these materials can be used to show the shape of a magnetic field when sprinkled around a bar magnet?

  A   Copper dust

  B   Plastic sprinkles

  C   Iron filings

  D   Gold flakes

2 The diagram shows the ends of two magnets placed close together. Which of the following statements is true?

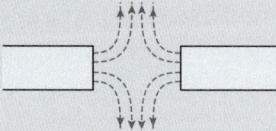

  A   One end is a south pole and one is a north pole; the magnets are attracting each other.

  B   Both ends are south poles and the magnets are attracting each other.

  C   Both ends are north poles and the magnets are repelling each other.

  D   Both ends are south poles and the magnets are repelling each other.

3 Which of the following is a description of magnetic induction?

  A   An unmagnetised iron bar acts as a magnet when placed near to a permanent magnet.

  B   A bar magnet attracts another bar magnet when they are placed close together.

  C   A bar magnet repels another bar magnet when they are placed close together.

  D   A compass points towards the magnetic poles of the Earth.

4 Which of the following explains why one magnet produces a force on another nearby magnet?

  A   The electric fields of the two magnets interact and produce a force.

  B   The magnetic fields of the two magnets interact and produce a force.

  C   The magnets are affected by each other's gravity.

  D   The poles of the magnets give off energy.

5 Which of these processes can be used to demagnetise a metal bar?

   A Place the bar inside a coil with direct current passing through it and slowly removing it.

   B Heat the bar strongly.

   C Place the bar next to another permanent magnet.

   D Expose the bar to ultraviolet radiation.

6 Simple bar magnets are used in a wide range of experiments in schools. Over time they become weaker magnets.

   (a) Suggest why the magnets become weaker over time. [1]

   (b) Copy this shape and draw the shape of the magnetic field which surrounds a weak bar magnet. [4]

S	N

   (c) Describe how your diagram would be different if the bar magnet was newer and much stronger. [2]

7 In a recycling plant, steel and aluminium cans are separated using a magnet above a conveyor belt. The steel cans are attracted to the magnet and are removed while the aluminium cans go on to be recycled. Some separators use strong permanent magnets while others use electromagnets.

   (a) Why aren't the aluminium cans attracted to the separator? [1]

   (b) Describe two advantages and two disadvantages of using an electromagnet instead of a permanent magnet. [4]

8 A student places an iron rod inside a coil of wire which has a d.c. current running through it. The iron rod becomes magnetic. The student repeats the process with a similarly sized steel rod which also becomes magnetic

   (a) Describe the changes which happen inside the steel or iron rod which causes them to become magnetic. [4]

The student bangs the two rods together several times. One of the rods loses most of its magnetic strength while the other does not.

   (b) State which rod becomes demagnetised quickly and explain why. [3]

9 Two identical bar magnets are placed side-by-side as below.

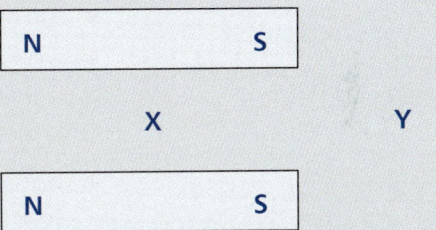

   (a) Draw the arrangement and the pattern of the magnetic field lines in the gap between the magnets. [3]

   (b) i) A plotting compass is placed in the gap at X which is equidistant from the poles of the magnets. On your drawing, show the plotting compass in this position and show the direction in which it points. [1]

      ii) The plotting compass is then moved to Y. Describe how the direction of the plotting compass changes as it moves from X to Y. [3]

11.1 Static electricity

EXAM TIP

You must learn the facts about the charge on protons, electrons and neutrons.

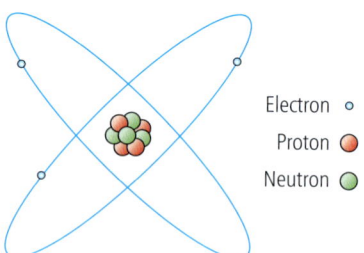

Electron ∘
Proton ●
Neutron ●

Figure 11.1.1 Inside an atom

EXAM TIP

Charged atoms are called <u>ions</u>. Atoms that gain electrons become negative ions. Atoms that lose electrons become positive ions. Any process that makes uncharged atoms into ions is called <u>ionisation</u>. You will meet ionisation in the next topic and in Chapter 15 on radioactivity.

Producing electric charge

To stick a balloon on a ceiling, all you need to do is to rub the balloon on your clothing before you touch it on the ceiling. The rubbing action gives the balloon **electric charge**. This stays on the balloon so we refer to it as **static electricity**. The charge on the balloon attracts it to the ceiling.

Inside the atom

To understand why certain materials can become charged, we need to consider what is inside an atom. Every atom has a positively charged nucleus composed of protons and neutrons. Electrons move about in the space around the nucleus.

- A proton has a positive charge.
- An electron has an equal negative charge.
- A neutron is uncharged.

An uncharged atom has equal numbers of electrons and protons. Only electrons can be transferred to or from an atom.

1   Adding electrons to an uncharged atom makes it negative (because the atom then has more electrons than protons).

2   Removing electrons from an uncharged atom makes it positive (because the atom has fewer electrons than protons).

Charge is measured in **coulombs**. We need over 6 million million million electrons to charge an object with 1 coulomb of charge. The magnitude of the charge of the electron is 1.6×10^{-19} coulombs. We will meet the coulomb again in Topic 11.4.

Charging by friction

Some insulators become charged by rubbing them with a dry cloth.

- Rubbing a perspex rod with a dry cloth transfers electrons from the surface atoms of the rod onto the cloth. So the perspex rod becomes positively charged.
- Rubbing a polythene rod with a dry cloth transfers electrons to the surface atoms of the rod from the cloth. So the polythene rod becomes negatively charged.

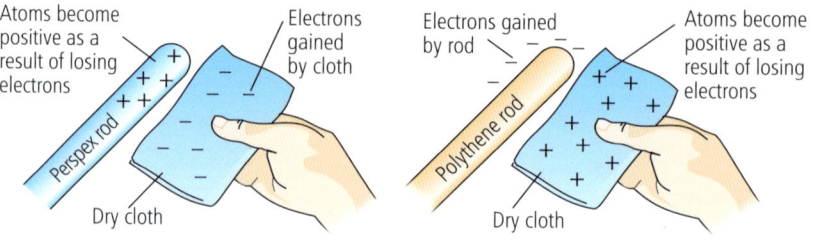

Atoms become positive as a result of losing electrons

Electrons gained by cloth

Electrons gained by rod

Atoms become positive as a result of losing electrons

Perspex rod

Polythene rod

Dry cloth

Dry cloth

Figure 11.1.2 Charging by friction

The force between two charged objects

Two charged objects exert a force on each other. Figure 11.1.3 shows how you can investigate this force. Your results should show that:

● two objects with the same type of charge (i.e. like charges) repel each other

● two objects with opposite types of charge (i.e. unlike charges) attract each other.

Like charges repel; unlike charges attract

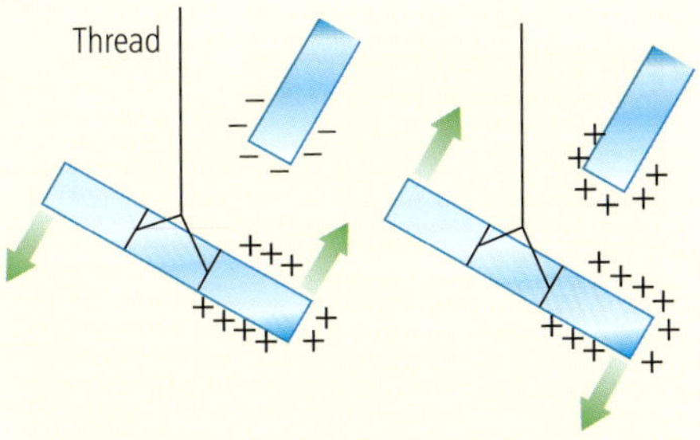

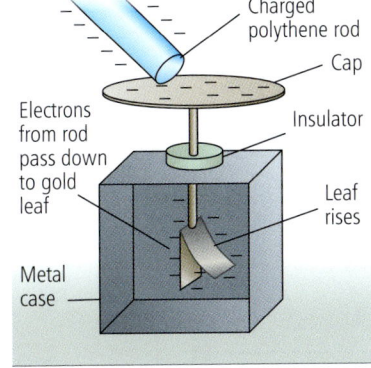

Figure 11.1.3 The law of force for charges

The electroscope

Figure 11.1.4 shows an electroscope, a device that detects charge. We can use an electroscope to see if an object is charged. In Figure 11.1.4, the electroscope is charged by direct contact with a charged rod. The gold leaf of the electroscope is repelled by the metal plate when the electroscope is charged. This happens because they both gain the same type of charge.

Figure 11.1.4 The electroscope

1 Choose words from the list to complete **a** and **b**:

to   from   loses   gains

a When a polythene rod is charged using a dry cloth, it becomes negative because it _____ electrons that transfer _____ it _____ the cloth.

b When a perspex rod is charged using a dry cloth, it becomes positive because it _____ electrons that transfer _____ it _____ the cloth.

2 **a** When rubbed with a dry cloth, perspex becomes positively charged. Polythene and ebonite become negatively charged. State whether or not attraction or repulsion occurs when:

   i a perspex rod is held near a polythene rod

   ii a perspex rod is held near an ebonite rod

   iii a polythene rod is held near an ebonite rod.

b Glass is charged when it is rubbed with a cloth. When a charged glass rod is held near a positively charged rod, it is repelled by the rod.

   i What type of charge, positive or negative, does a glass rod have when it is charged?

   ii Does glass gain or lose electrons when it is charged?

● Like charges repel; unlike charges attract.

● Insulating materials that lose electrons when rubbed become positively charged.

● Insulating materials that gain electrons when rubbed become negatively charged.

● Charge is measured in coulombs.

11.2 Electric fields

LEARNING OUTCOMES

- Describe what is meant by an electric field

S • Explain what is meant by a line of force in an electric field

- Describe simple electric field patterns

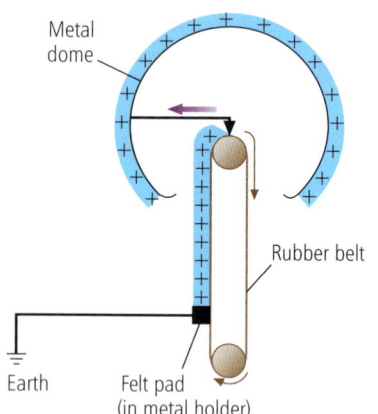

Figure 11.2.1 The Van de Graaff generator

Supplement

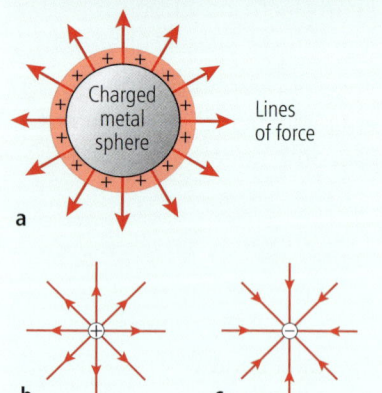

Figure 11.2.2 Lines of force near:
(a) a charged sphere
(b) a positive point charge
(c) a negative point charge

Lines of force

A **Van de Graaff generator** can make your hair stand on end. The dome charges up when the generator is switched on. Massive sparks are produced if the charge on the dome builds up too much.

The Van de Graaff generator charges up because:

- the belt rubs against a felt pad and becomes charged,
- the belt carries the charge onto an insulated metal dome,
- sparks are produced when the dome can no longer hold any more charge.

When the dome is charged, any other charged object brought near to it is repelled if it carries the same type of charge as the dome (or attracted if it carries the opposite type of charge). The charged dome is surrounded by an **electric field** due to its charge. Any other charged object near the dome is acted on (i.e. repelled or attracted) by the electric field of the dome.

A Van de Graaff generator in use

Supplement

Electric field patterns

We can picture an electric field using the idea of **lines of force**. A line of force is the path a free positive charge would take if released in the field. Figure 11.2.2 shows the lines of force around a charged sphere and around a positive and a negative 'point charge'. Note the direction of the lines of force in each case (i.e. the direction which a free positive charge would move in the field). In all three examples, the electric field is said to be **radial**.

You can see the pattern of an electric field using the apparatus shown in Figure 11.2.3. Two conductors, connected to a Van de Graaff generator are submerged in castor oil sprinkled with semolina powder. The powder grains line up along the lines of force of the field.

- In Figure 11.2.4a, the field between the two oppositely charged parallel plates is said to be uniform because the lines of force are parallel to one another.
- In Figure 11.2.4b, one of the conductors has a curved section. The lines of force are concentrated at the curve because this is where the charge is most concentrated.

Note the direction of the field lines is always from the positive conductor to the negative conductor. In general, the direction of an electric field at a point in the field is the direction of the force on a positive charge at that point

Supplement

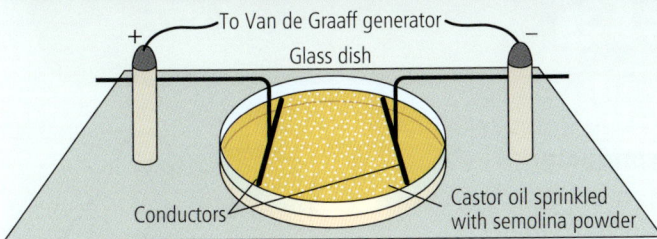

Figure 11.2.3 Demonstrating field patterns

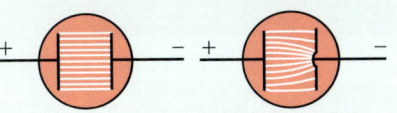

a A uniform field   **b** A concentrated field

Figure 11.2.4 Field patterns

Electric fields at work

The electrostatic paint sprayer is used to paint metal panels. The spray nozzle is connected to the positive terminal of an electrostatic generator. The negative terminal is connected to the metal panel. The panel attracts positively charged paint droplets from the spray.

The lightning conductor

A lightning strike is a massive flow of charge between a thundercloud and the ground. A lightning conductor on a tall building prevents lightning strikes by allowing the thundercloud to discharge gradually.

When a thundercloud is above a lightning conductor, a very strong electric field is created in the surrounding air. Air molecules near the conductor tip become **ionised** due to electrons being pulled off them. The ions then discharge the thundercloud so no lightning flash is produced. The conductor is joined to the ground by a thick copper strip. This allows charge to flow safely to earth.

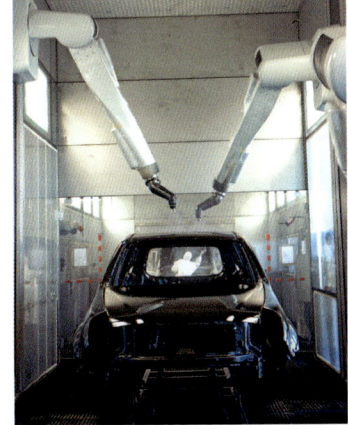

An electrostatic paint sprayer

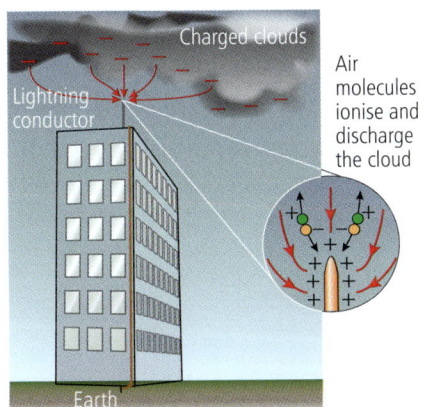

Figure 11.2.5 A lightning conductor

KEY POINTS

- An electric field is the region around a charged object in which a second charged object experiences a force due to its charge.

- A line of force is the path of a free positive charge released in an electric field.

- The lines of force surrounding a point charge or a charged sphere are radial.

- The lines of force between two oppositely charged parallel conductors are parallel to one another.

SUMMARY QUESTIONS

1 Copy and complete the following sentences using words from the list below.

   charge   field   force   ionisation

   a The region around a charged metal sphere is referred to as an electric _____.

   b If a particle in this region does not experience a _____, its _____ must be zero.

2 Draw a diagram to show the pattern of the electric field:

   a around a negative 'point charge',

   b between two oppositely charged flat conductors.

   In both cases show the direction of the lines of force of the field.

11.3 Conductors and insulators

A metal is a **conductor** of electricity. In Figure 11.3.1, the torch lamp lights up if a metal object is connected in the gap XY. The lamp will not light up if a plastic object is connected in the gap. Plastic is an **insulator**, not a conductor of electricity.

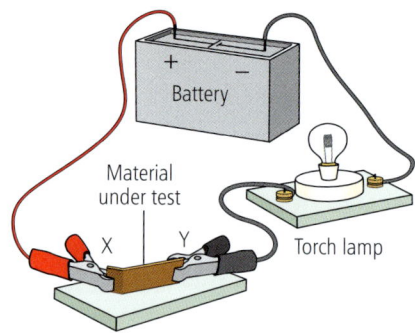

Figure 11.3.1 Conductors and insulators

Metals conduct electricity because they contain **conduction (delocalised) electrons**. These electrons have broken free from atoms in the metal and they move about freely inside the metal. When a metal conducts electricity, electrons transfer electric charge because they are forced to move through the metal. Insulators cannot conduct electricity because all the electrons are held in atoms.

Charging and discharging a conductor

A conductor can hold charge only if it is insulated from the ground. If it isn't insulated, it will not hold any charge because electrons transfer between the conductor and the ground.

A conductor cannot be charged by rubbing it with a cloth because a cloth is not a good insulator. So it won't hold charge and neither will the conductor.

To charge an insulated conductor, we can bring it into **direct contact** with a charged object as shown in the charging of an electroscope in Figure 11.3.3.

- If the object is negatively charged, electrons transfer to the conductor from the object. So the conductor becomes negative because it gains electrons.

- If the object is positively charged, electrons transfer from the conductor to the object. So the conductor becomes positive because it loses electrons.

To discharge a charged conductor safely, a conducting path (e.g. a wire) needs to be provided between the object and the ground. The conducting path allows electrons to transfer between the conductor and the ground (Figure 11.3.2). We say the object is then **earthed**.

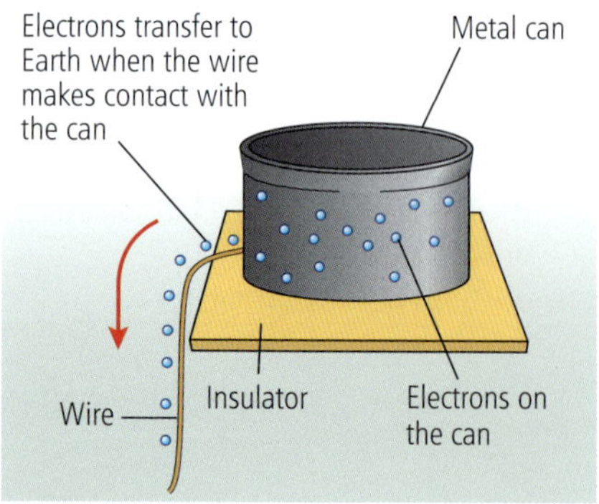

Electrons transfer to Earth when the wire makes contact with the can

Metal can

Wire — Insulator — Electrons on the can

Figure 11.3.2 Earthing a negatively charged conductor

Extension

Charging by induction

An insulated conductor can be charged from a charged body indirectly by **induction** as well as by direct contact. Figure 11.3.4 shows the process of charging an insulated metal sphere by induction using a negatively charged rod. Charging by induction gives the conductor the opposite type of charge to the charge on the rod. Also, no charge is gained or lost by the rod in this process.

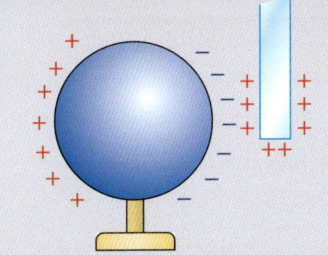

a The charged rod is held near the sphere

PRACTICAL

Charge and discharge tests

1 Place a metal can on an electroscope as shown in Figure 11.3.3. Use a wire to make sure the can is discharged.

2 Charge a polythene rod by rubbing it with a dry cloth and place the rod in direct contact with the can. You should see that the electroscope leaf rises, indicating the presence of charge on the can and the electroscope.

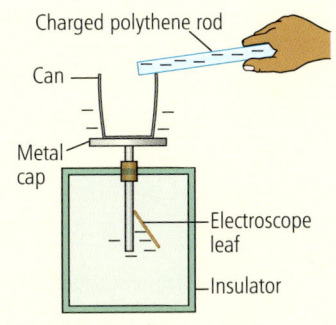

Charged polythene rod

Can

Metal cap

Electroscope leaf

Insulator

Figure 11.3.3 Charge tests

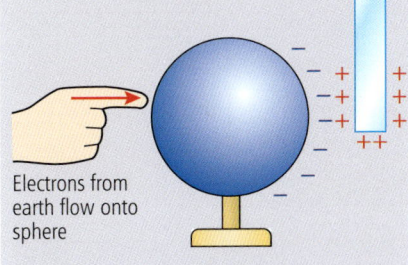

Electrons from earth flow onto sphere

b The sphere is earthed briefly

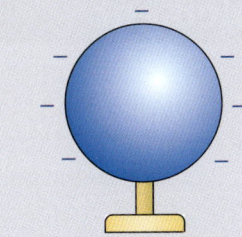

c The rod is removed. The sphere is left with an opposite charge to the rod

Figure 11.3.4 Charging by induction

SUMMARY QUESTIONS

1 **a** Which of the following substances conduct electricity?

   brass cork glass lead nylon

   polythene rubber wood

   b i Explain why a metal object can hold charge only if it is insulated from the ground.

   ii Describe how you would charge a metal sphere using a polythene rod.

2 **a** When a road tanker pumps oil or petrol into a storage tank, the connecting pipe must be earthed. If it is not, it could become charged. A build-up of charge would cause a spark which could make the fuel vapour explode.

   i Why is the rubber hose of a petrol pump made of conducting rubber?

   ii Why must the storage tank and the road tanker also be earthed before the fuel is pumped into the storage tank?

   b In a dry climate, road vehicles carry a metal chain that is attached to the back of the vehicle so it touches the ground.

   i Explain why the chain prevents the road vehicle from becoming charged?

   ii Why is such a chain unnecessary in damp climates?

KEY POINTS

- Any object that charge can pass through is a conductor.
- Any object that charge cannot pass through is an insulator.
- An insulated conductor can be charged by direct contact with a charged object.

11.4 Charge and current

LEARNING OUTCOMES

- State that an electric current is due to a flow of charge
- Recall the unit of electric current
- Describe the use of an ammeter
- **S** State that charge is measured in coulombs
- Recognise that a current is a rate of flow of charge and use the equation $I = \dfrac{Q}{t}$

Supplement

In Figure 11.4.1, each time the ball touches the left-hand plate, conduction electrons transfer to it from the plate so the ball becomes negatively charged. When the ball makes contact with the right-hand plate, electrons leave the ball so the ball becomes positively charged and is then attracted back to the left-hand plate. In this way, the ball transfers electrons, and therefore charge, across the gap.

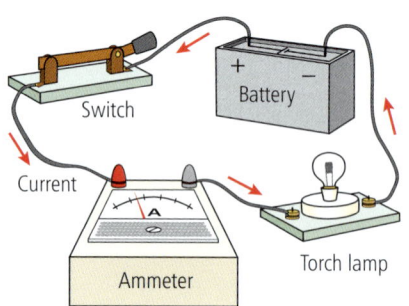

Figure 11.4.2 Using an ammeter

The shuttling ball experiment

An electric current is a flow of charge. Figure 11.4.1 shows a demonstration in which a ball bounces 'to and fro' between two oppositely charged metal plates when the electrostatic generator is on. The ball is a table-tennis ball coated with conducting paint.

- When the ball touches the negative plate, it becomes negatively charged itself and is then repelled by the plate.
- The ball is then attracted onto the positive plate where it loses its negative charge and becomes positively charged.
- It is then repelled by the positive plate and attracted back to the negative plate and then repeats the 'to and fro' motion for as long as the generator is on.

The ball transfers charge from one plate to the other each time it moves across the gap. The meter in Figure 11.4.1 registers a tiny **electric current** because charge passes through it each time charge is transferred across the gap.

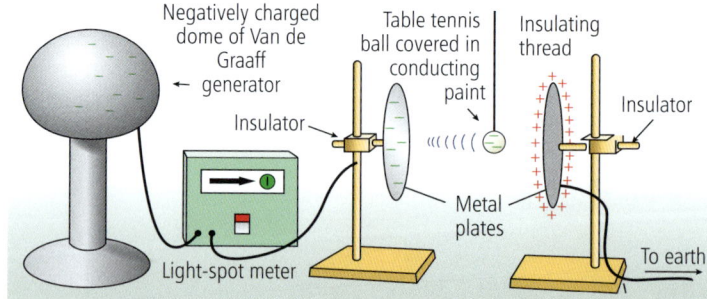

Figure 11.4.1 The shuttling ball experiment

If the plates are brought closer together, the ball shuttles between the plates at a faster rate. This makes the meter reading increase because the ball is 'ferrying' charge across the gap at a greater rate than before. So the electric current through the meter increases.

An electric current is due to a flow of electric charge.

Electric current

An **ammeter** is a meter designed to measure electric current. The unit we use to measure current is the **ampere (A)**. Figure 11.4.2 shows an electric circuit consisting of a torch lamp connected to a battery, a switch and an ammeter. We say the components are **in series** because the same amount of charge flows through each circuit component every second. The ammeter therefore measures the current in the torch lamp.

The direction of the current around an electric circuit is always shown on a circuit diagram from the **+** terminal of the battery to the **–** terminal, as in Figure 11.4.2.

Supplement

Electric charge is measured in coulombs (C). One coulomb is defined as the amount of charge passing a point in a circuit each second when the current is one ampere. In other words, a current of one ampere is equal to a rate of flow of charge of one coulomb per second.

1 For a steady current I in a circuit, if the charge flowing is Q in a certain time t, then

the current (in amperes) = $\dfrac{\text{charge flowing (in coulombs)}}{\text{time taken (in seconds)}}$

$$I = \frac{Q}{t}$$

Rearranging the above equation gives $\quad Q = It \quad$ or $\quad t = \dfrac{Q}{I}$

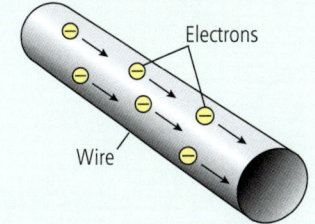

Figure 11.4.3 Current equals charge flow per second

2 **The direction in which electrons move around an electric circuit is from the – terminal of the battery to the + terminal.** This is opposite to the direction in which the current is marked on a circuit diagram. The reason is that the current direction convention was established as the direction of the flow of **positive charge** long before the discovery of electrons. Furthermore, an electric current can be due to the transfer of positive or negative charged particles. It just so happens that metals contain electrons that can move about freely inside the metal.

PRACTICAL

Measure the current in a torch lamp

1 Connect a torch lamp, a battery, a switch, an ammeter and a variable resistor in series with each other. Use the variable resistor to alter the brightness of the lamp.

2 Measure the lamp current when:

- the lamp is at normal brightness
- the lamp can just be seen to be emitting light.

SUMMARY QUESTIONS

1 Copy and complete the following sentences using words from the list below:

ampere ammeter charge coulomb current stopwatch

a The unit of charge is the _____.

b To measure the _____ flowing along a wire carrying a constant _____, we would need to use an _____ and a _____.

2 a In the 'shuttling ball' experiment shown in Figure 11.4.1, describe and explain the effect on the ball of moving the plates further apart.

b For each of the examples **i** to **iv** in the table below, calculate the missing quantities in the table below.

current / A	3.5	ii	5	iv
charge / C	i	200	500	0.50
time / s	60	40	iii	20

EXAM TIP

The unit for current is often called the amp (short for ampere).

KEY POINTS

- Electric current is due to a flow of charge.
- Electric current is measured in amperes (A).
- An ammeter is used to measure electric current.
- Current is the rate of flow of charge.
- Electric charge is measured in coulombs (C).
- charge flow = current × time (in seconds)

1 The diagram shows an atom. Name the parts labelled A, B and C.

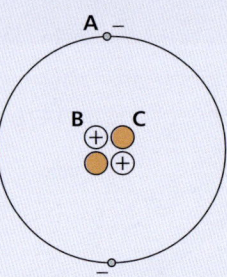

2 Copy and complete the following:

Rubbing a polythene strip with a dry cloth transfers _____ from the duster to the polythene. The polythene therefore becomes _____ charged, leaving the dry cloth _____ charged.

3 Describe a simple experiment using strips of insulating material hanging by insulating threads that will demonstrate the forces between charges. Describe the results that you would expect to observe and the conclusion that you would draw from the results.

4 The diagram shows a long-haired wig on top of the charged dome of a Van de Graaff generator.

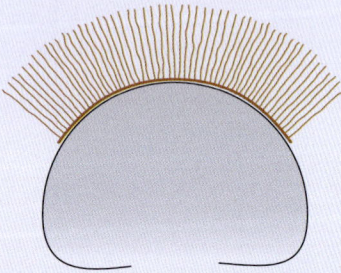

Explain why the hairs 'stand on end'.

5 Write a list of four materials that are electrical conductors and four materials that are insulators.

S 6 (a) Write down the equation that links charge, current and time.

(b) State the units for each of the quantities in part (a).

(c) A current of 0.8 A is switched on. Calculate the amount of charge that flows in a time of 5 minutes.

Practice Questions

1 Which row of the table gives the correct information about the electrical charges of protons, neutrons and electrons?

	proton	neutron	electron
A	negative charge	neutral	positive charge
B	neutral	neutral	negative charge
C	positive charge	neutral	negative charge
D	positive charge	positive charge	positive charge

2 A small piece of plastic has an overall negative charge. Which of the following statements must be true?

A The plastic has more electrons than protons.

B The plastic has more protons than electrons.

C The plastic has more neutrons than protons.

D The plastic has an equal number of electrons and protons.

3 What is an electrical current?

A The flow of electrical charge

B The force holding charges together in an atom

C The energy carried by electrical charges

D The movement of small masses in a wire

4 Which of the following is a measurement of electric charge?

A 4.0 volts

B 4.0 joules

C 4.0 coulombs

D 4.0 amperes

5 Why are metals good electrical conductors while most other materials are not?

   A Metals contain protons which can travel freely through them.

   B Metals contain ions which can travel freely through them.

   C Metals contain electrons which travel freely through them.

   D Non-metals do not contain electrons.

6 A polythene rod is charged negatively by rubbing it with a cloth. The rod is used to touch a metal cup which has been placed on top of an electroscope as shown in the diagram.

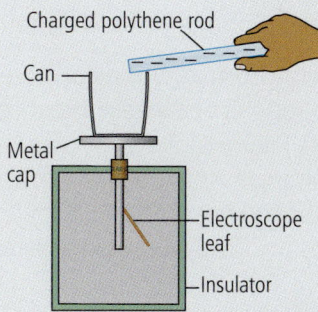

Charged polythene rod

Can

Metal cap

Electroscope leaf

Insulator

(a) Explain why the electroscope leaf rises in terms of the movement of charges and forces. [4]

(b) Why does the electroscope leaf need to be made of metal? [1]

The student discharges the electroscope and the leaf falls. They bring a strongly positively charged rod close to the metal can without touching it. The leaf rises slightly.

(c) Explain why the leaf rises. [2]

7 While working with sensitive electronic equipment, an engineer attaches a strap to their wrist. The strap contains electrical conductors and it is connected to the engineer's wrist at one end and a metal table at the other. This strap helps to reduce the build-up of electric charge.

(a) Name a material that could be used as an electrical conductor. [1]

(b) Which type of particles can travel through solid electrical conductors? [1]

(c) Name an electrical insulator. [1]

(d) Explain why the engineer should wear the strap and what could happen if they did not. [3]

8 The diagram shows an experiment used to demonstrate charge flow. When the Van de Graaff (VDG) generator is turned on, the ball begins to bounce back and forth between the two metal plates. The dot on the light-spot meter moves from side to side.

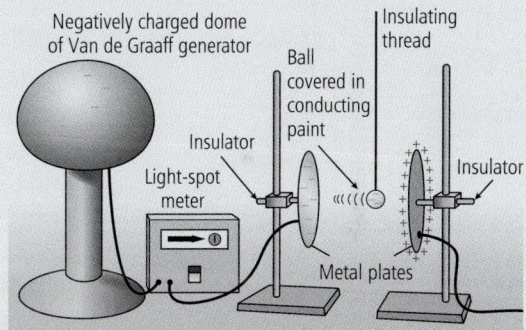

Negatively charged dome of Van de Graaff generator

Ball covered in conducting paint

Insulating thread

Insulator

Insulator

Light-spot meter

Metal plates

(a) Describe how the ball becomes charged when it touches the plate connected to the VDG generator. [2]

(b) Explain why the ball is then attracted towards the second metal plate. [1]

S

During the demonstration, the ball takes 1.5 s to move from the left plate to the right and back again. Each swing transfers 20 mC of charge.

(c) Calculate the total amount of charge which has been transferred during one minute of operation. [2]

(d) Calculate the average current provided by the device. [3]

12.1 Batteries and cells

Batteries in action

Calculators, smart watches, cameras, radios and mobile phones all use batteries. They are used in cars, hearing aids, torches, toys and many other items. Batteries vary in size from the tiny batteries used in smart watches to heavy duty car batteries.

Different types of battery

A battery consists of electric cells connected together. A cell (or a battery) is a source of energy. One terminal of a cell is positively charged and the other is negatively charged. For example:

- A chemical cell transfers energy from its chemical store of energy into electrical energy. This happens because substances inside the cell react with each other. Figure 12.1.1 shows a simple non-rechargeable cell containing a zinc rod and a copper rod in dilute sulfuric acid. The zinc rod becomes negatively charged and the copper rod becomes positively charged as a result of the reactions in the cell.

- A solar cell transfers energy from light into electrical energy. This happens because atoms in the solar cell release electrons as a result of absorbing light. This movement of electrons causes one terminal of the cell to become positively charged and the other negatively charged.

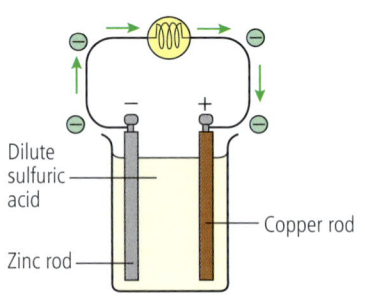
Dilute sulfuric acid
Copper rod
Zinc rod

Figure 12.1.1 A simple chemical cell

Electromotive force

When a cell is in a circuit, charge is forced to flow around the circuit by the cell. This flow of charge transfers energy from the cell to the circuit components then into the surroundings. In a chemical cell, this process continues until one of the substances in the cell has all reacted. In a solar cell, the process stops if there is no light reaching the cell.

The **electromotive force** (or 'emf') of a cell or a battery measured in **volts (V)** is a measure of how much 'push' the cell or battery can provide to force charge around the circuit. The greater the emf, the more energy the cell can deliver for every electron that passes through it.

The emf of a cell, sometimes referred to as its 'voltage', depends on the substances in the cell. Batteries and cells are usually marked clearly in volts. For any given battery-operated device, the battery in it must be of the correct voltage. If the battery emf is too low, the device is unlikely to work. If the battery emf is too high, the device is likely to be damaged.

A panel of solar cells

The electromotive force (emf) of a source of electrical energy is the work done per unit charge by the source in moving charge round a complete circuit.

S

emf, *E*, of a battery (in volts) = $\dfrac{\text{work done, } W, \text{ by the battery (in joules)}}{\text{charge (in coulombs)}}$

$$E = \frac{W}{Q}$$

The work done by a cell or battery is the amount of energy it transfers from its chemical store of energy. This depends on its emf. For every coulomb of charge that passes through it, a cell of emf 12 V transfers 12 J of electrical energy. Therefore, a 12 V battery can transfer twice as much energy per coulomb of charge as a 6 V battery.

EXAM TIP

It is useful to remember that 1 volt is equal to 1 joule per coulomb (ie. 1 V = 1 J/C).

PRACTICAL

Measuring the emf of a battery

1 The emf of a cell or battery can be measured using a **voltmeter**, as shown in Figure 12.1.2.

2 The voltmeter must be connected so its + terminal is connected to the + terminal of the battery and its − terminal is connected to the − terminal of the battery.

Figure 12.1.2 Measuring the emf of a battery

MORE ABOUT VOLTAGE

Batteries marked at more than 12 V can give you a nasty shock if you touch a terminal. Mains electricity is lethal because it is supplied at a much higher voltage. Never ever touch a mains circuit or terminal.

SUMMARY QUESTIONS

1 Copy and complete the following sentences using words from the list below:

an ammeter   charge   current   a voltmeter

a The emf of a battery is measured using _____.

b The _____ through a battery in a circuit is measured using _____.

c The longer a battery is in a circuit, the greater the _____ that flows through it.

2 Describe how you would use a voltmeter to measure the emf of a battery.

KEY POINTS

- Electromotive force (emf) is measured in volts.

- The 'emf' of a battery is a measure of its 'push' or charge.

- A voltmeter is used to measure the emf of a battery.

- The emf of the battery is the work done by the battery per unit charge in moving charge round a complete circuit.

12.2 Potential difference

LEARNING OUTCOMES

- Explain what is meant by potential difference
- State that potential difference (pd) is measured in volts
- Describe how to use a voltmeter to measure the pd across a component in a circuit
- Explain the significance in terms of energy of the emf of a battery and a pd in a circuit

EXAM TIP

A voltmeter is connected IN PARALLEL with a component to measure the pd ACROSS the component.

MORE ABOUT METERS

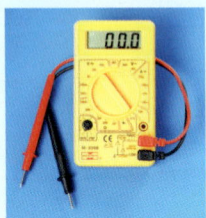

Figure 12.2.2 A digital meter

An ammeter or voltmeter with a pointer is called an **analogue** meter. An ammeter or a voltmeter with a digital read-out is called a **digital** meter. Whichever type of meter you use, always make sure it reads zero when it is not connected in a circuit or to a battery. To ensure an accurate reading of the analogue meter, the pointer must be viewed with the eye directly 'above' the pointer.

Energy transfer in a circuit

Figure 12.2.1 shows a lamp, a variable resistor and a battery connected in series. Charge from the battery has the **potential** to deliver energy to the circuit components. When it flows around the circuit, it transfers energy from the battery to the lamp and the variable resistor.

In Figure 12.2.1, a **voltmeter** is connected across the lamp. We say it is in **parallel** with the lamp. A voltmeter can be connected to any two points in a circuit. Its reading gives the **potential difference** (abbreviated to 'pd' and sometimes referred to as 'voltage') between those two points. This is a measure of the energy transferred by each electron as it passes between those two points.

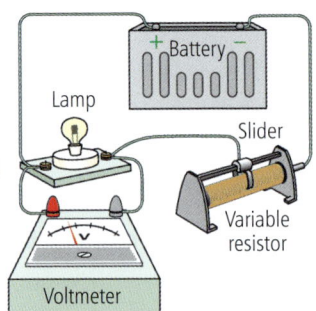

Figure 12.2.1 In series

- An electric circuit transfers energy from a source of energy such as an electrical cell or mains supply to the circuit components.
- If a voltmeter is connected across the terminals of the battery, it measures the emf of the battery in volts, provided no energy is 'wasted' as thermal energy inside the battery.

The potential difference (pd) across an electrical component is the work done per unit charge by the charge flowing through the component.

$$\text{potential difference, } V = \frac{\text{work done, } W, \text{ by the charge (in joules)}}{\text{charge, } Q \text{ (in coulombs)}}$$
(in volts)

$$V = \frac{W}{Q}$$

The work done by a battery is the energy transferred from its chemical energy store. Therefore, in Figure 12.2.1, in a given time:

the electrical energy supplied by the battery = the electrical energy supplied to the lamp + the electrical energy supplied to the variable resistor

- The emf of the battery is the electrical energy per coulomb of charge supplied by the battery.
- The potential difference across a component is the electric energy per coulomb of charge supplied to the component.

Therefore, from the above energy equation, we can say that:

emf of the battery = the pd across the lamp + the pd across the variable resistor

WORKED EXAMPLE

s

A 6 V lamp is connected in series with a 9 V battery and a variable resistor as shown in Figure 12.2.1. The variable resistor is adjusted until the lamp is at its normal brightness. The current through the lamp is then 2 A. Calculate

a the charge passing through the lamp in 10 s

b the electrical energy supplied to the lamp in this time

c the electrical energy supplied by the battery in this time.

Solution

a charge = current × time = 2 A × 10 s = 20 C.

b electrical energy supplied to the lamp = pd × charge
= 6 V × 20 C = 120 J

c electrical energy produced by
battery = emf × charge = 9 V × 20 C = 180 J

SUMMARY QUESTIONS

1 Copy and complete the following sentences using words from the list below:

charge   emf   energy   pd

a _____ and _____ are measured in volts.

b The bigger the _____ of a battery, the more _____ it can deliver for the same _____.

s

2 In a certain time interval, the battery in Figure 12.2.1 transfers 120 J of electrical energy and delivers 80 J of this energy to the lamp.

a Calculate the energy supplied to the variable resistor in this time.

b The charge passing around the circuit in this time is 10 C. Calculate:

i the emf of the battery

ii the pd across the lamp

iii the pd across the variable resistor.

c The current during this time was 0.5 A. Calculate the duration of the time interval in seconds.

PRACTICAL

Measuring potential differences in a circuit

1 Connect a lamp, a variable resistor and a battery in series.

2 Use the slider of the variable resistor to adjust the brightness of the lamp so it is at less than normal brightness.

3 Use the voltmeter as shown in Figure 12.2.1 to measure:

- the pd across the lamp.

- the pd across the variable resistor; you will need to reconnect the voltmeter across the variable resistor to do this.

- the emf of the battery; you will need to reconnect the voltmeter across the battery on its own to measure this.

4 Repeat the test for a different lamp brightness. Record your measurements.

In both cases, assuming there is negligible resistance inside the battery, your measurements should show that:

the emf of the battery = the pd across the lamp + the pd across the variable resistor

KEY POINTS

- Electromotive force (emf) and potential difference (pd) are both measured in volts.

- A voltmeter is used to measure a pd in a circuit.

- The pd across a component is the work done per unit charge by charge flowing through it.

12.3 Resistance

LEARNING OUTCOMES

- Recall the definition of resistance and its unit

- Recognise that the current through a component depends on its resistance and on the pd across it

- Describe an experiment to measure the resistance of a wire

- Calculate resistance or pd or current for a component given two of the three quantities

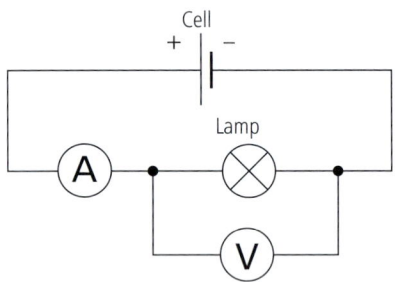

Figure 12.3.1 Using an ammeter and a voltmeter

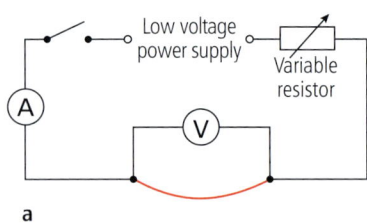

a

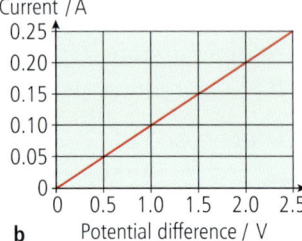

b

Figure 12.3.2 Investigating the resistance of a wire: (a) circuit diagram, (b) a current–potential difference graph for a wire

Ammeters and voltmeters

The current in and the pd across a lamp or any other component can be measured at the same time using the circuit represented in Figure 12.3.1. The standard symbol for each component is used in the circuit diagram.

Look at the ammeter and the voltmeter in the circuit in Figure 12.3.1.

- The ammeter measures the current in the torch lamp. It is connected in **series** with the lamp so the current in them is the same. The ammeter reading gives the current in amperes (A) or milliamperes (mA) for small currents where 1 mA = 0.001 A.

- The voltmeter measures the potential difference across the torch lamp. It is connected in **parallel** with the torch lamp so it measures the pd across it. The voltmeter reading gives the pd in volts (V).

Electrons passing through a torch lamp have to push their way through lots of vibrating atoms. The atoms resist the passage of electrons through the torch lamp.

We define the **resistance** of an electrical component:

$$\text{resistance (in ohms)} = \frac{\text{potential difference (in volts)}}{\text{current (in amperes)}}$$

The unit of resistance is the **ohm**. The symbol for the ohm is the Greek letter Ω ('omega').

We can write the definition above as:

$$R = \frac{V}{I}$$

where V = potential difference in volts

I = current in amperes

R = resistance in ohms.

Note Rearranging the above equation $R = \dfrac{V}{I}$ gives $V = IR$ or $I = \dfrac{V}{R}$

PRACTICAL

Investigating the resistance of a wire

Does the resistance of a wire change when the current in it is changed? Figure 12.3.2a shows how we can use a variable resistor to change the current in a wire. Make your own measurements and use them to plot a current–potential difference graph like the one in Figure 12.3.2b.

current / A	0	0.05	0.10	0.15	0.20	0.25
potential difference / V	0	0.50	1.00	1.50	2.00	2.50

a Discuss how your measurements compare with the ones from the table used to plot the graph in Figure 12.3.2b.

b Calculate the resistance of the wire you tested and the resistance of the wire that gave the results in Figure 12.3.2b.

Current–potential difference graphs

For a wire, the graph in Figure 12.3.2b and your own graph should show:

- a straight line through the origin
- that the current is proportional to the potential difference.

Reversing the potential difference makes no difference to the shape of the line. The resistance is the same whichever direction the current is in.

The graph shows that the resistance (= potential difference/current) is constant. This is known as **Ohm's law:**

The current in a resistor at constant temperature is proportional to the potential difference across the resistor.

Supplement

If a filament lamp is tested instead of a wire, the measurements give a graph which curves, as shown in Figure 12.3.3. This is because the resistance of the filament lamp increases as the current in it increases and makes it hotter. This is because its atoms vibrate more, so they oppose the flow of electrons more.

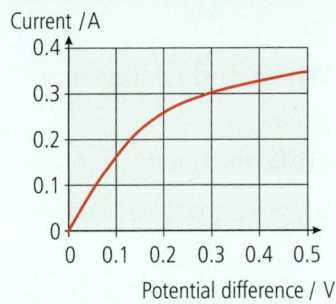

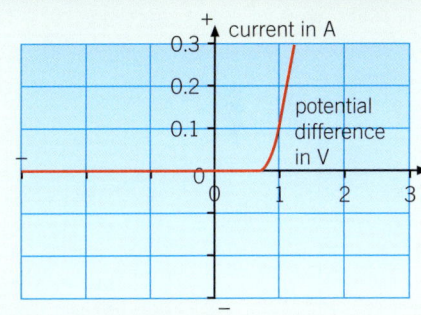

Figure 12.3.3 A current–potential difference graph for a filament lamp

Figure 12.3.4 A current–potential difference graph for a diode

For a diode, the measurements depend on which way round it is connected in the circuit because it only conducts in one direction (called its 'forward' direction). Figure 12.3.4 shows a graph of the measurements for each direction. Its resistance in the forward direction is much lower than in the 'reverse' direction.

SUMMARY QUESTIONS

1 a Draw a circuit diagram to show how you would use an ammeter and a voltmeter to measure the current in and the potential difference across a wire.

  b The potential difference across a resistor was 3.0 V when the current in it was 0.5 A. Calculate the resistance of the resistor.

2 Calculate the missing values in each line of the table.

resistor	current / A	potential difference / V	resistance / Ω
W	2.0	12.0	?
X	4.0	?	20
Y	?	6.0	3.0
Z	0.5	12.0	?

KEY POINTS

- Resistance R (in ohms) =

$$\frac{\text{potential difference } V \text{ (in volts)}}{\text{current } I \text{ (in amperes)}}$$

- $R = \dfrac{V}{I}$ rearranged gives

  $V = IR$ or $I = \dfrac{V}{R}$

- The current in a resistor at constant temperature is proportional to the potential difference across the resistor.

12.4 More about resistance

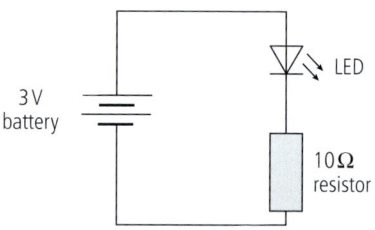

Figure 12.4.1 Using a resistor

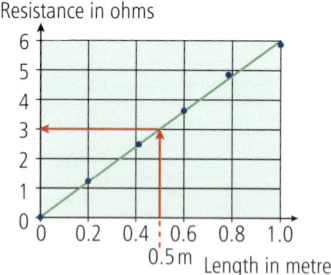

Figure 12.4.2 A graph of resistance against length for a uniform wire

Resistors

A **resistor** is a component designed to have a certain resistance. Resistors are usually made from metal wires or carbon. The symbol for a resistor is shown in the circuit diagram in Figure 12.4.1. Resistors are used in circuits to limit the current in the circuit.

The circuit diagram in Figure 12.4.1 shows a circuit with a 3.0 V battery in series with a light-emitting diode (LED) and a 10 Ω resistor. An LED emits light when a current passes through it. It is often used as an indicator in an electronic circuit. However, if too much current passes through it, it overheats and no longer conducts charge or emits light.

Because the pd across the resistor can never be more than the battery emf of 3.0 V, the current in the resistor can never exceed 0.3 A (= 3.0 V/10 Ω).

Wires and resistance

A wire that has the same diameter all along its length is said to be **uniform**. The resistance of a uniform wire depends on:

- its length – the longer the length of the wire, the greater its resistance

- its diameter – the smaller a wire's diameter, the greater its resistance

- the material it is made from – for example, a metal such as copper conducts more readily than a metal such as steel. So a copper wire of the same diameter and length as a steel wire has less resistance.

PRACTICAL

Make a wire-wound resistor

1 A wire-wound resistor consists of a wire wrapped around a suitable insulating tube. Constantan wire is often used because constantan, a metal alloy, does not rust or oxidise. Use the circuit in Figure 12.3.2 of Topic 12.3 to measure the resistance of different measured lengths of constantan wire (or a wire made of other suitable material) up to 1 m in length. For each length,

- use the variable resistor to adjust the current to the same value and measure the pd across the wire for this current

- use the equation resistance = $\dfrac{\text{pd}}{\text{current}}$

length of wire / m	0	0.20	0.41	0.60	0.79	1.00
current / A	1.5	1.5	1.5	1.5	1.5	1.5
pd / V	0	1.8	3.7	5.4	7.2	8.9
resistance / Ω	0	1.2	2.5	3.6	4.8	5.9

2 Record your measurements and the results of your calculations in a table as shown above.

3 Use your results to plot a graph of resistance against length as shown in Figure 12.4.2.

The length of wire needed to make a resistor of any value less than $6\,\Omega$ can then be found from the graph. For example, we can see from Figure 12.4.2 that a length of 0.50 m of the wire tested would be needed to make a resistor of resistance $3.0\,\Omega$. From your graph, find the length of wire needed to make a resistance of $3.0\,\Omega$. It will not necessarily be the same as that given by Figure 12.4.2 because your wire might have a different diameter and/or be made of a different material. Because the line on the graph is straight throught the origin, the graph shows that the resistance of the wire is directly proportional to its length.

SUMMARY QUESTIONS

1 Copy and complete the following sentences using words from the list below.

greater   the same as   smaller

a A wire W in a circuit has a fixed pd across it. The length of wire W in a circuit is reduced. As a result of this change:

  i the resistance of wire W in the circuit will become _____,

  ii the current in wire W will become _____.

b A wire X in a different circuit is replaced by a thicker wire Y of the same length and the same material.

  i The resistance of wire Y is _____ than that of wire X.

  ii The current in wire Y will be _____ than the current through wire X.

2 a Describe how you would make measurements on wires of different known diameters and of the same material to show that the narrower a wire is, the greater its resistance.

b A wire of resistance $5.0\,\Omega$ and length 1.0 m in a circuit is replaced with a wire of the same material and of length 0.5 m which has half the diameter of the first wire.

  i A student suggests the second wire has a resistance of $2.5\,\Omega$. Explain why this suggestion is incorrect.

  ii Calculate the resistance of the second wire. Explain the steps in your calculation.

Supplement

Tests show that the resistance of a uniform conductor is:

- proportional to its length. For example, if the length is doubled, the resistance is also doubled. The graph in Figure 12.4.2 shows the resistance is proportional to the length because the line is straight and passes through the origin.

- inversely proportional to the area of cross-section of the wire. For example, if the wire is replaced by wire of the same length and of the same material but with twice the cross-sectional area, the resistance is halved.

Wire X of resistance R

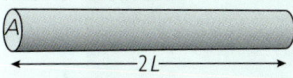

Wire Y of resistance $2R$

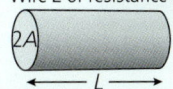

Wire Z of resistance $0.5R$

Figure 12.4.3 Wires of the same material

- The resistance of a uniform wire increases if its length is increased or if its diameter is reduced.

- The resistance of a wire depends on the material it is made from and is:

  i proportional to its length

  ii inversely proportional to its area of cross-section.

12.5 Electrical power

EXAM TIP

Remember than 1 watt = 1 joule per second. See Topic 4.9.

THE HUMAN HEART

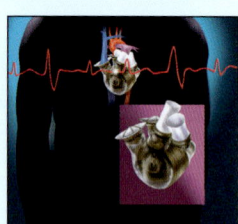

An artificial heart

The power of a human heart varies with activity and is about 1 watt on average. In 1 year, a human heart is supplied with about 30 million joules of energy.

MATHS NOTE

The electrical power equation may be written as

Power $P = IV$

where I = current and V = potential difference

Energy and power

When you use an appliance, it transfers energy from one store to other stores of energy. The **power** of an appliance, in watts, is the rate at which it transfers energy, in joules per second.

$$\text{power (in watts)} = \frac{\text{energy transferred (in joules)}}{\text{time taken (in seconds)}}$$

WORKED EXAMPLE

A light bulb transfers 30 000 J of electrical energy when it is on for 300 s. Calculate its power.

Solution

$$\text{Power} = \frac{\text{energy transferred}}{\text{time taken}} = \frac{30\,000\,\text{J}}{300\,\text{s}} = 100\,\text{W}$$

Calculating power

Millions of millions of electrons pass through the circuit of an artificial heart every second. Each electron transfers a small amount of energy to it from the power supply. So the energy transferred to it each second is large enough to enable the device to work.

For any electrical appliance:

- the current in it is a measure of the number of electrons passing through it each second (i.e. the charge flow per second)
- the potential difference across it is a measure of how much energy each electron passing through it transfers to it (i.e. the electrical energy transferred per unit charge)
- the power supplied to it is the energy transferred to it each second.

Therefore,

$$\begin{matrix} \text{the energy transfer to} \\ \text{the device each second} \end{matrix} = \begin{matrix} \text{the charge flow} \\ \text{per second} \end{matrix} \times \begin{matrix} \text{the energy transfer} \\ \text{per unit charge} \end{matrix}$$

In other words,

$$\begin{matrix} \textbf{power supplied} \\ \text{(in watts)} \end{matrix} = \begin{matrix} \textbf{current} \\ \text{(in amperes)} \end{matrix} \times \begin{matrix} \textbf{potential difference} \\ \text{(in volts)} \end{matrix}$$

For example, the power supplied to:

- a 4 A, 12 V electric motor is 48 W (= 4 A × 12 V),
- a 0.1 A, 3 V torch lamp is 0.3 W (= 0.1 A × 3.0 V).

Electrical energy and potential difference

When a resistor is in a circuit, electrons are forced through the resistor by the power supply. Each electron repeatedly collides with the vibrating atoms of the resistor, transferring energy to them (Figure 12.5.1). The atoms of the resistor therefore gain kinetic energy and vibrate even more. The resistor becomes hotter.

When charge flows through a resistor, electrical energy is transferred into thermal energy.

The electrical energy supplied to a resistor (or any other electrical component) in a given time can be calculated if we know the pd across the resistor and the current in it. This is because:

- energy = power × time
- electrical power = current × potential difference

Therefore, we can calculate the electrical energy supplied using the equation:

electrical energy supplied	=	**current**	×	**potential difference**	×	**time**
(in joules)		(in amperes)		(in volts)		(in seconds)

$$E = IVt \qquad \text{where} \quad I = \text{current}$$
$$V = \text{potential difference}$$
$$t = \text{time taken}$$

Notes

1 For a resistor of resistance R, the pd across it, $V = IR$ when the current in it is I.

In time t, the electrical energy supplied, $E = IVt = I^2Rt$

Therefore the energy per second or power supplied to the resistor = $IV = I^2R$

2 The **kilowatt-hour** (kWh) is defined as the energy transferred by a 1 kilowatt appliance in one hour. Mains electricity meters measure energy supplied in kilowatt hours. 1 kWh = 1000 W × 3600 s = 36 MJ.

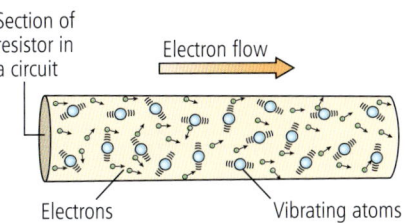

Section of resistor in a circuit

Electron flow

Electrons   Vibrating atoms

Figure 12.5.1 Electrons moving through a resistor

SUMMARY QUESTIONS

1 Copy and complete the sentences **a** to **c** using words from the list below.

current   energy   power potential difference

  a When an electrical appliance is on, _____ is supplied to it as a result of _____ passing through it.

  b When an electrical appliance is on, a _____ is applied to it which causes _____ to pass through it.

  c Charge flowing through a resistor transfers _____ to the resistor.

2 a Calculate the power supplied to each of the following devices in normal use:

    i a 12 V, 3 A light bulb

    ii a 230 V, 2 A heater

  b Calculate the energy transfer:

    i for a charge flow of 20 C when the potential difference is 6.0 V

    ii in 20 s for a current of 3 A in a resistor when the potential difference is 5 V.

KEY POINTS

- Electrical power supplied (in watts) = current (in amperes) × potential difference (in volts)

- Energy transferred

  = power × time (in seconds)

  = current × pd × time

1 Draw a diagram using standard circuit symbols of a circuit designed to show the relationship between the potential difference across a lamp and the current in the lamp at a variety of different potential differences.

2 **(a)** Explain the meaning of the term electromotive force (emf).

(b) Explain how you would measure the emf of a cell.

3 Write down the equation linking potential difference, current and resistance. State the units of each of the three quantities.

4 Use the readings in the table to plot a graph of V / V (y-axis) against I / A (x-axis).

V / V	I / A
0	0
1.8	0.33
4.0	0.68
6.2	1.00
7.9	1.33
10.0	1.70
12.1	2.00

Use the graph to calculate the resistance of the component in which the current is being measured.

5 **(a)** State two factors that affect the resistance of a metal wire.

(b) State whether an increase in the factor increases or decreases the resistance of the wire.

6 **(a)** Calculate the power of a 240 V lamp that carries a current of 1.6 A.

(b) Calculate the amount of energy supplied if the lamp is left switched on for 10 min.

S 7 **(a)** Write down the equation that links the emf of a battery with the electrical energy it supplies and the amount of charge that passes through it.

(b) (i) A 12 V battery passes 48 C of charge. Calculate how much energy it supplies.

(ii) Which other quantity would you need to have measured in order to calculate the power?

Practice Questions

1 The unit of electrical resistance is the
A amp
B ohm
C volt
D joule

2 A length of wire has a current of 0.50 A passing through it when the potential difference across it is 6.0 V. What is the resistance of the wire?
A 3.0 Ω
B 12 Ω
C 6.5 Ω
D 0.08 Ω

3 A mobile phone is charged for 30 minutes using a charger with a power rating of 50 W. How much energy is transferred to the phone?
A 80 J
B 1.6 J
C 1.5 kJ
D 90 kJ

4 Which of these changes will decrease the resistance of a metal wire?
A Increasing the length of the wire
B Increasing the cross-sectional area of the wire
C Increasing the temperature of the wire
D Increasing the current in the wire

S 5 How much energy is transferred when a charge of 80 C passes through a potential difference of 12 V?
A 960 J
B 6.6 J
C 92 K
D 0.15 J

6 During an investigation into how the resistance of a diode changes, a researcher measures the current and potential differences with the results shown in table below.

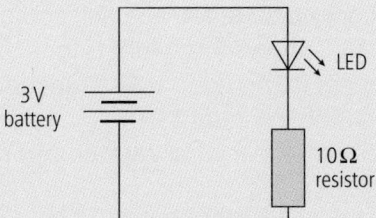

V / V	0	0.2	0.4	0.6	0.8	1.0	1.2
I / A	0	0.0	0.0	0.1	0.3	0.6	0.9

(a) Name the instruments which are used to measure **(i)** current and **(ii)** potential difference. [2]

(b) Plot a graph of V / V against I / A to show the current–potential difference characteristics of the diode. [4]

(c) Calculate the resistance of the diode when the current through it is 0.3 A. [2]

7 A group of students is investigating how the current in a lamp varies with the potential difference across it.

(a) Draw a circuit which could be used in this investigation. [4]

(b) Describe how you would carry out the experiment to see the relationship. [3]

(c) Sketch the shape of the graph that would be produced. [2]

8 A battery-powered electric screwdriver uses a motor which has a power rating of 180 W and operates with a voltage of 12 V.

(a) What is the current in the motor when it is operating? [2]

(b) How much energy is transferred by the motor if it operates for 5.0 s? [2]

(c) Calculate the resistance of the electric motor. [2]

(d) Explain why the electric motor heats up as it is operated. [3]

9 A 12 V, 36 W bulb is connected to a 12 V supply.

(a) Calculate:

(i) the current through the bulb [2]

(ii) the charge flow through the bulb in 200 s. [2]

(b) (i) Show that 7200 J of electrical energy is delivered to the bulb in 200 s. [2]

(ii) Calculate the energy delivered to the bulb by each coulomb of charge that passes through it. [2]

10 An electrician has the job of connecting a 6.6 kW electric oven to the 230 V mains supply in a house.

(a) Calculate the current needed to supply 6.6 kW of electrical power at 230 V. [2]

(b) The table below shows the maximum current that can pass safely through five different mains cables. For each cable the cross-sectional area of each conductor is given in square millimetres (mm²).

	Cross-sectional area of conductor (mm²)	Maximum safe current (A)
A	1.0	14
B	1.5	18
C	2.5	28
D	4.0	36
E	6.0	46

(i) To connect the oven to the mains supply, which cable should the electrician choose? Assume the cost per metre of each cable is proportional to its area of cross-section. Give a reason for your answer. [3]

(ii) State and explain what would happen if she chose a cable with thinner conductors. [2]

13.1 Circuit components

LEARNING OUTCOMES

- Recall commonly-used circuit symbols
- Draw and interpret circuit diagrams
- **s** Draw and interpret circuit diagrams containing diodes

PRACTICAL

Circuit tests

1 Connect a variable resistor in series with the torch lamp and a battery, as shown in Figure 13.1.1. Adjusting the slider of the variable resistor alters the amount of current flowing through the bulb and therefore affects its brightness.

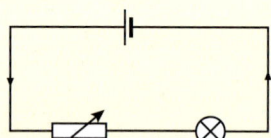

Figure 13.1.1 Using a variable resistor

2 Move the slider one way so the torch lamp becomes dim. This happens because more of the variable resistor is included in the circuit when the slider is moved this way. So the current decreases.

s 3 Replace the variable resistor with a **diode** in its 'forward' direction so the torch lamp is on. Reverse the diode in the circuit and you should find the lamp is off. Sketch and label the circuit diagram with the diode in its forward direction.

Components and their symbols

A circuit diagram is a very helpful way of showing how the components in a circuit are connected together. Each component has its own symbol. Figure 13.1.2 shows the symbols for some of the components you will meet in this course. The function of each component is also described in Figure 13.1.2. You need to recognise these symbols and remember what each component is used for.

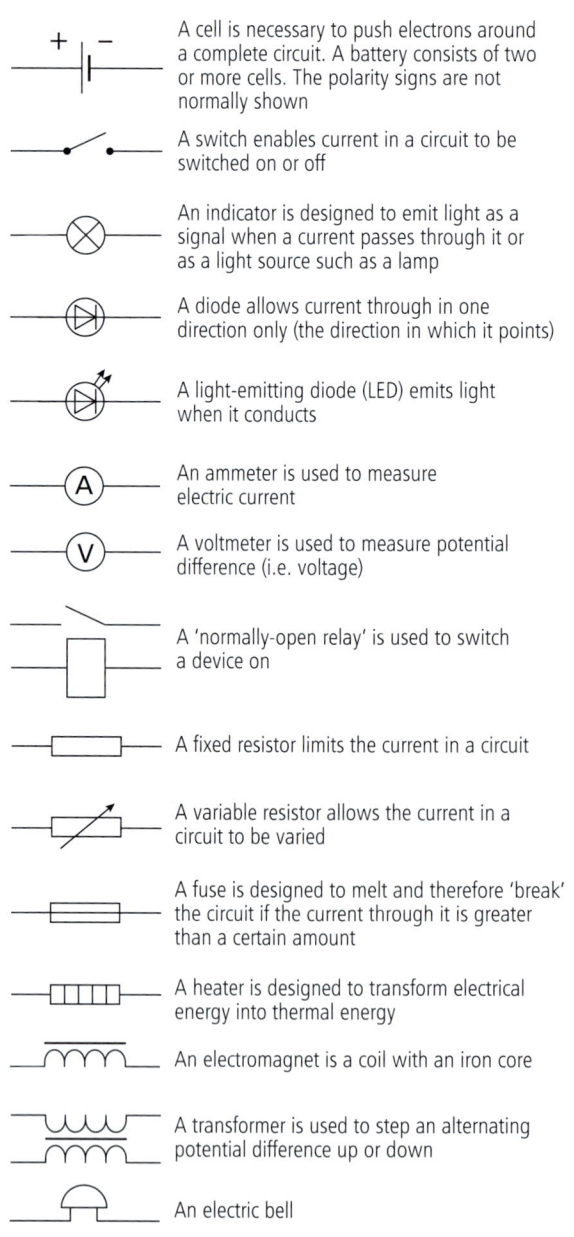

Figure 13.1.2 Components and symbols

Fuses

A fuse is designed to protect an appliance. It must be connected in series with the appliance. A fuse contains a thin wire that melts if too much current passes through it. If a fault occurs in an appliance and too much current passes through it, the fuse wire melts and the current through both it and the appliance is cut off.

The fuse for a given appliance must be chosen so it melts only when the current is greater than the 'normal' current that passes through the appliance.

Electricity costs

An electricity meter in your home measures how much electrical energy your family uses. It records the total energy supplied, no matter how many appliances you all use. It shows how many kilowatt-hours (kWh) of energy has been supplied by the mains.

In most houses, the meter is read every three months. The difference between the two readings is the number of kilowatt-hours supplied since the last bill.

The cost of the electricity supplied = number of kWh supplied × cost per kWh

SUMMARY QUESTIONS

1 Name the numbered components in the circuit diagram in Figure 13.1.4 and state the function of each one.

2 Figure 13.1.5 shows the construction of a torch.

   a Draw the circuit diagram for Figure 13.1.5.

   b i What is the purpose of the spring in Figure 13.1.5?

   ii Why is the case of the torch made of plastic rather than metal?

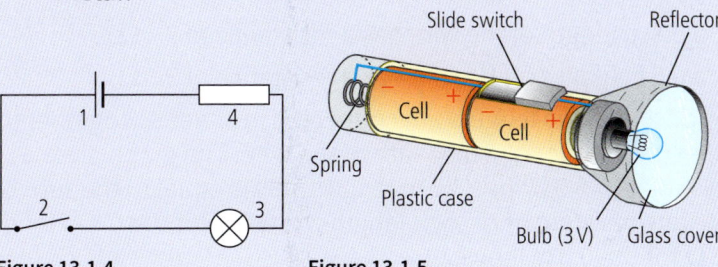

Figure 13.1.4   Figure 13.1.5

EXAM TIP

Make sure that you learn the symbols and do not confuse them with each other. A common mistake is drawing a fuse instead of a resistor.

Supplement

DIODES IN ACTION

A portable radio would be damaged if the batteries were put in the wrong way around unless a diode is in series with the battery. The diode allows current through only when it is connected as shown in Figure 13.1.3. If the battery is reversed in the circuit, the diode stops electrons passing around the circuit.

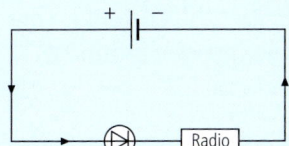

Figure 13.1.3 Using a diode

KEY POINTS

- Every component has its own agreed symbol.
- A circuit diagram shows how components are connected together.
- A diode allows current through in its 'forward' direction only.

13.2 Series circuits

LEARNING OUTCOMES

- Recognise that the same current passes through each component in series
- Calculate the combined resistance of resistors in series
- **S** • Recall and use the pd rule for components in series

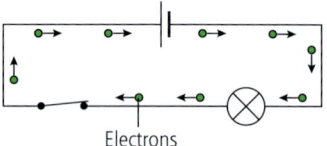

Figure 13.2.1 A torch lamp circuit

Electrons

EXAM TIP

Remember the current is the same all the way around a series circuit.

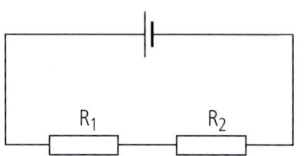

Figure 13.2.3 Resistors in series

Circuit rules

1 The same current passes through components in series with each other.

In the torch circuit in Figure 13.2.1, the lamp, the cell and the switch are connected in series with each other. Each electron moving around the circuit passes through every component in the circuit. The same number of electrons passes through each component every second, so the same current is in each component. This is true for any series circuit.

PRACTICAL

Measuring the current in a series circuit

1 Figure 13.2.2 shows a lamp, a cell, a variable resistor and two ammeters A_1 and A_2 in series with each other.

2 Set up this circuit and use the variable resistor to change the current in the circuit.

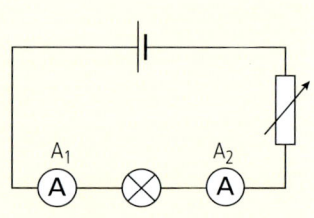

Figure 13.2.2 A current test

You should find that the two ammeters always show the same current as each other. The current is the same in every part of the circuit because the components are in series.

2 The total resistance of two or more resistors in series is equal to the sum of their separate resistances.

In Topic 12.4, we saw that the resistance of a uniform wire depends on its length. We can use the graph in Figure 12.4.2 to find the resistance of any length of wire up to 1.0 m. For example:

- a 0.4 m length of the wire has a resistance of 2.4 Ω,
- a 0.6 m length of the wire has a resistance of 3.6 Ω.

Suppose the two lengths are connected in series so together they are equivalent to a 1.0 m length of the wire. The graph gives a resistance of 6.0 Ω for a 1.0 m length. This is equal to the sum of the separate resistances of each length (i.e. 2.4 Ω for the 0.4 m length + 3.6 Ω for the 0.6 m length).

In general, for two resistors of resistances R_1 and R_2 in series, their:

$$\textbf{combined resistance} = \textbf{\textit{R}}_\textbf{1} + \textbf{\textit{R}}_\textbf{2}$$

Supplement

3 The total potential difference across a voltage supply in a series circuit is shared between the components.

In Figure 13.2.4, each electron is pushed through the lamps by the cell. The potential difference (or voltage) of the cell is a measure of the energy transferred from the cell by each electron that passes through it. Since each electron in the circuit in Figure 13.2.4 goes through both lamps, the potential difference of the cell is shared between the lamps. In other words, **in any series circuit, the total potential difference across the components is equal to the sum of potential differences across the individual components.**

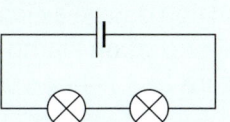

Figure 13.2.4 Lamps in series

KEY POINTS

- For components in series:
  - the current is the same in each component
  - the resistances add up to give the total resistance.
- **S** The potential differences across each component add up to give the total potential difference.

SUMMARY QUESTIONS

1 Complete the following sentences using words from the list.

greater than   less than   the same as

For the circuit in Figure 13.2.6:

a the current in the battery is _____ the current in resistor P.

b the potential difference across resistor Q is _____ the potential difference across the battery.

c The pd across resistor Q is _____ the pd across resistor P.

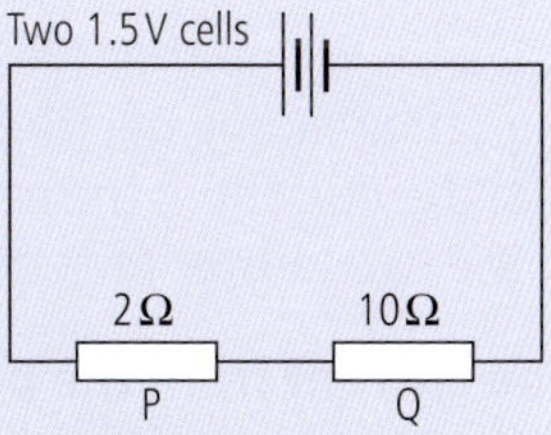

Two 1.5 V cells

$2\,\Omega$ P $10\,\Omega$ Q

Figure 13.2.6

2 For the circuit in question 1, the battery has a pd of 3.0 V.

a Calculate the total resistance of the two resistors.

b Show that current in the battery is 0.25 A.

S c Calculate the potential difference across each resistor.

PRACTICAL

Investigating potential differences in a series circuit

Set up the circuit shown in Figure 13.2.5 to test the potential difference rule for a series circuit. The circuit consists of a filament lamp in series with a variable resistor and a cell. Use the variable resistor to see how the voltmeter readings change when the current is changed. Make your own measurements and see how they compare with the ones in the table.

filament lamp	voltmeter V_1 / volts	voltmeter V_2 / volts
normal	1.5	0.0
dim	0.9	0.6
very dim	0.5	1.0

Your measurements should show that the voltmeter readings always add up to 1.5 V. This is the potential difference across the cell.

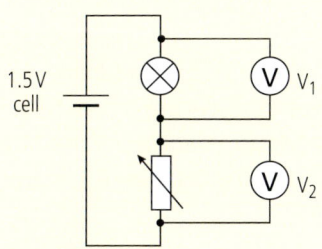

Figure 13.2.5 Voltage tests

EXAM TIP

Are your own readings correct 'within the limits of experimental accuracy'?

13.3 Parallel circuits

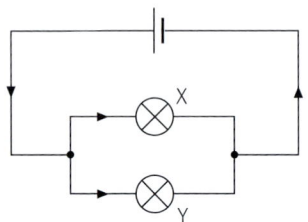

Figure 13.3.1 Torch lamps in parallel

Lamps in parallel

Can you design a circuit with two torch lamps, X and Y, and a cell so that one lamp stays on if the other one is disconnected? Figure 13.3.1 shows such a circuit. The two torch lamps are said to be in parallel because the current through X does not pass through Y and the current through Y does not pass through X. If X is disconnected, Y stays lit. If Y is disconnected instead, X stays lit. The circuit in Figure 13.3.1 has two parallel branches, one with lamp X in and the other with lamp Y in.

Investigating parallel circuits

Figure 13.3.2 shows how you can investigate the current in two lamps in parallel with each other. Ammeters in series with the lamps and the cell are used to measure the current in the lamp. Set up your own circuit and see how the measurements compare with the ones in the table for different settings of the variable resistor.

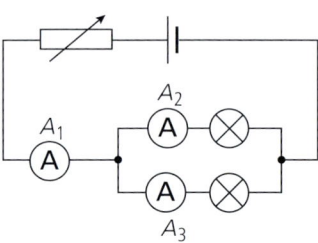

Figure 13.3.2 At a junction

ammeter A_1 / A	ammeter A_2 / A	ammeter A_3 / A
0.50	0.30	0.20
0.30	0.20	0.10
0.18	0.12	0.06

You should find that the reading of ammeter A_1 is always greater than the reading of either of the other two ammeters. This means that the current in the cell is always larger than the current in either of the two lamps. In general, for a parallel circuit:

the current from the power supply is larger than the current in each branch.

Supplement

Your results from the circuit in Figure 13.3.2 should show, as in the table, that the reading of ammeter A_1 is equal to the sum of the readings of ammeters A_2 and A_3. In general, for any parallel circuit:

the current from the power supply is the sum of the currents in the separate branches of the circuit.

A household lighting circuit

In a building, each lamp is switched on or off by a separate switch. The lamps can be switched on or off independently. When the lighting circuit is installed, the lamps and switches must be correctly connected to the lighting circuit. Otherwise, the lamps might switch on and off together or the fuses in the circuit might blow.

Figure 13.3.3 shows the circuit diagram for a household lighting circuit supplied with **mains electricity** via two wires referred to as the **live wire** and the **neutral wire**.

Each lamp is in series with a switch that switches the lamp on or off.

- Each lamp and its switch is a parallel branch of the lighting circuit connected between the live wire and the neutral wire.

- A fuse in series with the live wire cuts the mains supply off if too much current passes along the live wire. For example, if a fault occurred in one of the lamps and the current through it increased considerably, the fuse would melt and cut the lamps off from the circuit.

KEY POINTS

- The current in each branch of a parallel circuit is less than the current from the power supply.

- The current from the power supply is the sum of the currents in the branches of a parallel circuit.

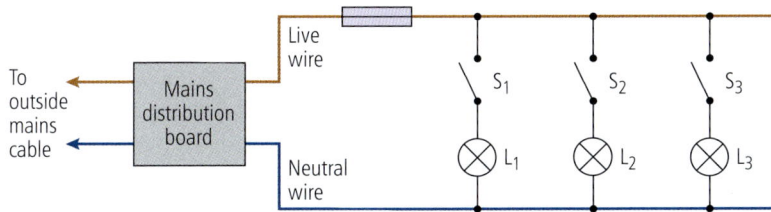

Figure 13.3.3 A household lighting circuit

SUMMARY QUESTIONS

1 Figure 13.3.4 shows two circuits with identical lamps.

   a Which circuit allows each torch lamp to be switched on and off without affecting the other torch lamp?

   b In circuit 1, how could both lamps be switched on?

   c Both switches are closed in each circuit. State and explain whether the current in the cell in circuit 1 is greater than, less than or the same as the cell current in the other circuit.

2 The circuit diagram in Figure 13.3.5 shows three resistors $R_1 = 2\,\Omega$, $R_2 = 3\,\Omega$ and $R_3 = 6\,\Omega$ connected to each other in parallel and to a 6 V battery.

   Calculate:

   a the current in each resistor

   b the current in the battery.

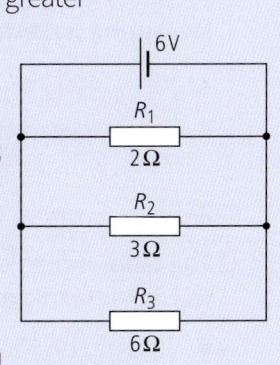

Figure 13.3.5

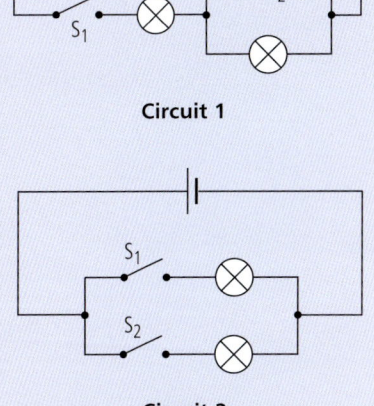

Circuit 1

Circuit 2

Figure 13.3.4

13.4 More about series and parallel circuits

LEARNING OUTCOMES

- Relate the current through a parallel component to its resistance

s • Recognise that components in parallel have the same pd

- Calculate the combined resistance of two resistances in parallel

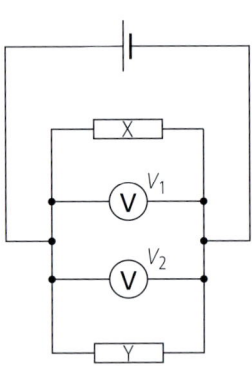

Figure 13.4.1 Components in parallel

EXAM TIP

Many students do not realise the pd across parallel components is the same and therefore misunderstand much basic electricity!

Potential differences in a parallel circuit

Figure 13.4.1 shows two resistors X and Y in parallel with each other. A voltmeter is connected across each resistor. The voltmeter across resistor X shows the same reading as the voltmeter across resistor Y. This is because each electron from the cell either passes through X or through Y. So it delivers the same amount of energy from the cell whichever resistor it goes through. In other words, for components in parallel:

s the potential difference across each component is the same.

Cells in series

What happens if we use two or more cells in series in a circuit (Figure 13.4.2). Provided we connect the cells so they act in the 'same direction', each electron gets a push from each cell. So an electron would get the same push and therefore the same energy from a battery of three 1.5 V cells in series as it would from a single 4.5 V cell.

1.5 V   3.0 V   4.5 V

Figure 13.4.2 Cells in series

In other words:

the total emf of cells in series acting in the same direction is the sum of the individual emfs of the cells.

Note Provided no energy is wasted as thermal energy inside a cell (i.e. the internal resistance of a cell is zero when current passes through it), the pd across the terminals of a cell is equal to its emf. In examination questions for this specification, assume cells, batteries and other power supplies have no internal resistance unless told otherwise in a question.

PRACTICAL

Measuring potential differences in a parallel circuit

1 Set up and test the circuit in Figure 13.4.1.

2 Repeat the test with two cells in series, then with three cells in series, all acting in the same direction.

3 Record your readings in a table.

The table should show that the voltmeter readings:

	V_1 / V	V_2 / V
one cell	1.5	1.5
two cells		
three cells		

- depend on the number of cells in the circuit

- are the same in each of the three tests.

Rules for resistors in parallel

Resistors in parallel have the same potential difference. For two or more resistors in parallel:

- The current in any of the resistors depends on the resistance of the resistor. The bigger the resistance of the resistor, the smaller the current in it. The resistor which has the largest resistance passes the smallest current.

- The current in the battery is larger than the current in any of the resistors. Therefore, the combined resistance of the resistors is less than the smallest individual resistance.

For example, Figure 13.4.3 shows a $3\,\Omega$ resistor and a $6\,\Omega$ resistor connected in parallel with a 12 V battery.

- The current in the $3\,\Omega$ resistor is larger than the current in the $6\,\Omega$ resistor.

- The current in the battery is larger than the current in either resistor.

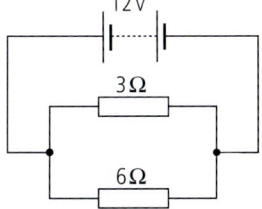

Figure 13.4.3 Resistors in parallel

SUMMARY QUESTIONS

1 Copy and complete the sentences **a** to **c** using words from the list below.

   current   potential difference   circuit

   a Components in parallel with each other have the same _____.

   b For components in parallel, the _____ is greatest for the component with the least resistance.

   c For two unequal resistors in parallel connected to a cell, the _____ is not the same for each resistor.

2 A circuit consists of a $4.0\,\Omega$ resistor in parallel with a $12.0\,\Omega$ resistor connected to a 12.0 V battery.

   a **i** Draw the circuit diagram.

   ii State and explain which resistor current has the greater current passing through it.

   b Calculate:

   i the effective resistance of the two resistors

   ii the current through each resistor

   iii the current through the battery.

Supplement

In Figure 13.4.3:

the current in the

$3\,\Omega$ resistor $= \dfrac{12\,V}{3\,\Omega} = 4\,A$

the current through the

$6\,\Omega$ resistor $= \dfrac{12\,V}{3\,\Omega} = 2\,A$

Therefore, the current in the cell $= 4\,A + 2\,A = 6\,A$

For a current of 6 A to pass through a single resistor connected to a 12 V battery, the resistance would need to be $2\,\Omega$ ($=$ pd/current $= 12\,V/6\,A$). Therefore, the parallel combination has a combined resistance of $2\,\Omega$.

Using the same approach as above, it can be shown that the combined resistance R of two or more resistors of resistances R_1 and R_2 in parallel is given by:

$$\frac{1}{R} = \frac{1}{R_1} + \frac{1}{R_2}$$

Prove for yourself that the combined resistance of a $3\,\Omega$ resistor in parallel with a $6\,\Omega$ resistor is $2\,\Omega$.

KEY POINTS

For components in parallel:

- The potential difference is the same in each component.

- The total current is the sum of the currents through each component.

- The bigger the resistance of a component, the smaller the current.

- The combined resistance R of two resistances R_1 and R_2 in parallel is given by:

$$\frac{1}{R} = \frac{1}{R_1} + \frac{1}{R_2}$$

13.5 Sensor circuits

LEARNING OUTCOMES

- Explain how a thermistor works in a temperature sensor circuit
- Explain how an LDR thermistor works in a light sensor circuit
- **S** Describe a potential divider
- Describe the action of a variable potential divider

EXAM TIP

The potential divider seems difficult at first. But persevere – it's not as difficult as it appears!

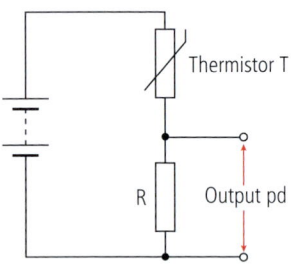

Figure 13.5.1 Using a thermistor

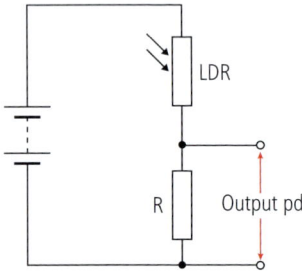

Figure 13.5.2 Using an LDR

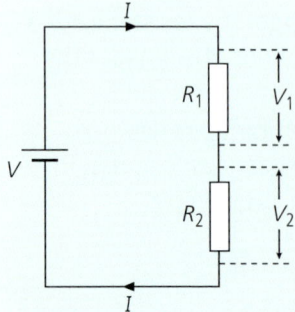

Figure 13.5.3 The potential divider

Input transducers

We use thermistors and light-dependent resistors (LDRs) in **sensor circuits**. A change of the output pd of a sensor circuit can be used to activate an alarm or to control a device. A sensor circuit is sometimes described as an **input transducer**.

A **thermistor** is a temperature-dependent resistor. Its resistance decreases the hotter it becomes. The symbol for a thermistor is shown in Figure 13.5.1.

Figure 13.5.1 shows a sensor circuit in which a thermistor T and a resistor R form a potential divider connected to a battery. When the temperature of the thermistor increases, its resistance decreases so its share of the battery pd decreases. As a result, the share of the battery pd across resistor R increases so the output pd increases.

An **LDR** has a resistance that decreases if the brightness of the incident light increases. The symbol for an LDR is shown in Figure 13.5.2.

When the brightness of the incident light is increased, its resistance decreases so the output pd increases as explained above.

In both of the sensor circuits, the resistor may be replaced with a variable resistor which is adjusted to give a certain output pd for a certain temperature or light brightness. The output pd would then change if the temperature or the light brightness changes.

Supplement

The potential divider

Sensor circuits are designed to respond to an external change such as a change of temperature or light intensity. Many sensor circuits consist of a **potential divider** which includes a resistor that is sensitive to a change in the surroundings.

A potential divider consists of two or more resistors in series. A fixed pd is connected across the combination, as shown in Figure 13.5.3. The pd across the combination is shared or 'divided' between the resistors.

Figure 13.5.3 shows two resistors of resistances R_1 and R_2 connected to a source of fixed pd V. The same current I passes through the two resistors. Therefore,

- the pd, V_1, across resistor R_1 is given by the equation $V_1 = I R_1$
- the pd, V_2, across resistor R_2 is given by the equation $V_2 = I R_2$

So $$\frac{V_1}{V_2} = \frac{I R_1}{I R_2} = \frac{R_1}{R_2}$$

(because I cancels out in the middle equation).

In other words, the ratio of the resistor pds = the ratio of their resistances

$$\frac{V_1}{V_2} = \frac{R_1}{R_2}$$

For example, for two resistors $R_1 = 10\,\Omega$ and $R_2 = 40\,\Omega$ in series and a current $I = 0.1\,A$:

- the pd, V_1, across the $10\,\Omega$ resistor $= IR_1 = 0.1\,A \times 10\,\Omega = 1.0\,V$
- the pd, V_2, across the $40\,\Omega$ resistor $= IR_2 = 0.1\,A \times 40\,\Omega = 4.0\,V$.

Because the resistors are in series with each other, the pd across the combination $= 1.0\,V + 4.0\,V = 5.0\,V$. Therefore, the fixed pd of $5.0\,V$ across the combination is shared between the two resistors as stated above.

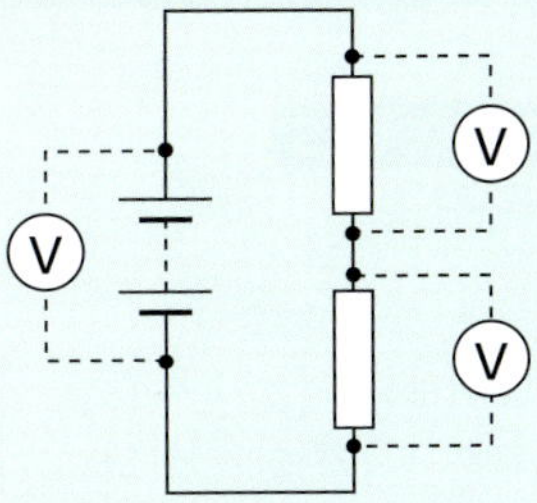

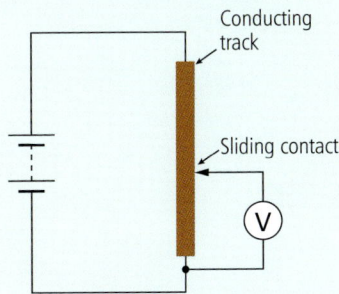

Figure 13.5.4 Investigating the potential divider

Investigating the potential divider

1. Connect a battery to two resistors of known resistance in series with each other, as shown in Figure 13.5.4. Use a voltmeter to measure the battery pd and the pd across each resistor. You should find that the sum of the pds across the resistors is equal to the battery pd.

2. Replace one of the resistors with a variable resistor and measure the pd across the other resistor for several different settings of the variable resistor. For each setting, you should find the pd across the other resistor is between zero and the battery pd and is different for each setting.

A **potentiometer** is a variable potential divider consisting of a track of a suitable resistive material with a fixed contact at each end and a sliding contact on the track, as shown in Figure 13.5.5. The track is straight in a linear potentiometer and circular in a rotary potentiometer.

In Figure 13.5.5, the voltmeter reading:

- increases when the sliding contact is moved up
- decreases when the sliding contact is moved down.

The pd between the sliding contact and either fixed contact depends on the position of the sliding contact.

Figure 13.5.5 A potentiometer

Conducting track

Sliding contact

KEY POINTS

- A thermistor is a resistor which has a resistance that decreases when its temperature is increased.

- An LDR is a resistor which has a resistance that decreases when the incident light is made brighter.

- A potential divider consists of two resistors in series connected to a fixed pd.

SUMMARY QUESTIONS

1. **a** In Figure 13.5.2, explain why the output pd decreases when the LDR is covered completely so it is in darkness.

   b The thermistor and the resistor in Figure 13.5.1 are swapped over in the circuit so the output pd is across the thermistor.

   i Draw the new circuit diagram.

   ii State and explain how the output pd in the new circuit changes when the temperature of the thermistor is reduced.

2. A potential divider consisting of a $20\,\Omega$ resistor in series with a $30\,\Omega$ resistor is connected to a $5.0\,V$ battery.

   a Sketch the circuit diagram and explain why the pd across the $30\,\Omega$ resistor is greater than the pd across the $20\,\Omega$ resistor.

   b Calculate the current through the battery and the pd across the $30\,\Omega$ resistor.

13.6 Switching circuits

The relay

Relays are used in control circuits to switch machines on or off. Figure 13.6.1 shows the construction of a relay.

- When a current passes through the coil of the electromagnet, the iron armature is attracted on to the electromagnet.

- The armature turns about a pivot and closes the switch gap.

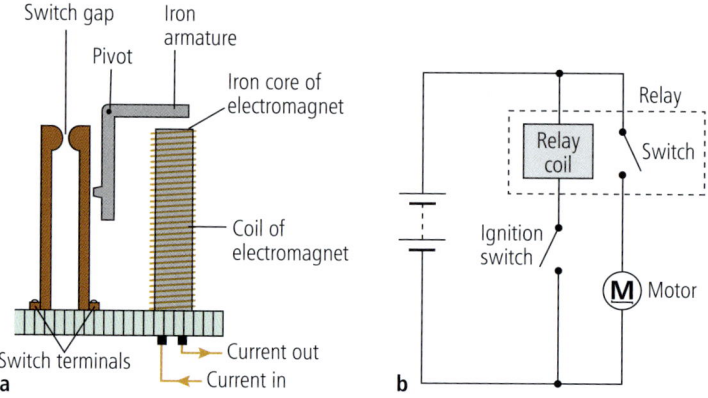

Figure 13.6.1 A normally open relay: (a) construction, (b) in a circuit

In this way a small current can be used to switch on a much greater current. For example, when the ignition switch of a car is turned on, a small current passes through the electromagnet coil so the relay switch closes. This allows a much greater current to pass through the starter motor.

Supplement

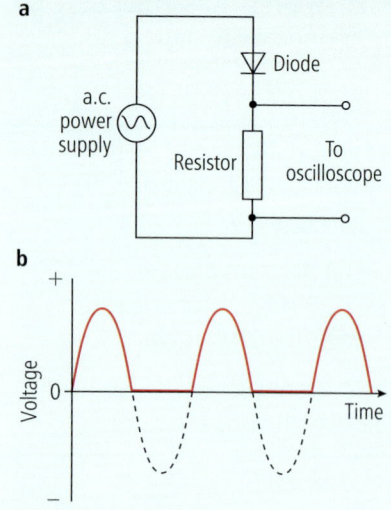

Figure 13.6.2 (a) Half-wave rectification (b) a half-wave

Converting alternating to direct current

- **Rectifier circuits** are circuits that convert alternating current to direct current. An alternating current is one that reverses its direction repeatedly. Such a circuit includes one or more diodes. Figure 13.6.2a shows a half-wave rectifier circuit.

- The diode allows current to pass around the circuit only when the pd across it is in its 'forward' direction. The diode is said to 'rectify' the alternating current. The 'half-wave' variation of the pd across the resistor may be displayed on an oscilloscope as shown in Figure 13.6.2b.

Supplement

A temperature-operated fan

Figure 13.6.3 shows a circuit which is used to switch an electric fan on if the temperature of a thermistor increases above a certain level.

When the thermistor is cold, its resistance is high so the pd across the relay coil is too small to switch the relay on.

1   Suppose the variable resistor is then adjusted so the relay is only just switched off.

2   If the thermistor temperature then increases, its resistance decreases and the pd across the relay coil therefore increases, thus switching on the relay.

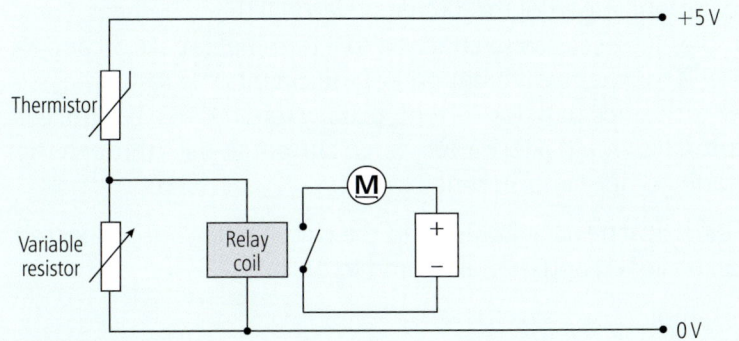

Figure 13.6.3 A temperature-operated fan

KEY POINTS

- A relay is a switch operated by an electromagnet.
- **s** A potential divider containing a thermistor or an LDR can be used with a relay to switch a device on and off.

SUMMARY QUESTIONS

1   Copy and complete the sentences **a** and **b** using words from the list below.

armature   coil

electromagnet   switch

a   When a current passes through the _____ of a relay, the _____ attracts the _____.

b   The movement of the _____ closes the _____ of the relay.

2 a   In Figure 13.6.3, explain the function of:

   i   the variable resistor

   ii   the relay coil.

b   In Figure 13.6.3, the fan motor is replaced by a lamp.

   i   What further changes would need to be made to the circuit in order to switch the lamp on automatically at night?

   ii   Draw a circuit diagram to show the changes.

13.7 Electrical safety

LEARNING OUTCOMES

- Explain why electrical circuits can be dangerous
- Describe common electrical faults and their causes
- Describe the features of electrical wires and fittings that make them safe to use

Brass is used to conduct electricity in plugs. Tap water conducts electricity. Never use an electrical appliance if you have wet or damp hands. The contact resistance between your hand and a metal terminal is much lower if your hand is wet or damp. Less resistance means more current.

A mains appliance with a plastic case has a **'double insulation'** case made of tough, stiff plastic carrying the double insulation symbol (the double square) shown in the photograph.

Electrical hazards

High voltage circuits are dangerous for obvious reasons. Anyone touching a bare wire or terminal at high voltage would receive an electric shock that is likely to be fatal. An electric current through the body would stop the heart. A current of no more than 0.02 A (= 20 mA) is enough to electrocute someone. The resistance of the human body is no more than about 1000 ohms. For a current of 0.02 A and a resistance of 1000 ohms, using '$V = IR$' gives a potential difference of 20 V. **Never** touch a wire or terminal at a voltage of about 20 V or more!

Low voltage circuits can be dangerous due to overheating leading to a fire. This might happen if a fault develops and causes a very large current in part of the circuit, which will then overheat. For example, suppose a wire attached to a terminal breaks off and its bare end touches a different part of the circuit. The result may be a low-resistance path for current, usually referred to as a **short-circuit**. This would allow a very large current to pass through the wire and components in series with it.

The electrical devices we use and the circuits they are connected to are all designed and manufactured to be safe.

For example:

- all wires are made of copper inside flexible hard-wearing plastic insulation. Copper is used because it is a very good electrical conductor, it does not rust and it is flexible so it doesn't snap.
- a cable used to connect a device such as an electric kettle to a socket has an outer layer of insulation surrounding the separate insulated wires inside the cable.
- plugs and sockets are made of stiff heat-resistant plastic materials shaped to hold the wires and terminals sealed firmly inside so they cannot make contact with each other.
- terminals used in plugs, sockets and other electrical fittings are made of brass because brass is a good conductor. It is more hard-wearing than copper and it does not rust or oxidise.

Common faults that develop in electrical devices and circuits have many causes including wear and tear, overloading a socket, fitting an incorrect fuse and ignoring manufacturers' instructions. Electrical faults may result in an exposed bare wire which would be lethal to touch or a short circuit if two bare wires in a cable touch each other.

Common faults include the following:

- **damaged insulation**

  The layer of insulation around a wire or a cable may wear away.

  A plug or socket may become chipped or broken, exposing a bare wire or terminal in the plug.

- **overheating of cables**

  Too much current through the wires of a cable may cause overheating which could make the insulation of the wires soften or melt, exposing a bare wire in the cable. The wires inside the cable might also short-circuit causing a fire.

  A coiled cable in use could overheat because it retains heat instead of losing thermal energy to its surroundings (as it would if it was not coiled up).

- **damp or wet conditions**

  Dampness in a device, a socket or a plug could cause a short-circuit inside the device or it could provide a conducting path to the outer surface. Anyone touching the device could then suffer an electric shock.

  Water in a device would also cause a short-circuit and a conducting path to outside the device, as outlined above.

- **unsuitable cables and sockets**

  A cable that is too long is a hazard because someone could trip over it. In addition, it may become a hazard if part of it is coiled up when it is in use.

  A socket with more than one appliance connected to it may be overloaded if too much current passes through it and the cables connected to it.

An overloaded socket

SUMMARY QUESTIONS

1. **a** Match the list of parts 1–4 below in a electrical plug with the list of materials A–D.

1 cable insulation	A brass
2 case	B copper
3 pin	C rubber
4 wire	D stiff plastic

2. **a** Why is a frayed or worn mains cable dangerous?

   b **i** Why do the wires of an electrical cable become warm if too much current passes through them?

   ii Why could a cable become dangerous if too much current passes through it?

   c Why should you never use a mains appliance if you have damp or wet hands?

KEY POINTS

- Electric circuits are dangerous because of the risk of electrocution (in high-voltage circuits) and/or overheating.

- Common electrical hazards include damaged insulation, overheating of cables, damp conditions and overloaded sockets.

13.8 More about electrical safety

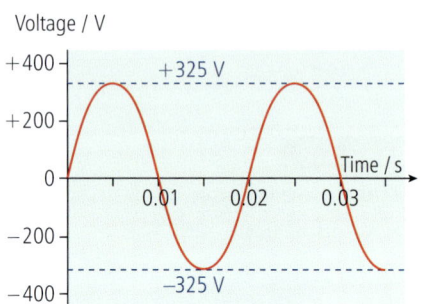

Figure 13.8.1 Mains voltage v. time

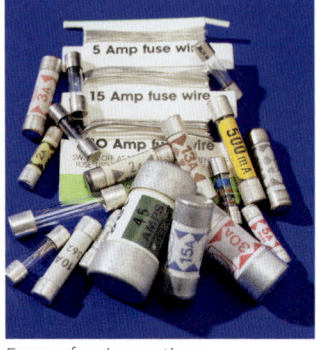

Fuses of various ratings

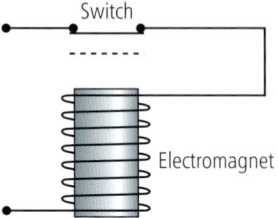

Figure 13.8.2 A circuit breaker

Mains circuits

Every mains circuit has:

- a live wire which alternates from positive to negative voltage and back every cycle, and
- a neutral wire which stays at a potential or voltage close to zero because it is earthed (i.e. connected to ground) at the local sub-station. The current in a mains circuit is an **alternating current** because it reverses its direction each time the voltage reverses.

Figure 13.8.1 shows how the voltage (relative to the Earth) of the live wire varies with time.

The maximum voltage of the live wire is different in different parts of the world. In many parts of the world, it is about 325 V so the voltage of the live wire alternates between + 325 V and – 325 V. In terms of electrical power, this is equivalent to a direct voltage of 230 V. So we say the 'voltage' of the mains is 230 V.

The frequency of the mains supply (i.e. the number of cycles per second) is 50 Hz in most countries except those in North America, parts of South America and a few other parts of the world. At 50 Hz, one complete cycle takes 0.02 s (= 1/50th of a second) as shown in Figure 13.8.1.

Fuses and circuit breakers

Fuses and circuit breakers are used to protect appliances and cables.

A **fuse** contains a thin wire that heats up and melts if too much current passes through it. The rating of a fuse is the maximum current that can pass through it without melting the fuse wire. If the rating is too large, the fuse will not blow when it should. The heating effect of the current could make the appliance catch fire. If the rating of the fuse is too low, the fuse will blow every time the appliance is switched on.

A **circuit breaker** is an electromagnetically operated switch in series with an electromagnet (Figure 13.8.2). The switch is held closed by a spring. If the current exceeds a certain value, the electromagnet pulls the switch open (i.e. the switch 'trips') so the current is cut off. The switch stays open until it can be reset once the fault that made it trip has been put right. They work faster than fuses and can be reset quicker.

Earthing

Mains appliances with metal frames and panels are made safer by earthing the frame. This is done automatically when the appliance is plugged in to a socket if:

- a cable is used that includes a third wire, the **earth wire**, and a suitable plug is used, and

- the circuit wiring from the distribution board includes an 'earth' wire in addition to the live and the neutral wire. The earth wire is connected to the ground outside the building.

Figure 13.8.3 shows why an electric heater is made safer by earthing its frame.

In **a**, the heater works normally and its frame is earthed. The frame is safe to touch.

In **b**, the earth wire is broken. The heater element has touched the unearthed frame so the frame is live. Anyone touching it would be electrocuted. The fuse provides no protection to the user because a current of no more than 20 mA can be lethal.

Suppose in **b** that the earth wire has been repaired but the heater element still touches the frame. The current is greater than normal and goes to earth and should blow the fuse. Because the frame is earthed, anyone touching it would not be electrocuted. But the heater could still be dangerous because the current might not be enough to blow the fuse. The appliance might therefore overheat.

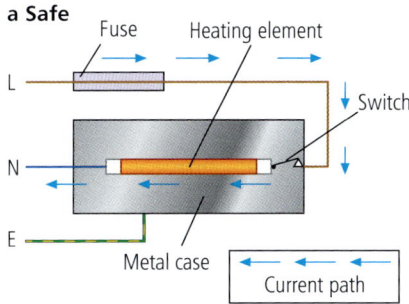

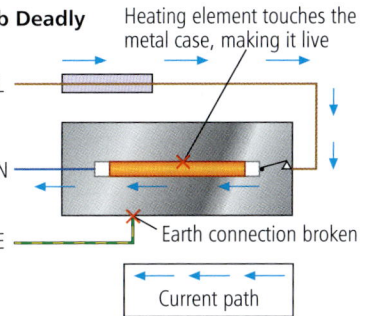

Figure 13.8.3 Earthing

KEY POINTS

- Mains electricity is an alternating current supply. A mains circuit has a live wire, which is alternately positive and negative every cycle, and a neutral wire at zero volts.

- A fuse contains a thin wire that heats up and melts and so cuts off the current if too much current passes through it.

- A circuit breaker is an electromagnetic switch that opens (i.e. 'trips') and cuts the current off if too much current passes through it.

SUMMARY QUESTIONS

1 **a** What is the purpose of a fuse in a mains circuit?

   b Why is the fuse of an appliance always on the live side?

   c What advantage does a circuit breaker have compared with a fuse?

2 Figure 13.8.4 shows the circuit of an electric heater that has been wired incorrectly.

   a Does the heater work when the switch is closed?

   b When the switch is open, why is it dangerous to touch the element?

   c Redraw the circuit correctly wired.

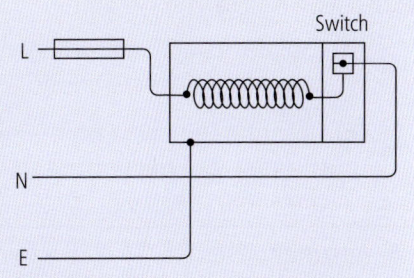

Figure 13.8.4 An incorrectly wired heater

1 Draw the circuit symbols for:

(a) a buzzer

(b) a thermistor

(c) a diode

(d) an LDR

(e) a normally open relay

(f) a variable resistor.

2 Draw a circuit diagram showing a potential divider consisting of a 50 Ω resistor and a thermistor connected to a 9.0 V battery.

3 (a) List four common faults in electrical circuits that are a safety hazard. Explain the possible danger of each one.

(b) Describe the action of a fuse.

(c) State the importance of an earth wire in an electrical circuit.

(d) Explain how an earth wire protects a user if the metal casing of an appliance becomes live.

S 4 The circuit diagram shows a light-operated buzzer.

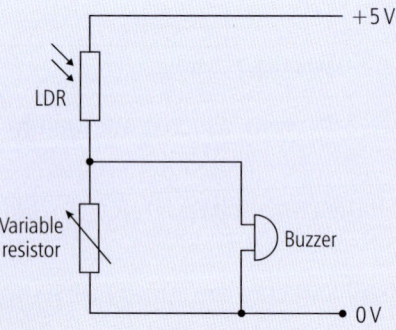

(a) Describe the effect on the resistance of the LDR if the light shining on it changes from very dim to bright.

(b) Explain how changing light conditions cause the buzzer to sound.

(c) Describe the effect on the operation of the circuit of decreasing the resistance of the variable resistor.

Practice Questions

1 Which of the following units is used to measure potential difference?

   A amps   C volts

   B ohms   D joules

2 Which of the following is the correct symbol for a variable resistor?

   A

   B

   C

   D

3 Which of the following statements about series circuits are correct?

   1. The same current passes through components in series with each other.

   2. The total resistance of two or more resistors in series is equal to the sum of their separate resistances.

   3. The total potential difference across a voltage supply in a series circuit is shared between the components.

   A 1 only

   B 1 and 2 only

   C 2 and 3 only

   D 1, 2 and 3

4 What is the total resistance of a 30 Ω and a 20 Ω resistor connected in parallel with each other?

   A 50 Ω

   B 10 Ω

   C 12 Ω

   D 1/12 Ω

5 A student uses a 4.0 A, 230 V microwave oven for 10 minutes every day and a 2500 W electric kettle three times a day for 4 minutes each time.

(a) Calculate the total number of kilowatt-hours supplied to each of these appliances for 7 days.

(b) Calculate the total cost of the energy supplied in (a) if the unit cost of electricity is 24p per kWh.

5 The figure shows a light-emitting diode (LED) in series with a resistor and a 3.0 V battery.

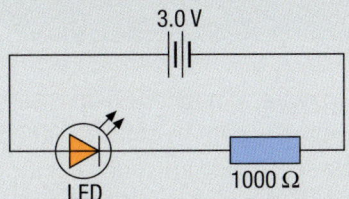

3.0 V

LED   1000 Ω

(a) The LED in the circuit emits light. The potential difference across it when it emits light is 0.6 V.

   (i) Explain why the potential difference across the 1000 Ω resistor is 2.4 V.   [2]

   (ii) Calculate the current in the circuit. [2]

(b) The current through the LED must not exceed 15 mA or else it will be damaged. If the resistor in the figure is replaced by a different resistor R, what should be the minimum resistance of R?   [2]

(c) If the LED in the circuit is reversed, what would be the current in the circuit? Give a reason for your answer.   [3]

6 A fixed value resistor (R_1 with resistance 10 Ω) and a variable resistor (R_2) are placed in series in a circuit and connected to a battery which produces an emf of 6.0 V.

The variable resistor is adjusted so that the total resistance of the circuit is 25 Ω.

(a) What is the resistance of resistor R_2?   [1]

(b) Calculate the current in the circuit.   [2]

(c) What is the potential difference across resistor R_1?   [2]

(d) What is the potential difference across resistor R_2?   [1]

(e) What name is given to a circuit containing two or more resistors designed to control an output potential difference?   [1]

7 Most mains powered electrical devices contain a fuse for safety and some contain a circuit breaker.

(a) Describe what a fuse is and how a fuse protects an electrical device.   [3]

(b) Describe the advantages a circuit breaker has when compared to a fuse.   [2]

Many electrical devices are protected by being connected to an earth wire.

(c) Explain how the earth wire prevents the user being electrocuted by a faulty device.   [2]

(d) Why don't some mains powered devices need an earth connection?   [1]

8 An electrical circuit contains two resistors in parallel as shown in the figure below. The battery consists of 3 identical cells, each with an emf of 2.0 V. The cells are connected the same way around.

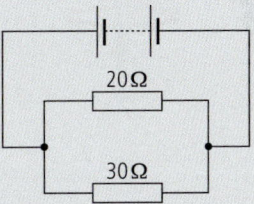

20 Ω

30 Ω

(a) State the potential difference across both the resistors in this arrangement.   [1]

(b) Calculate the total resistance of the circuit.   [3]

(c) Calculate the total current from the battery.   [2]

(d) Which of the two resistors will have the greatest current in it?   [1]

One of the cells is removed and replaced so that its emf acts in the opposite direction.

(e) Describe what happens to the size of the current in the cells for the new circuit compared to the original.   [2]

14.1 Magnetic field patterns

LEARNING OUTCOMES

- Describe the magnetic field pattern around a wire and a solenoid carrying a current
- Describe applications of the solenoid
- S State how the strength and direction of the above fields vary with position

When an electric current passes along a wire, a magnetic field is set up around the wire. Figure 14.1.1 shows how the pattern of the magnetic field around a long straight wire can be seen using iron filings or a plotting compass. The lines of force due to a straight current-carrying wire are circles, centred on the wire. The field is strongest near the wire.

The direction of the field is reversed if the direction of the current is reversed. We can use the corkscrew rule to remember the direction of the magnetic field for each direction of the current. If the current is downwards, the field turns in the same direction as a downward-moving corkscrew or screw. Increasing the current increases the strength of the field everywhere in the field.

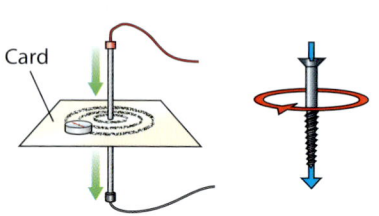

Figure 14.1.1 The magnetic field near a long straight wire

PRACTICAL

Plotting the magnetic field near a current-carrying wire

1. Set up the arrangement as shown in Figure 14.1.2. Make sure a suitable lamp or resistor is in series with the wire so as to limit the current from the power supply. Use a wooden stand (or any non-ferrous object) to support the card so it is horizontal.
2. Use the plotting compass to plot magnetic field lines near the wire as explained in Topic 10.2. You should find that the field lines are circles (in the plane of the card) centred on the wire.

S

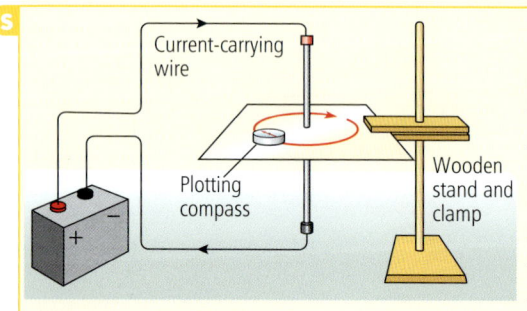

Figure 14.1.2 Magnetic field tests

Use the arrangement shown to observe the effect of:

Reversing the current: You should find that the direction of the plotting compass reverses. This shows that the magnetic field lines reverse direction when the direction of the current is reversed.

Moving the plotting compass away from the wire: You should find it points more towards 'magnetic north'. This is because the magnetic field due to the wire decreases in strength further from the wire so the Earth's magnetic field has more effect. Remember that the direction of a magnetic field line at any point is the direction of the force on the N pole of a magnet at that point.

Figure 14.1.3 The magnetic field near a solenoid

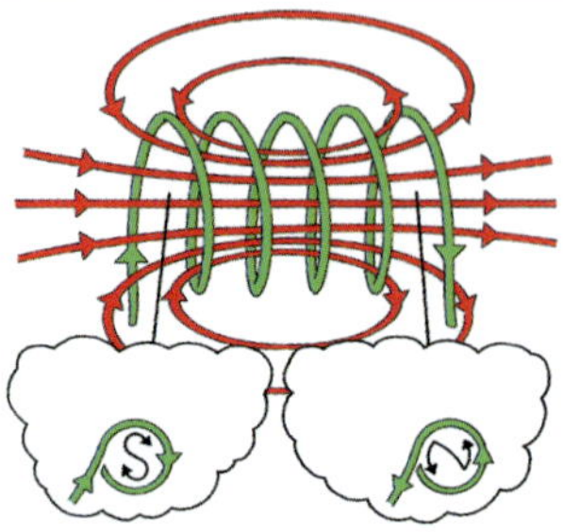

The magnetic field of a current-carrying solenoid

A **solenoid** is a long coil of wire. When there is a current in it, there is a magnetic field in and around the solenoid. The magnetic field pattern is shown in Figure 14.1.3. Reversing the direction of current reverses the direction of the magnetic field lines. Increasing the current increases the strength of the field everywhere in the field.

We can use iron filings or a plotting compass to plot the field lines. Notice that each field line is a continuous loop through the inside of the solenoid and around the outside. The field lines are:

- concentrated at the ends of the solenoid
- parallel to each other inside the solenoid
- spread out beyond the ends of the solenoid.

Outside the solenoid, the magnetic field pattern is like that of a bar magnet. Using this comparison, we can say that the end of the solenoid from which the field lines emerge is like the north pole of a bar magnet and the other end is like the south pole. The end-view diagrams in Figure 14.1.3 show the 'solenoid rule', which is a simple way to relate the polarity of each end to the current direction.

Inside the solenoid, the magnetic field lines are parallel to each other so the field is uniform except near the ends. This means the magnetic field has the same strength and direction everywhere inside the solenoid (except near the ends where it is weaker).

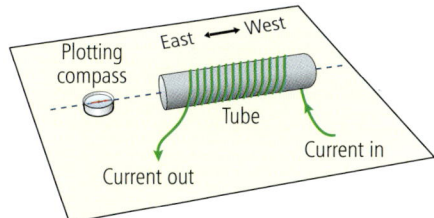

Figure 14.1.4 Investigating a solenoid

PRACTICAL

Investigating the magnetic field of a solenoid

1 Make a solenoid using insulated wire wrapped around a cardboard or plastic tube. Fix the solenoid with its axis horizontal and aligned along an east-west line, as shown in Figure 14.1.4.

2 Place a plotting compass near one end of the solenoid then switch the solenoid current on. You should observe that the plotting compass points along the axis of the solenoid (instead of pointing to magnetic north when the current is off).

3 Reverse the current direction and you should see the plotting compass direction reverse.

4 Move the plotting compass through the inside of the solenoid. You should find the direction of the plotting compass is unchanged as it is moved along.

S 5 With the plotting compass on the solenoid axis just outside one end, move the plotting compass away from the solenoid along the axis. You should observe the plotting compass points more and more towards magnetic north as it is moved away from the solenoid. This shows that the strength of the magnetic field decreases with increased distance from the solenoid.

Applications

1 The uniform magnetic field inside a solenoid is used in applications such as the 'magnetic resonance' (MR) brain scanner.

2 The magnetic field outside a solenoid is used in many applications from large electromagnets used in scrap yards (see Topic 10.3) to small electromagnets used in relays (see Topic 13.6). Their iron core makes the magnetic field much stronger.

KEY POINTS

- The magnetic field lines around a wire are circles centred on the wire in a plane perpendicular to the wire.

- The magnetic field of a solenoid is uniform inside the solenoid and like that of a bar magnet outside. **S**

- Increasing the current increases the strength of the magnetic fields above; reversing the current reverses the magnetic field lines.

SUMMARY QUESTIONS

1 Sketch the pattern of the magnetic field lines for each of the following:

   a vertical wire carrying current upwards,

   b an air-filled solenoid.

2 A bar magnet is held near the end of an unfilled solenoid to repel it. Describe and explain the effect of:

   a increasing the current in the solenoid

   b switching the current off

   c reversing the current

   d reversing the bar magnet.

14.2 The motor effect

LEARNING OUTCOMES

- Describe the motor effect
- Relate the force in the motor effect to the current and the magnetic field
- **S** Describe an experiment to show the effect of a magnetic field on an electron beam

Magnetic fields at work

How often do you use an electric motor? If you think you rarely use an electric motor, think again. Electric motors are used in many devices you use every day including computer disc drives, air-conditioning systems, fridges and electrically operated doors and windows.

An electric motor works because a force can be exerted on a wire in a magnetic field when a current passes through the wire. This effect, shown in Figure 14.2.1, is known as the **motor effect**. In addition to electric motors, the motor effect is made use of in other devices such as the loudspeaker (see Topic 10.3) and the analogue meter (see Topic 12.2). In the motor effect, the wire and the magnet (or electromagnet) exert equal and opposite forces on each other because their magnetic fields interact.

PRACTICAL

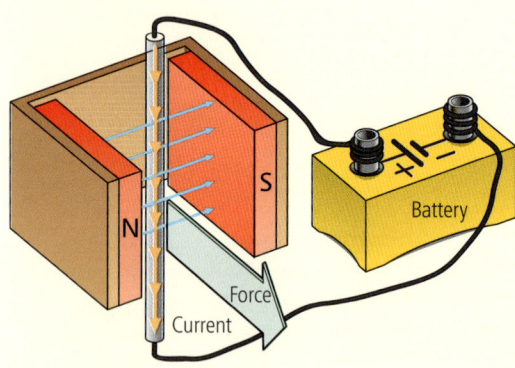

Figure 14.2.1 Investigating the motor effect

Investigating the motor effect

1   Figure 14.2.1 shows one way to investigate the motor effect. You should find that a force acts on the wire unless it is parallel to the magnetic field lines. Show that:

   - increasing the current (by using two cells) increases the force on the wire
   - reversing the current (by reversing the cell in the circuit) reverses the direction of the force on the wire.

2   In addition, test the effect with the wire in different directions in the magnetic field. As explained below, you should find this affects the force in a big way!

Force factors

Your investigations should show that:

1   **the force can be increased** by:
   - increasing the current
   - using a stronger magnet.

2   **the direction of the force is reversed** if the direction of either the current or the magnetic field is reversed. If the current and the magnetic field are both reversed, the direction of the force is unchanged.

3   **the force depends on the angle between the wire and the magnetic field lines.** The force is:
   - greatest when the wire is perpendicular to the magnetic field
   - zero when the wire is parallel to the magnetic field lines.

4   **the direction of the force is always at right angles to the wire and the field lines.** The direction of the force can be worked out from the direction of the current and the direction of the magnetic field using Fleming's left hand rule shown in Figure 14.2.2.

EXAM TIP

You must learn these factors that affect the size and direction of the force.

Supplement

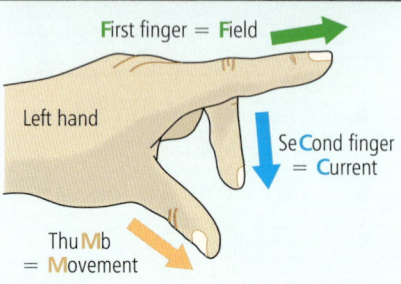

First finger = Field

Left hand

SeCond finger = Current

ThuMb = Movement

Figure 14.2.2 Fleming's left hand rule

The combined magnetic field

Figure 14.2.3 shows the pattern of the magnetic field lines of a current-carrying wire in a magnetic field. The magnetic field of the magnet and that of the wire cancel each other out below the wire and reinforce each other above it. The result is that the wire is pushed down by the magnet.

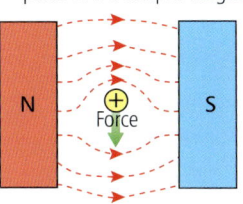

a Magnetic field between the poles of a U-shaped magnet

b Wire carrying current downwards

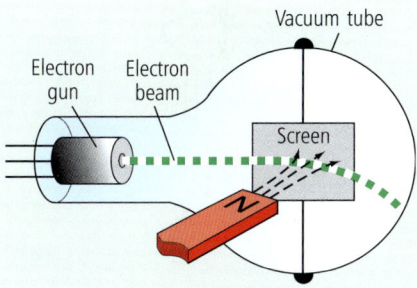

c Combined magnetic field

Figure 14.2.3 A current-carrying wire in a magnetic field of a U-shaped magnet

Supplement

The effect of a magnetic field on an electron beam

A magnetic field applied to a beam of electrons pushes the beam down, as shown in Figure 14.2.4. Each electron in the beam experiences a downward force due to the magnetic field. The same happens to electrons moving along a current-carrying wire. The force on the wire and hence the motor effect is because the moving electrons are pushed sideways by the field.

In Figure 14.2.4: **the force on each electron is at right angles to the direction of the magnetic field and to the direction in which the electron is moving.**

If the magnet is turned around, the magnetic field direction is reversed so the beam is pushed upwards instead of downwards.

Using a current-carrying electromagnet instead of a permanent magnet would have the same effect. Increasing the current in the electromagnet increases the magnetic field strength of the electromagnet which increases the force on the electron beam and pushes it down further. Fleming's left hand rule gives the direction of the force on each electron in the beam provided the direction of the conventional current is used. See Topic 11.4.

Figure 14.2.4 An electron beam in a magnetic field

SUMMARY QUESTIONS

1 Copy and complete the following sentences using words from the list below.

**horizontal   reversed   perpendicular
unchanged   vertical**

A vertical wire carrying an electric current is in a uniform horizontal magnetic field.

a The direction of the force on the wire is _____ and _____ to the direction of the magnetic field lines.

b When the magnetic field is reversed, the direction of the force on the wire is _____.

c When the current in the wire is reversed and the magnetic field is reversed, the direction of the force on the wire is _____.

S 2 In Figure 14.2.1, describe how the force on the wire changes if the wire is gradually turned until it is parallel to the magnetic field.

KEY POINTS

- In the motor effect, the force:

  – is increased if the current or the strength of the magnetic field is increased

  – is reversed if the direction of the current or the magnetic field is reversed.

  – is at right angles to the direction of the magnetic field and to the wire **S**

- Electrons in a beam are pushed sideways when a magnetic field is applied at right angles to the beam.

14.3 The electric motor

LEARNING OUTCOMES

- Explain why a current-carrying coil in a magnetic field turns
- Describe how an electric motor works
- Describe the effect of increasing or reversing the current in an electric motor or increasing the strength of the magnetic field

Using the motor effect

A coil in a magnetic field can be forced to turn by passing a current through it. Figure 14.3.1 shows a rectangular coil between the poles of a U-shaped magnet. When a current passes around the coil, a force acts as shown on each of the two long sides of the coil. As these forces are in opposite directions, they act to make the coil turn.

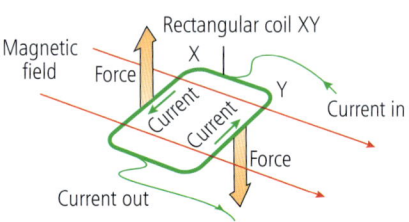

Figure 14.3.1 Turning a coil

PRACTICAL

Test a coil

1 Make a rectangular coil to fit between the poles of a U-shaped magnet. Use suitable insulated wire and a wooden block. Tape the wires to the block.

2 Place the coil in the field of a U-shaped magnet as shown in Figure 14.3.1. Use a suitable low voltage supply to pass a current through the wires of the coil. You should observe that the coil tries to turn.

- Increase the current and you should observe the turning effect increases.
- Reverse the current and you should observe the coil tries to turn in the opposite direction.
- Use a stronger magnet or wind more turns on the coil and you should observe that the turning effect increases.

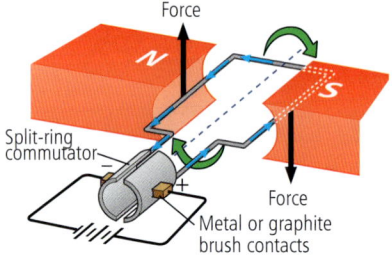

Figure 14.3.2 The electric motor

EXAM TIP

Study the diagrams with care and do your best to understand how the motor works.

Supplement

The simple electric motor

An electric motor is designed to use the motor effect.

The simple motor in Figure 14.3.2 consists of a rectangular coil of insulated wire (the armature coil) that is forced to rotate. The coil is connected via two metal or graphite 'brushes' to the battery. The brushes press onto a metal **'split ring' commutator** fixed to the coil.

When a current is passed through the coil, the coil spins because:

- a force acts on each side of the coil due to the motor effect
- the force on one side is in the opposite direction to the force on the other side.

The split ring commutator reverses the current around the coil every half-turn of the coil. Because the sides also swap over each half-turn, the coil is pushed in the same direction every half-turn.

We can control the speed of an electric motor by changing the current. Also, we can reverse its turning direction by reversing the current.

Make and test an electric motor

Make and test a simple electric motor like the one in Figure 14.3.2.

1 The coil could be wound on a wooden block with a narrow hollow tube (not shown) through it. With a thin rod through the tube as the spindle, the wooden block and the tube should be able to spin freely.

2 To make the coil, wrap 'single-core' insulated wire around the sides of the coil so that the two bare ends of the wire can be laid on insulating tape to either side of one end of the tube.

3 Bend two small pieces of metal foil so they fit on the tube over the bare ends of the wire to form the split-ring commutator, as shown in Figure 14.3.2. Use two narrow strips of tape or small elastic bands to hold the metal foil in place.

4 Use two suitable metal pins fixed to a board as the pivots to hold the coil in place between opposite poles of a U-shaped magnet. If necessary, make each pivot using a paper clip.

Use the bare ends of the two wires from the battery as 'brushes'. With practice, when these are held in contact with opposite sides of the split-ring commutator, the coil should spin.

Practical electric motors

In a practical electric motor, the rotating part of the motor or 'armature' consists of several evenly-spaced coils wound on an iron core as shown in Figure 14.3.3a. The iron core makes the field much stronger so the turning effect is much greater.

● Each coil is connected to its own section of the commutator. The result is that each coil in sequence experiences a turning effect when it is connected to the voltage supply, so the armature is repeatedly pushed around.

● Because the iron core makes the field radial as shown in Figure 14.3.3b, each coil is in the field for most of the time. As a result, the overall turning effect is much steadier than in a simple electric motor with one coil only, so the motor runs more smoothly.

Graphite is a form of carbon which conducts electricity and is very slippery. It therefore causes very little friction when it is in contact with the rotating commutator.

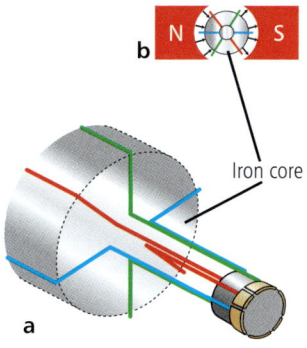

Figure 14.3.3 Inside a practical motor

● A simple electric motor has a rectangular coil of wire that spins in a magnetic field when a current passes through the coil.

● The speed of an electric motor is:

  – increased if the current or the strength of the magnetic field is increased

  – reversed if the current is reversed.

1 Copy and complete the following sentences using words from the list below.

   coil   current   force   magnet

   a When a _____ passes through the _____ of an electric motor, a _____ due to the _____ acts on each side of the _____.

   b The _____ along each side is in opposite directions so the _____ is in opposite directions and the _____ turns.

2 a Explain why a simple electric motor connected to a battery reverses if the battery connections are reversed.

   b Discuss whether or not an electric motor would run faster if the coil was wound on:

   i a plastic block

   ii an iron block instead of a wooden block.

14.4 Electromagnetic induction

A hospital has its own electricity generator always 'on standby' in case the mains electricity supply fails. Patients' lives would be put at risk if the mains power failed and there was no standby generator.

A generator contains coils of wire that spin in a magnetic field. A potential difference (pd) is created or **induced** in the wire when it cuts across the magnetic field lines. We refer to this source of pd as an **induced electromotive force** (emf) or an induced voltage. If the wire is part of a complete circuit, the induced emf makes an electric current pass around the circuit.

PRACTICAL

Investigating a simple generator

1 Connect some insulated wire to an ammeter as shown in Figure 14.4.1. Move the wire between the poles of a U-shaped magnet and observe the ammeter. You should discover the ammeter pointer deflects as a current is generated when the wire is moved across the magnetic field. The current is generated because an emf is induced in the wire when it cuts across the magnetic field lines. Make the wire into a coil or use a stronger magnet and you should find the current is much greater.

2 Carry out tests to investigate the above effects.

You should find that:

- there is no current generated when the wire is stationary
- a current is generated when the magnet instead of the wire is moved
- a larger current is generated when the wire moves faster or more turns are added to the coil
- the current is reversed when the direction of motion is reversed.

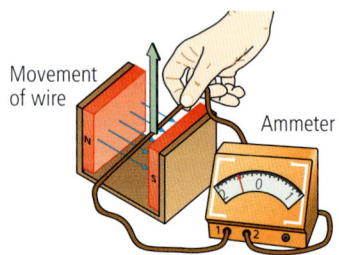

Figure 14.4.1 The generator effect

A generator test

Figure 14.4.2 shows a coil of insulated wire connected to a centre-reading ammeter. When one end of a bar magnet is pushed into the coil, the ammeter pointer deflects. This is because:

- the movement of the bar magnet causes an induced emf in the coil
- the induced emf causes a current because the coil is part of a complete circuit.

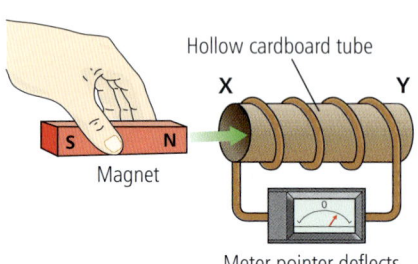

Figure 14.4.2 Testing electromagnetic induction

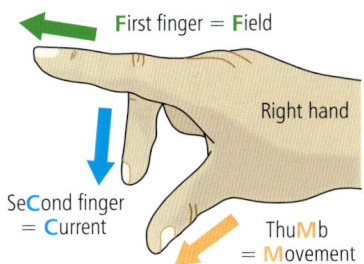

First finger = **F**ield

Right hand

Se**C**ond finger = **C**urrent

Thu**M**b = **M**ovement

Figure 14.4.3 Fleming's right hand rule for induced current

Supplement

In Figure 14.4.2, if the bar magnet is then withdrawn from the coil, the ammeter pointer deflects in the opposite direction. This is because the induced emf acts in the opposite direction so the induced current is in the opposite direction. The direction of the induced current also depends on which way around the polarity of the magnet is.

The table below shows the results of testing each direction of motion of the magnet with the magnet each way around. The table gives the current direction as seen by someone viewing end X of the coil.

magnetic pole entering or leaving the coil	pushed in or pulled out	current direction	induced polarity of X	magnet and coil
north pole	in	anticlockwise	north pole	repel
north pole	out	clockwise	south pole	attract
south pole	in	clockwise	south pole	attract
south pole	out	anticlockwise	north pole	repel

The induced current passing through the wires creates a magnetic field as long as the coil is moving. We can use the solenoid rule shown in Figure 14.4.4 to work out whether end X of the coil is like the north or the south pole of a bar magnet. The result of using the solenoid rule is shown in the 'induced polarity' column of the table. These results show that the induced polarity of end X always acts against the change that causes it. The electrical energy generated is the result of work done by the person moving the magnet to overcome the magnetic field of the induced current.

The induced current always acts in such a direction as to oppose the change that causes it.

FLEMING'S RIGHT HAND RULE

When a wire cuts across a magnetic field, the direction of the induced current in the wire can be worked out using this rule as shown in Figure 14.4.3. Notice that the direction of the induced current is at right angles to the direction of motion of the wire and to the direction of the magnetic field.

Figure 14.4.4 The solenoid rule

KEY POINTS

- When a wire cuts the lines of a magnetic field, an emf is induced in a wire.

- If the wire is part of a complete circuit, the induced emf causes a current in the circuit.

- The current is increased if the wire moves faster or a stronger magnet is used.

- The direction of an induced current opposes the change that causes it.

SUMMARY QUESTIONS

1 A coil of wire is connected to a centre-reading ammeter. A bar magnet is inserted into the coil, making the ammeter pointer flick briefly to the right.

What would you observe:

a if the magnet had been inserted more slowly into the coil?

b if the magnet was then held at rest in the coil?

c if the magnet is withdrawn rapidly from the coil?

S 2 Describe how you would use a coil, a bar magnet and an ammeter to demonstrate that the direction of an induced current opposes the change that causes it.

14.5 The alternating current generator

Supplement

The simple alternating current generator

A simple **ac generator** consists of a rectangular coil which is forced to spin in a magnetic field, as shown in Figure 14.5.1. The coil is connected to a centre-reading meter via metal 'brushes' that press on two metal slip rings. The slip rings and brushes provide a continuous connection between the coil and the meter.

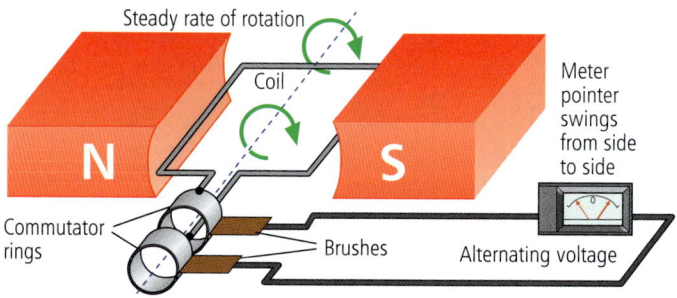

Figure 14.5.1 The construction of a simple a.c. generator

When the coil turns steadily in one direction, the meter pointer deflects first one way then the opposite way then back again. This carries on as long as the coil keeps turning in the same direction. The current in the circuit repeatedly changes its direction through the meter because the induced emf in the coil repeatedly changes its direction. This carries on as long as the coil keeps turning. The induced emf and the current are said to alternate because they repeatedly change direction.

The induced emf varies as the coil rotates, as shown in Figure 14.5.2. In one complete rotation of the coil (or 'one full cycle'), the induced emf increases from zero to a maximum value then decreases to zero, reverses and increases to a negative maximum and then becomes zero again. We refer to both the positive and negative maximum values as the **peak value**.

1 **The magnitude of the induced emf is greatest** when the plane of the coil is parallel to the direction of the magnetic field. At this position, the sides of the coil parallel to the axis of rotation (labelled X and Y in Figure 14.5.2) cut directly across the magnetic field lines. As a result, the induced emf is at its peak value.

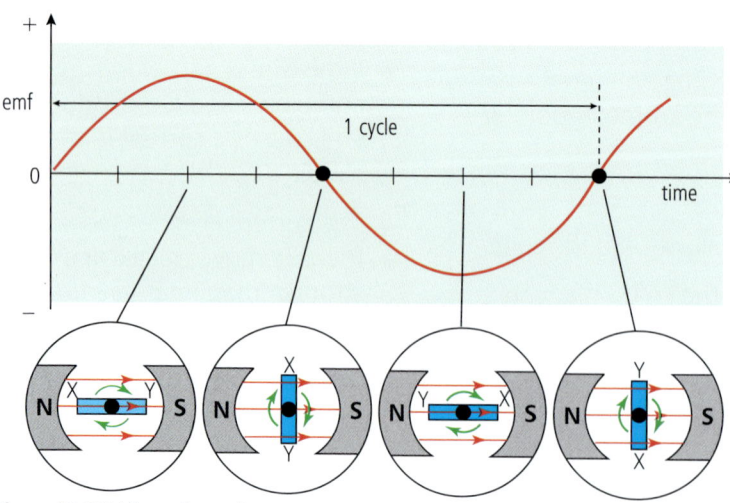

Figure 14.5.2 Alternating voltage

2 The magnitude of the induced emf is zero when the plane of the coil is perpendicular to the magnetic field lines. At this position, the sides of the coil move parallel to the field lines and do not cut them. So the induced emf is zero.

In one complete rotation of the coil starting with its plane **perpendicular** to the field, the induced emf therefore:

- increases from zero to a maximum as the coil turns through 90° to where its plane is parallel to the field
- decreases from the maximum to zero as the coil turns through a further 90° to where its plane is perpendicular to the field (and is upside down compared with its starting position)
- decreases from zero to a negative maximum as the coil turns through a further 90° to where its plane is once again parallel to the field (and is upside down compared with its starting position)
- increases from the negative maximum to zero as the coil turns through a further 90° to the position it was in at the start of the cycle.

Therefore each full cycle of the alternating emf takes the same time as one full rotation of the coil.

The faster the coil rotates:

1 the greater the frequency (i.e. the number of cycles per second) of the alternating current. This is because each full cycle of the alternating emf takes the same time as one full rotation of the coil.

2 the larger the peak value of the alternating current. This is because the sides of the coil move faster and therefore cut the field lines at a faster rate, so the induced emf is greater.

An alternating voltage can be displayed on an oscilloscope screen. If the induced emf from an ac generator is displayed on an oscilloscope and the generator is rotated faster, the screen display will show more waves on the screen (because the frequency of the induced emf will be greater). The waves will also be taller (because the peak value of the induced emf will be greater).

KEY POINTS

- The simple ac generator consists of a coil that spins in a uniform magnetic field.
- The slip rings and brush contacts enable the coil to stay connected to the external circuit.
- The peak value of the induced emf is when the sides of the coil cut directly across the magnetic field lines.
- When the sides of the coil move parallel to the field lines, the induced emf is zero.

SUMMARY QUESTIONS

1 Copy and complete the following sentences **a** to **c** using words from the list below:

alternates   reverses   spins   increases

a An emf is induced in the coil of an ac generator when the coil _____.

b The induced emf _____ which means its polarity repeatedly _____.

c If the coil _____ faster, the peak value of the induced emf _____ as well as its frequency.

2 Figure 14.5.2 shows how the alternating voltage produced by an ac generator changes with time.

a How would the graph in Figure 14.5.2 differ if the coil was rotated more slowly?

b Give reasons for your answer in **a**.

14.6 Transformers

LEARNING OUTCOMES

- Describe the construction of a transformer
- Recall and use the transformer equation relating the voltage ratio to the turns ratio
- **S** Describe how a transformer works

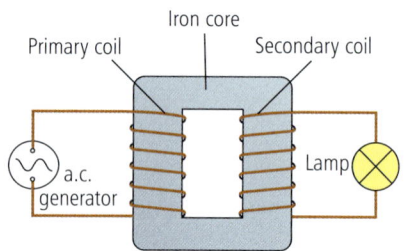

Figure 14.6.1 Transformer in action

EXAM TIP

Make sure you know the difference between a step-up and a step-down transformer and that you can draw the appropriate diagrams.

A typical power station generator produces an alternating voltage of about 25 000 V. The alternating voltage of the cables from a power station is typically 132 000 V. Mains electricity to homes and offices is much lower than this, for example 230 V in many countries.

Transformers are used to change an alternating voltage. We use them in low voltage supply units such as mobile phone chargers to step the alternating voltage from the mains down to the required lower voltage.

The construction of a basic transformer

A **transformer** has two coils of insulated wire, both wound around a soft iron core as shown in Figure 14.6.1. The two coils are referred to as the primary coil and the secondary coil. When an alternating voltage from a suitable source is applied to the primary coil, an alternating voltage is induced in the secondary coil. If the secondary coil is part of a complete circuit, the alternating voltage induced in the secondary coil causes an induced current in the circuit containing the secondary coil.

The voltage applied to the primary coil must be an alternating voltage. A transformer will not work with a constant voltage applied to the primary coil. The alternating voltage causes an alternating current in the primary coil which induces an alternating voltage in the secondary coil. A constant voltage applied to the primary coil would cause a constant current in the primary coil and a constant current cannot cause electromagnetic induction.

A transformer is described as:

- **a step-up transformer** if the transformer 'steps up' the primary voltage so the secondary voltage is greater than the primary voltage
- **a step-down transformer** if the transformer 'steps down' the primary voltage so the secondary voltage is less than the primary voltage.

The transformer equation

In terms of peak values, the alternating voltage in the secondary coil, V_S, depends on:

1   the alternating voltage applied to the primary coil, V_P
2   the number of turns of wire in the primary coil, N_P
3   the number of turns of wire in the secondary coil, N_S.

The above quantities are related to each other through the following equation which we can use to calculate any one of these quantities if we know the other three:

$$\frac{\text{voltage across primary, } V_P}{\text{voltage across secondary, } V_S} = \frac{\text{number of turns on primary, } N_P}{\text{number of turns on secondary, } N_S}$$

- **For a step-up transformer**, the number of secondary turns N_s is greater than the number of primary turns N_p, so V_s is greater than V_p.
- **For a step-down transformer**, the number of secondary turns N_s is less than the number of primary turns N_p, so V_s is less than V_p.

Make a model transformer

1. Wrap a coil of insulated wire around the iron core of a model transformer as the primary coil. Connect the coil to a 1 V ac supply unit and connect a second length of insulated wire to a 1.5 V torch lamp. When you wrap the second wire around the iron core, the lamp should light up.

2. Observe the effect of wrapping more turns of wire around the iron core. You should find the lamp is brighter. This is because the secondary voltage is greater.

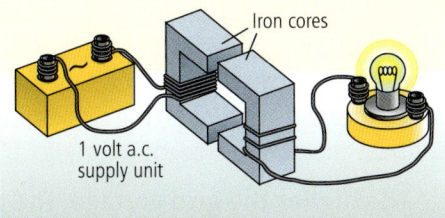

Figure 14.6.2 A model transformer

Supplement

How a transformer works

A transformer has two coils of insulated wire, both wound around the same iron core as shown in Figure 14.6.1. When alternating current is in the primary coil, an alternating voltage is induced in the secondary coil. This happens because:

- alternating current in the primary coil produces an alternating magnetic field
- the lines of the alternating magnetic field pass through the secondary coil and induce an alternating voltage in it.

If a lamp is connected across the secondary coil, the induced voltage causes a current in the secondary circuit. So the lamp lights up. Electrical energy is therefore transferred from the primary to the secondary coil. This happens even though they are not electrically connected in the same circuit.

WORKED EXAMPLE

A transformer is used to step a voltage of 230 V down to 10 V. The secondary coil has 60 turns. Calculate the number of turns of the primary coil.

Solution

$V_P = 230\,V$, $V_S = 10\,V$, $N_S = 60$ turns

Using $\dfrac{V_P}{V_S} = \dfrac{N_P}{N_S}$

gives $\dfrac{230}{10} = \dfrac{N_P}{60}$

Therefore $N_P = \dfrac{230 \times 60}{10}$

$= 1380$ turns

KEY POINTS

- A transformer consists of a primary coil and a secondary coil wrapped on the same iron core.

- Transformers only work using alternating current.

  The transformer equation is:

  $$\frac{V_P}{V_S} = \frac{N_P}{N_S}$$

- The alternating current in the primary coil creates an alternating magnetic field in the iron core which induces an alternating voltage in the secondary coil.

SUMMARY QUESTIONS

1. Copy and complete the following sentences about a transformer using words from the list below.

   current   voltage   primary   secondary

   In a transformer, an alternating _____ is passed through the _____ coil. As a result, an alternating _____ is induced in the _____ coil.

2. **a i** Why does a transformer not work with direct current?

   ii Why is it important that the coil wires of a transformer are insulated?

   b A transformer with 1200 turns in the primary coil is used to step a voltage of 120 V down to 6 V. Calculate the number of turns on the secondary coil.

14.7 High-voltage transmission of electricity

LEARNING OUTCOMES

- Describe the use of transformers in the high-voltage transmission of electricity

- State the advantages of high voltage transmission

S - Recall and use the power equation relating the current and pd in each coil for a transformer that is 100% efficient

- Explain why energy loss in cables is lower when the voltage is high

The **electricity grid** is the network of cables and transformers used to distribute mains electricity to our homes and other buildings from power stations.

The cables of the grid waste energy because the cables have resistance. Current in them has a heating effect. As a result, the cables do waste some energy. However, by operating the grid at a high voltage, we can reduce the current through the cables and transfer the same amount of electrical energy every second through them.

The higher the grid voltage, the greater the efficiency of transferring electrical power through the grid.

The **efficiency** may be expressed as the percentage of the electrical energy supplied to the grid that is used by the appliances and machines connected to it. The greater the efficiency, the lower the energy wasted in the cables. For example, if an electricity grid is supplied with 1000 million joules of electrical energy every second and 950 million joules is transferred to 'users', the efficiency of the grid is 95%. In other words, 5% of the electrical energy supplied every second to this grid is wasted.

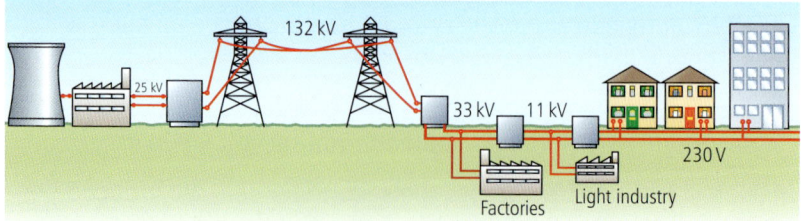

Figure 14.7.1 The electricity grid

Supplement

Transformer efficiency

Transformers are almost 100% efficient. This means that almost all the electrical energy supplied each second (i.e. electrical power) to the primary coil of a transformer is transferred to the device or devices connected to the secondary coil.

Since power = current × voltage:

- power supplied to the transformer = primary current, I_p × primary voltage, V_p

- power delivered by the transformer = secondary current, I_s × secondary voltage, V_s

> **EXAM TIP**
>
> You must know that high voltage transmission is used because it is efficient – much less energy is wasted in the form of thermal energy.

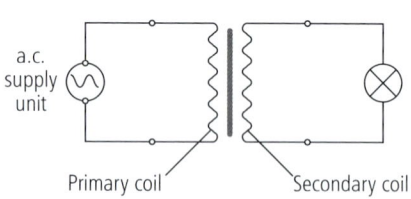

Figure 14.7.2 Transformer efficiency

Therefore, for 100% efficiency:

$$\frac{\text{primary}}{\text{current}} \times \frac{\text{primary}}{\text{voltage}} = \frac{\text{secondary}}{\text{current}} \times \frac{\text{secondary}}{\text{voltage}}$$

$$I_p V_p = I_s V_s$$

WORKED EXAMPLE

A 120 V, 60 W lamp lights normally when it is connected to the secondary coil of a transformer and a 10 V ac supply is connected to the primary coil. The transformer is 100% efficient.

Calculate **a** the primary current, **b** the lamp current.

Solution

a Rearranging   power supplied to the primary coil

= the primary current × primary voltage,

$$\text{primary current} = \frac{\text{power supplied to the primary coil}}{\text{primary voltage}}$$

$$= \frac{60\,\text{W}}{10\,\text{V}} = 6.0\,\text{A}$$

b For 100% efficiency, power supplied to the lamp by the secondary coil = 60 W.

Rearranging   power supplied by the secondary coil

= the secondary current × secondary voltage,

$$\text{lamp current} = \frac{\text{power supplied to the lamp}}{\text{secondary voltage}} = \frac{60\,\text{W}}{120\,\text{V}} = 0.5\,\text{A}$$

Transformers and the grid

Transformers are used to step up the alternating voltage from a power station to the grid voltage and to step the grid voltage down to the mains voltage. The grid voltage is at least 132 000 V. What difference would it make if the grid voltage was much lower? Much more current would be needed to deliver the same amount of power. Because of their resistance, the grid cables would therefore heat up more and waste more power.

The power P wasted in the cables can be calculated using the equation $P = I^2R$ from Topic 12.5. For example, to transfer 100 000 W of electrical power at 1000 V along a 0.5 Ω cable, the current passing along the cable would be 100 A (= 100 000 W/1000 V). So the power wasted in the cable would be 5000 W (= I^2R = (100 A)^2 × 0.5 Ω). In other words, 5% of the power supplied to the cable is wasted as thermal energy.

If the same amount of power was transferred along the same cable at 100 000 V instead of 1000 V, the current would be 1.0 A (= 100 000 W/100 000 V). The voltage drop along the cable would be 0.5 V (= 1.0 A × 0.5 Ω).

So the power wasted in the cable would be only 0.5 W (= I^2R = (1.0 A)^2 × 0.5 Ω). In other words, the power wasted is negligible compared with the power transferred.

KEY POINTS

- Transformers are used to step voltages up or down.

- High voltage transmission of electricity is much more efficient than transmission at much lower voltages.

- For a transformer that is 100% efficient:

  the primary current × the primary voltage = the secondary current × secondary voltage

SUMMARY QUESTIONS

1 Copy and complete the following sentences using words from the list below:

   **down   primary
   secondary   up**

   a In a step-up transformer, the voltage across the _____ coil is greater than the voltage across the _____ coil.

   b The voltage from a power station is stepped _____ so the same amount of power can be delivered through the cables as a result of stepping the current _____.

2 A transformer with a secondary coil of 100 turns is to be used to step a voltage down from 240 V to 12 V.

   a Calculate the number of turns on the primary coil of this transformer.

   b A 12 V, 36 W lamp is connected to the secondary coil. Calculate the current in **i** the lamp, **ii** the primary coil. Assume the transformer is 100% efficient.

1 (a) Describe an experiment to plot the shape of the magnetic field near a straight current-carrying wire. Include a diagram in your answer to show the apparatus used and to show the shape of the magnetic field produced.

(b) How would you find the direction of the magnetic field?

(c) What is the effect of reversing the direction of the current?

2 (a) Describe an experiment to plot the shape of the magnetic field in and around a long current-carrying coil (solenoid). Include a diagram in your answer to show the apparatus used and to show the shape of the magnetic field produced.

(b) How would you find the direction of the magnetic field?

(c) What is the effect of reversing the direction of the current?

3 Explain, in simple terms, the difference between an electric motor and a generator.

4 (a) Describe the construction of a basic step-up transformer. Include a diagram in your answer.

(b) Explain how the transformer works.

(c) Suggest a use for a step-up transformer.

5 A transformer is used to convert 240 V ac to 12 V ac. The primary coil has 48 000 turns.

(a) Calculate the number of turns on the secondary coil.

(b) State whether this is a step-up or step-down transformer.

S (c) A 12 V, 5.0 A lamp is connected to the secondary coil. Calculate the current in the primary coil. Assume the transformer is 100% efficient.

1 Which of the following factors will increase the size of the force produced on a current-carrying metal wire in a magnetic field?

A Reversing the direction of the current

B Reversing the direction of the magnetic field

C Increasing the size of the electric current

D Decreasing the strength of the magnetic field

Practice Questions

2 The diagram shows the structure of a simple transformer. What is the output voltage of the transformer?

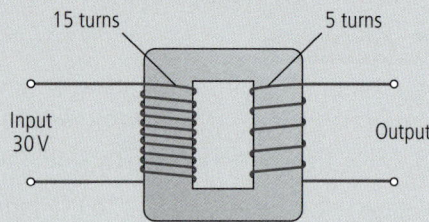

A 60 V

B 45 V

C 10 V

D 5 V

3 Which of the following statements about transformers is NOT true?

A Transformers only operate with alternating currents.

B A step up transformer has more turns on the primary coil than the secondary coil.

C The iron core of the transformer allows an alternating magnetic field to pass between the two coils efficiently.

D A step down has fewer turns on the secondary coil than the primary coil.

4 A magnet is moved slowly into a coil of wire as shown in the figure, and the meter pointer moves slightly to the right as the magnet moves. What will happen to the pointer while the magnet is pulled out of the coil quickly?

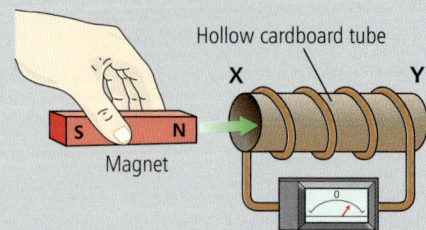

A It will move the same amount to the right.

B It will move the same amount to the left.

C It will move further to the right.

D It will move further to the left.

5 A transformer used in a mobile phone charger has a primary current of 0.10 A and a primary voltage of 110 V. The secondary voltage is 5.0 V. Assuming that the transformer is 100% efficient, what is the current in the secondary coil?

A 5.0 A

B 2.2 A

C 55 kA

D 4.5 A

6 A transformer can be connected to a mains supply to power a low voltage lamp.

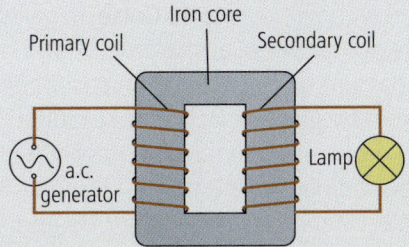

Iron core
Primary coil   Secondary coil
a.c. generator   Lamp

(a) Write the following statements in the correct order to explain how a transformer works: [3]

A. The alternating potential difference produces an alternating current in the lamp.

B. The alternating magnetic field passes through the iron core.

C. An alternating current in a primary coil produces an alternating magnetic field in the primary coil.

D. The alternating magnetic field induces a potential difference in the secondary coil.

(b) Explain why a transformer requires an alternating current source to operate. [2]

(c) The lamp operates with a voltage of 12 V. The transformer is connected to an electrical mains supply of 240 V. What is the ratio of the number of turns on the primary and secondary coils? [2]

(d) State what type of transformer this is. [1]

7 In a demonstration of electromagnetism, a teacher places a stiff wire between two magnets held inside a frame as shown in the diagram. The wire is connected to a battery and it experiences a force in the direction shown.

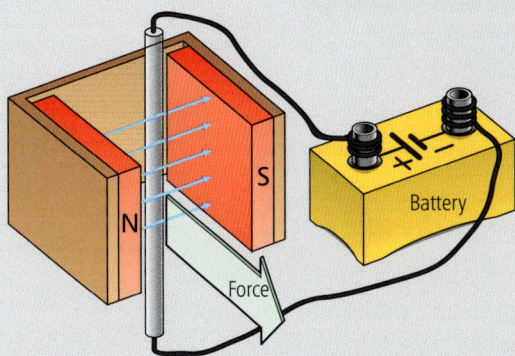

S
N
Battery
Force

(a) Name this effect. [1]

(b) State two ways that the direction of the motion of the wire can be reversed. [2]

(c) (i) State two factors which affect the size of the force acting on the wire. [2]

(ii) Describe how the effect of these factors could be demonstrated. [3]

(d) Describe the cause of this effect in terms of magnetic fields. [3]

8 A simple alternating current generator can be made as shown in the diagram.

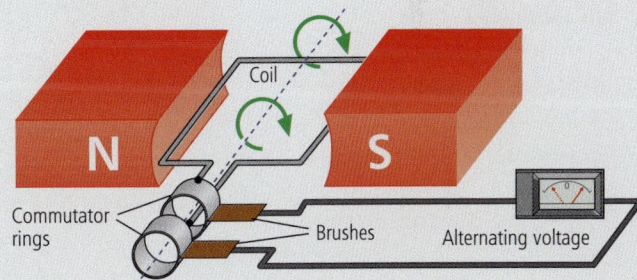

Coil
N   S
Commutator rings   Brushes   Alternating voltage

(a) Sketch a graph of the output emf produced by this generator for one complete rotation of the coil starting from the coil position shown in the diagram. [2]

(b) Describe what happens to **(i)** the frequency and **(ii)** peak voltage produced by the generator when it is spun faster. [2]

(c) Describe the purpose of the commutator rings and brushes in this generator. [2]

15.1 Observing nuclear radiation

LEARNING OUTCOMES

- State that radioactive substances emit radiation all the time
- State the three different types of radiation emitted by radioactive substances
- Describe how the different types of radiation from radioactive substances can be detected

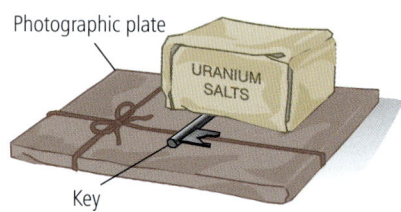

Figure 15.1.1 Becquerel's key

EXAM TIP

The word radiation is used in other contexts in Physics. Nuclear radiation has a particular meaning. Make sure you come to understand this as you work through this chapter.

A key discovery

If your photos showed a mysterious image, what would you think? In 1896, the French physicist **Henri Becquerel** discovered the image of a key on a film he developed. He remembered the film had been in a drawer under a key – with a packet of uranium salts on top (Figure 15.1.1). The uranium salts must have sent out some form of radiation that passed through paper (i.e. the film wrapper) but not through metal (i.e. the key).

Becquerel asked a young research worker, **Marie Curie**, to investigate. She found that the salts gave out radiation all the time. It happened spontaneously no matter what was done to them. She used the word **radioactivity** to describe this strange new property of uranium. She and her husband, Pierre, did more research into this new branch of science. They discovered new radioactive elements. They named one of the elements polonium, after Marie's native country, Poland.

MARIE CURIE (1867–1934)

Becquerel and the Curies were awarded the Nobel prize for the discovery of radioactivity. Pierre died in a road accident. Marie went on with their work. She was awarded a second Nobel prize in 1911 for the discovery of polonium and radium. She died in middle-age from leukaemia in 1934. This is a disease of the blood cells and was caused by the radioactive materials she worked with.

Marie Curie (1867–1934)

PRACTICAL

Investigating radioactivity

We can use a **Geiger counter** to detect radioactivity. Look at Figure 15.1.2. The counter clicks each time a particle of radiation from a radioactive substance enters the Geiger tube.

Figure 15.1.2 Using a Geiger counter

What stops the radiation? Ernest Rutherford carried out tests to answer this question about a century ago. He put different materials between the radioactive substance and a 'detector'. He discovered two types of radiation.

- One type (**alpha radiation**; symbol α) was stopped by paper.

- The other type (**beta radiation**; symbol β) went through it.

- Scientists later discovered a third type, **gamma radiation** (symbol γ), even more penetrating than beta radiation.

Radioactivity around us

When we use a Geiger counter, it clicks even without a radioactive source near it. This is due to **background radiation** from radioactive substances found naturally all around us.

Background radiation is ionising radiation from space (cosmic rays), from devices such as X-ray tubes and from radioactive substances in the environment. As explained in the next topic, the radiation from radioactive substances is harmful. Some of these substances are present because of nuclear weapons testing and nuclear power stations. But most of it is from substances in the Earth. For example, radon gas is radioactive and is a product of the decay of uranium in the ground. Radiation from substances in the Earth varies with location.

A radioactive puzzle

Why are some substances radioactive? Every atom has a nucleus made up of protons and neutrons. Electrons move about in the space surrounding the nucleus.

Most atoms each have a stable nucleus that doesn't change. But the atoms of a radioactive substance each have a nucleus that is unstable. An unstable nucleus becomes stable by emitting alpha, beta or gamma radiation. We say an unstable nucleus **decays** when it emits radiation. We can't tell when an unstable nucleus will decay or in which direction the radiation from it will be emitted. It is a **random** event that happens **spontaneously** without anything being done to the nucleus.

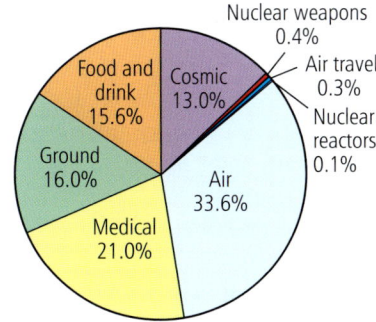

Figure 15.1.3 The origins of background radioactivity

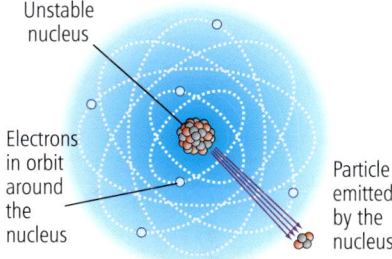

Figure 15.1.4 Radioactive decay

KEY POINTS

- Radioactive substances give out radiation all the time.

- Radioactive decay is a random event – we cannot predict or influence when it will happen.

- Background radiation occurs naturally from traces of radioactive substances in buildings, in the atmosphere and in the ground.

- There are three types of radiation from radioactive substances: alpha, beta and gamma radiation.

SUMMARY QUESTIONS

1 **a** The radiation from a radioactive source is stopped by paper. What type of radiation does the source emit?

 b The radiation from a different source goes through paper. What can you say about this radiation?

2 Look at the pie chart in Figure 15.1.3.

 a What is the biggest source of background radioactivity?

 b List the sources in the chart and say which ones could be avoided.

15.2 Alpha, beta and gamma radiation

LEARNING OUTCOMES

- Describe the nature and main properties of α-, β- and γ-radiation

S
- Explain the relative ionising effect of α-, β- and γ-radiation

- Describe the deflection of α-, β- and γ-radiation using an electric field and using a magnetic field

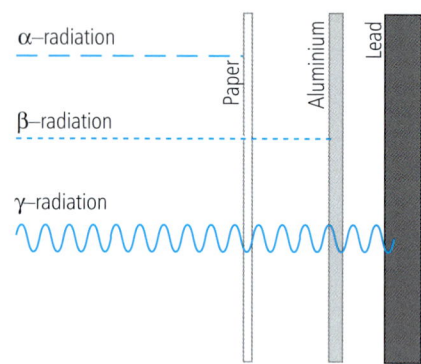

Figure 15.2.1 The penetrating powers of α-, β- and γ-radiation

EXAM TIP

The electrons that make up beta radiation come from the nucleus. They are NOT the electrons that orbit the nucleus.

Figure 15.2.3 Radioactive warning

Investigating the properties of alpha, beta and gamma radiation

Alpha radiation cannot penetrate paper. What stops beta and gamma radiation? How far can each type of radiation travel through air? The properties of each type of radiation can be demonstrated using a Geiger counter.

1 **To test different materials**, each type of material is placed between the tube and the radioactive source. We can add more layers of material until the radiation is stopped.

2 **To test the range in air**, the Geiger tube is moved away from the source. When the tube is beyond the range of the radiation, it cannot detect it.

The table below shows the results of the above tests.

type of radiation	absorber materials	range in air
alpha	paper	about 10 cm
beta	aluminium sheet (1 cm thick) lead sheet (2–3 mm thick)	about 1 m
gamma	thick lead sheet (several cm thick) concrete (more than 1 m thick)	unlimited see note below

Note Gamma radiation spreads out in air without being absorbed. It gets weaker as it spreads out.

The nature of alpha, beta and gamma radiation

What are these mysterious radiations? Experiments carried out by Ernest Rutherford and other scientists showed that:

- alpha radiation consists of positively charged particles, each consisting of two protons and two neutrons,

- beta radiation consists of electrons so it is negatively charged,

- gamma radiation is electromagnetic radiation so it is uncharged.

Supplement

Deflection tests using electric and magnetic fields

Alpha, beta and gamma radiation in a narrow beam can be separated using a magnetic field, as shown in Figure 15.2.2.

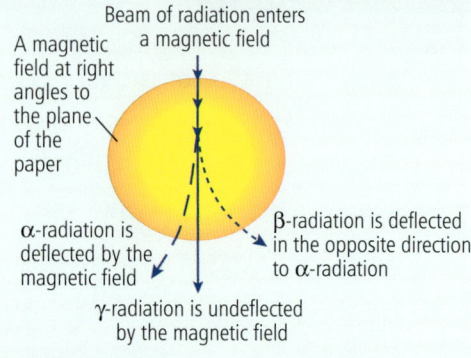

A magnetic field at right angles to the plane of the paper

Beam of radiation enters a magnetic field

α-radiation is deflected by the magnetic field

β-radiation is deflected in the opposite direction to α-radiation

γ-radiation is undeflected by the magnetic field

Figure 15.2.2 Radiation in a magnetic field

1 β-radiation is easily deflected because it consists of electrons. See Topic 14.2.

2 α-radiation is deflected in the opposite direction to β-radiation because an α-particle has a positive charge. α-radiation is harder to deflect than β-radiation because an α-particle has much more mass than a β-particle, so much more force is needed to change its direction of motion.

3 γ-radiation is not deflected by a magnetic field. This is because gamma radiation is electromagnetic radiation so it is uncharged.

We can also use an electric field as shown in Figure 15.2.4 to deflect α- and β-radiation. Note that they are deflected in opposite directions.

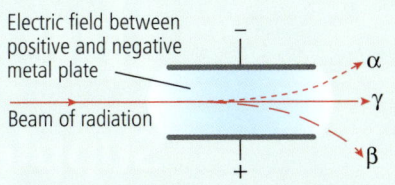

Electric field between positive and negative metal plate

Beam of radiation

Note that α- and β-particles passing through an electric field are deflected in opposite directions

Figure 15.2.4 Radiation in an electric field

Ionisation

The radiation from radioactive substances knocks electrons out of atoms. The atoms become charged as a result. Atoms can also be charged by adding electrons to them. The process of creating a charged atom is called **ionisation**. X-rays also cause ionisation. Alpha radiation has a much greater ionising effect than beta radiation, which has a much greater ionising effect than gamma radiation.

Ionisation in a living cell can damage or kill the cell. Damage to the genes in a cell can be passed on if the cell generates more cells. Strict rules must always be followed when radioactive substances are used.

S
- Alpha radiation has a much greater ionising effect than beta radiation. This is because an α-particle has more charge and much more mass than a β-particle and moves more slowly, so it has much more effect on the atoms it encounters.

- Beta radiation has a greater ionising effect than gamma radiation because β-particles are charged whereas gamma radiation is uncharged.

KEY POINTS

- A radioactive substance contains unstable nuclei.

- An unstable nucleus becomes stable by emitting radiation.

type of radiation	charge
alpha	positive
beta	negative
gamma	uncharged

SUMMARY QUESTIONS

1 Copy and complete the following sentences using words from the list:

protons   neutrons   nucleus   radiation

a The _____ of an atom is made up of _____ and _____.

b When an unstable _____ decays, it emits _____.

2 a Copy and complete the following sentences using words from the list:

alpha   beta   gamma.

i Electromagnetic radiation from a radioactive substance is called _____ radiation.

ii A thick metal plate will stop _____ and _____ radiation but not _____ radiation.

b Which type of radiation is **i** uncharged, **ii** positively charged, **iii** negatively charged?

c i Why is a radioactive source stored in a lead-lined box?

ii Why should long-handled tongs be used to move a radioactive source?

3 Look at Figure 15.2.4. **S**

Explain why the α-particles are deflected upwards.

15.3 The discovery of the nucleus

Supplement

LEARNING OUTCOMES

- Describe Rutherford's alpha-particle scattering experiment and its results

- Explain how the results provide evidence for the nuclear model of the atom

Alpha particle scattering

Ernest Rutherford had already made important discoveries about radioactivity when he decided to use alpha particles to probe the atom. He asked two of his research workers, Hans Geiger and Ernest Marsden, to investigate the scattering of alpha particles by a thin metal foil. Figure 15.3.1 shows the arrangement they used.

The radioactive source they used had to decay slowly enough so its activity effectively stayed the same during the experiment. They measured the number of alpha particles deflected per second through different angles. The results showed that:

- most of the alpha particles passed straight through the metal foil

- the number of alpha particles deflected per minute decreased as the angle of deflection increased

- about 1 in 10 000 alpha particles were deflected by more than 90°

- occasionally an alpha particle bounced off the foil back towards the source.

Rutherford was astonished by the results. He said it was like firing 'naval shells' at cardboard and discovering the occasional shell rebounds. He knew that alpha particles are positively charged. He deduced from the results that there is a **nucleus** at the centre of every atom. This nucleus is:

1 positively charged because it repels alpha particles (remember that particles with like charges repel)

2 much smaller than the atom because most alpha particles pass through without deflection

3 where most of the mass of the atom is located.

Using this model, Rutherford worked out the proportion of alpha particles that would be deflected for a given angle. He found an exact agreement with Geiger and Marsden's measurements. He used his theory to estimate the diameter of the nucleus and found it was about 100 000 times smaller than the atom itself.

Rutherford's nuclear model of the atom was quickly accepted by other scientists because:

- it agreed exactly with Geiger and Marsden's measurements
- it explains radioactivity in terms of changes that happen to an unstable nucleus when it emits radiation
- it predicted the existence of the neutron, which was later discovered.

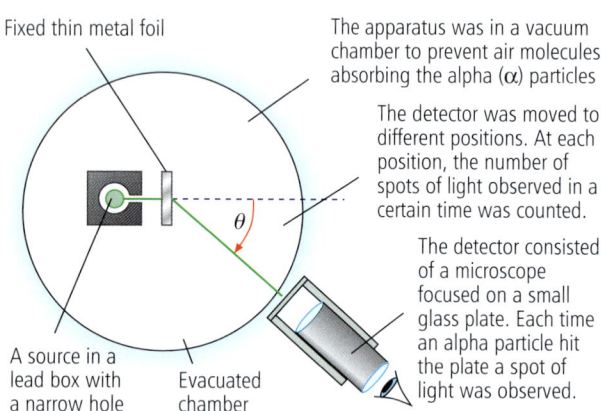

Fixed thin metal foil

The apparatus was in a vacuum chamber to prevent air molecules absorbing the alpha (α) particles

The detector was moved to different positions. At each position, the number of spots of light observed in a certain time was counted.

The detector consisted of a microscope focused on a small glass plate. Each time an alpha particle hit the plate a spot of light was observed.

A source in a lead box with a narrow hole

Evacuated chamber

θ

Figure 15.3.1 Alpha particle scattering

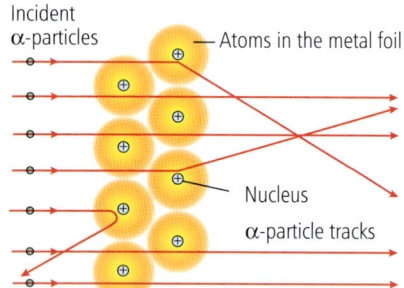

Incident α-particles

Atoms in the metal foil

Nucleus

α-particle tracks

Figure 15.3.2 Alpha particle paths

PRACTICAL

An α-scattering model

Fix a small metal disc about 2 cm thick at the centre of a table. Hide the disc under a cardboard disc about 20 cm in diameter. See if you can hit the metal disc with a rolling marble.

The scientist who split the atom!

When radioactivity was discovered, scientists couldn't work out what alpha radiation was – until Ernest Rutherford found the answer. He collected alpha particles in a tube. Then he made the 'alpha gas' light up by passing an electric current through it at high voltage. He found that light from the tube was the same as if helium gas was in the tube. So he concluded that alpha particles must be helium atoms without electrons.

After his discovery that every atom contains a nucleus, he went on to discover how to 'split the atom' and he found that nuclear reactions release much more energy then chemical reactions. He hoped no one would find out how to do this on a practical scale until the human race had learned to live in peace. He was made Lord Rutherford of Nelson in 1931. After his death in 1937, his ashes were placed close to Newton's tomb in Westminster Abbey in London.

Goodbye to the plum pudding atom!

Before the nucleus was discovered in 1914, scientists didn't know what the structure of the atom was. They did know it contained electrons and they knew these are tiny negatively charged particles. But they didn't know how the positive charge was arranged in an atom, although there were different models suggested. Some scientists thought the atom was like a 'plum pudding' with the positively charged matter in the atom evenly spread (like in a pudding) and the electrons buried inside (like plums in the pudding). Rutherford proved that the atom consists of a tiny nucleus with its electrons moving about in the empty space around the nucleus.

Lord Rutherford of Nelson (1871–1937)

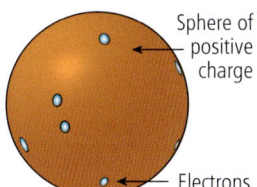
Figure 15.3.3 The plum pudding model

SUMMARY QUESTIONS

1 Copy and complete the following sentences using words from the list:

charge   diameter   mass

a A nucleus has the same type of _____ as an alpha particle.

b A nucleus has a much smaller _____ than the atom.

c Most of the _____ of the atom is in the nucleus.

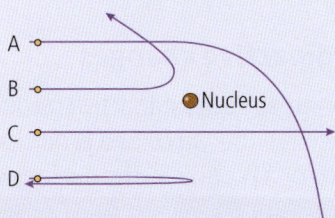
Figure 15.3.4

2 a Figure 15.3.4 shows four possible paths labelled A,B,C and D of an alpha particle deflected by a nucleus. Which path would the alpha particle travel along?

b Explain why each of the other paths in **a** is not possible.

15.4 More about the nucleus

LEARNING OUTCOMES

- Recall and use the symbols for a nuclide

S
- Describe the changes to an unstable nucleus that take place due to a radioactive emission

EXAM TIP

Isotopes have the same number of protons but different numbers of neutrons.

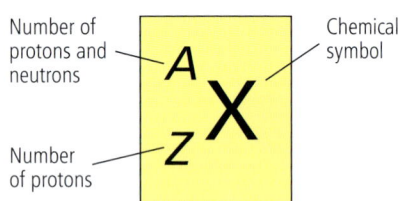

Number of protons and neutrons — A

Chemical symbol — X

Number of protons — Z

Figure 15.4.1 The symbol for a nuclide

The Table below lists the relative charge and mass of the particles in an atom.

	charge	mass
proton	1	1
neutron	0	1
electron	−1	0.00055

Nuclides

The nucleus of an atom consists of protons which are positively charged and neutrons which are uncharged. Every atom of any particular element always has the same number of protons in its nucleus. However, the number of neutrons in the nucleus can differ. Each type of atom is called a **nuclide**. For example:

- Every atom of uranium has 92 protons in its nucleus. Most of these atoms contain 146 neutrons but some have 3 fewer neutrons.

- Every atom of carbon has 6 protons in its nucleus. Almost all these atoms contain 6 neutrons as well but some do not. For example, carbon atoms with 8 neutrons in the nucleus form a very tiny proportion of any sample of carbon.

Different nuclides with the same number of protons are the same element because every atom of an element has the same number of protons in its nucleus. The different nuclides of the same element are referred to as **isotopes** of the element.

The proton number (or atomic number), Z, of a nuclide is the number of protons in a nucleus of any atom of the element. The **S** charge of a nucleus is therefore $Z \times$ the charge of a proton.

The nucleon number (or mass number), A, of a nuclide is the total number of protons and neutrons in any atom of that nuclide.

S As the mass of a neutron is almost the same as the mass of a proton, the mass of a nucleus is therefore almost equal to $A \times$ the mass of a proton (or a neutron).

Figure 15.4.1 shows the symbol for a nuclide. The values of A and Z always precede the chemical symbol with Z as the subscript. For example: for a uranium nuclide with 146 neutrons and 92 protons, $Z = 92$ and $A = 238$ ($= 92 + 146$), thus the symbol is $^{238}_{92}U$

Radioactive changes

In general, each radioactive nuclide emits one type of radiation only which is either an α- or a β-particle or γ-radiation. When a nucleus emits an α- or a β-particle, the nucleus changes to that of a different element.

Supplement

Alpha radiation consists of particles, each composed of two protons and two neutrons. We use the symbol $^{4}_{2}\alpha$ for an α-particle, as $Z = 2$ (because it contains 2 protons) and $A = 4$ (because it consists of 2 protons and 2 neutrons).

An α particle is emitted from a nucleus that is unstable because it has too many neutrons and protons. Such a heavy nucleus loses 2 protons and 2 neutrons, as shown in Figure 15.4.2. Therefore:

- the proton number of the nucleus decreases by 2 and its mass number decreases by 4

- it becomes a nuclide of a different element because it loses 2 protons.

We can represent this change by the equation shown under Figure 15.4.2:

1 The total number of protons on each side are equal: 90 = 88 + 2

2 The total number of protons and neutrons on each side are equal, i.e. 228 = 224 + 4

Beta radiation consists of electrons. We use the symbol $_{-1}^{0}\beta$ for a β-particle where $Z = -1$ (as its charge is equal and opposite to that of a proton) and $A = 0$ (because it does not contain any protons or neutrons).

A β-particle is emitted from a nucleus that is unstable due to an excess of neutrons. Such a nucleus becomes more stable when one of its neutrons changes into a proton, creating and emitting the β-particle at the same time. Figure 15.4.3 shows the process. Therefore:

- the proton number of the nucleus increases by 1 and its mass number is unchanged because the total number of protons and neutrons is unchanged.

- it becomes a nuclide of a different element because it has one more proton and one less neutron than it started with.

We can represent this change by the equation shown under Figure 15.4.3:

1 The nucleus has 19 protons and 21 (= 40 − 19) neutrons before the change. After the change, it has 20 protons and 20 neutrons.

2 The total number of protons and neutrons on each side are equal. There is a total of 40 protons and neutrons in the nucleus before the change and after the change.

Gamma radiation is high-energy electromagnetic radiation. After an unstable nucleus has emitted an α-particle or a β-particle, it sometimes has surplus energy. It emits this energy as γ-radiation.

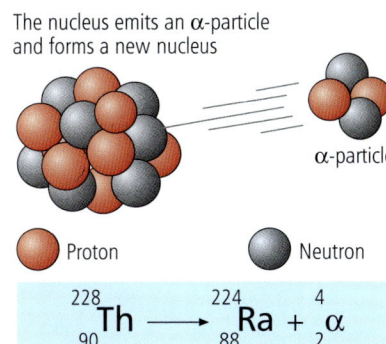

The nucleus emits an α-particle and forms a new nucleus

α-particle

● Proton   ● Neutron

$$_{90}^{228}\text{Th} \longrightarrow {}_{88}^{224}\text{Ra} + {}_{2}^{4}\alpha$$

Figure 15.4.2 α-emission

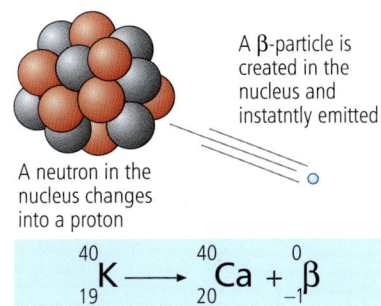

A β-particle is created in the nucleus and instatntly emitted

A neutron in the nucleus changes into a proton

$$_{19}^{40}\text{K} \longrightarrow {}_{20}^{40}\text{Ca} + {}_{-1}^{0}\beta$$

Figure 15.4.3 β-emission

EXAM TIP

Practise balancing nuclear equations until you are confident with them.

KEY POINTS

- A nuclide $_{Z}^{A}X$ contains Z protons and $(A - Z)$ neutrons.

- When a nucleus emits:

  - **an α-particle, it loses 2 protons and 2 neutrons**

  - **a β-particle, a neutron in the nucleus changes into a proton**

  - **γ radiation, the number of neutrons and protons it contains is unchanged.**

SUMMARY QUESTIONS

1 Work out the number of protons and the number of neutrons in each of the following nuclides.

   a $_{90}^{228}$Th   b $_{91}^{234}$Pa   b $_{89}^{227}$Ac

2 a $_{92}^{238}$U is an α-emitter.

   When such a nucleus emits an α-particle, which one of the following nuclides is formed?

   $_{92}^{234}$U $_{90}^{233}$Th $_{90}^{234}$Th $_{90}^{228}$Th $_{90}^{231}$Th

   b $_{91}^{234}$Pa is a β-emitter.

   i From the list of nuclides above in **a**, identify the nucleus formed when this change happens.

   ii Write an equation for the above change.

15.5 Half-life

LEARNING OUTCOMES

- Explain what is meant by the term *half-life*
- Use half-life in calculations or in graphs
- Describe how radioactive materials are stored and used safely

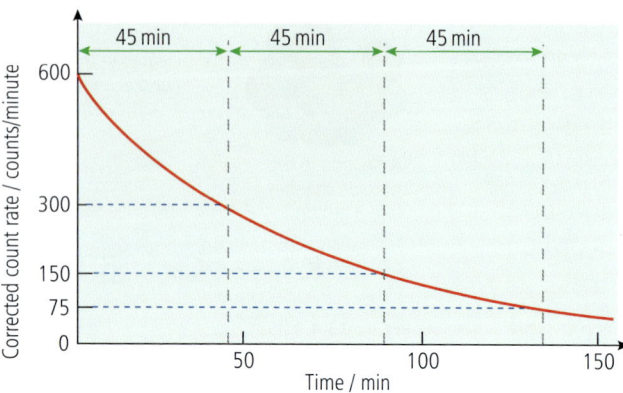

Figure 15.5.1 Radioactive decay: a graph of count rate versus time

Supplement

If a graph is plotted without subtracting the background rate, the curve flattens out above zero at the level of the background rate. This level can be estimated from the graph to give the background rate. Subtracting this estimate from the measured count rate then gives the corrected count rate due to the sample.

PRACTICAL

Modelling radioactive decay

Do an investigation modelling radioactive decay using dice or a computer simulation.

Radioactive decay

The **activity** of a radioactive substance is the number of nuclei that decay per second. Each unstable nucleus (the 'parent' nucleus) changes into a nucleus of a different isotope (the 'daughter' nucleus) when it decays. Because the number of parent nuclei goes down, the activity of the sample decreases. We say the sample **decays**.

We can use a Geiger counter to monitor the activity of a radioactive sample. We need to measure the **corrected count rate** due to the sample. This is the number of counts per second (or per minute) due to the sample only. The Geiger counter measures counts due to background radiation and due to the sample. So we need to subtract the background count rate from the measured count rate to obtain the count rate due to the sample.

Figure 15.5.1 shows how the count rate of a sample decreases with time. It falls from:

- 600 counts per minute (cpm) to 300 cpm in the first 45 minutes
- 300 counts per minute (cpm) to 150 cpm in the next 45 minutes.

The time taken for the count rate (and therefore the number of parent atoms) to fall by half is always the same. This time is called the **half-life**. The half-life in Figure 15.5.1 is 45 minutes.

The half-life of a radioactive isotope is the time it takes:

- **for the number (and therefore the mass) of parent nuclei in a sample of the isotope to halve**
- **for the count rate from the original isotope to fall to half its initial level.**

The random nature of radioactive decay

We can't predict *when* an individual nucleus will suddenly decay. But we *can* predict how many nuclei will decay in a certain time – because there are so many of them. This is a bit like throwing dice. If you threw 1000 dice, you would expect one-sixth to land at random on a particular number.

Suppose we start with 1000 unstable nuclei. If 10% disintegrate every hour, we would expect:

100 nuclei to decay in the first hour, leaving 900

90 nuclei (= 10% of 900) to decay in the second hour, leaving 810

81 nuclei (= 10% of 810) to decay in the third hour, leaving 729.

Continue the calculations to find how many unstable nuclei are left at the end of each hour for the next four hours. Check your results with the box above Figure 15.5.2. All the results are plotted in Figure 15.5.2. Check that the graph here has a 'half-life' of about 6.5 hours.

Using radioactive materials

Radioactive materials are used in nuclear power stations and in many hospitals and laboratories and in certain factories. When a radioactive source is not in use, it must be stored in a thick lead container in a secure location that is not accessible to unauthorised users.

S In use, to minimise exposure of users to ionising radiation, the radioactive source should be:

- kept at a safe distance from the user (or users)
- separated from users by thick lead screens if necessary
- in use for the shortest possible time
- only moved using long-handled tools so it is not near to anybody.

After use, it should be immediately returned to its container and its storage location.

Radiation workers are required to carry a radiation monitoring device, usually a film badge attached to clothing. The film badge is sealed from light and contains a strip of photographic film which is covered in different parts by different filters. Ionising radiation darkens the covered parts of the film by different amounts according to the filter in each part. In this way, the film can be used to find out the total exposure the wearer has been subjected to.

Unwanted radioactive substances, such as radioactive waste from nuclear power stations and unwanted radioactive sources from hospitals, factories, laboratories, schools and other users, must be disposed of safely and securely in strict accordance with legally enforced safety regulations.

CHECK YOUR CALCULATIONS

After 4, 5, 6 and 7 hours respectively, 656, 590, 530 and 477 unstable nuclei are left.

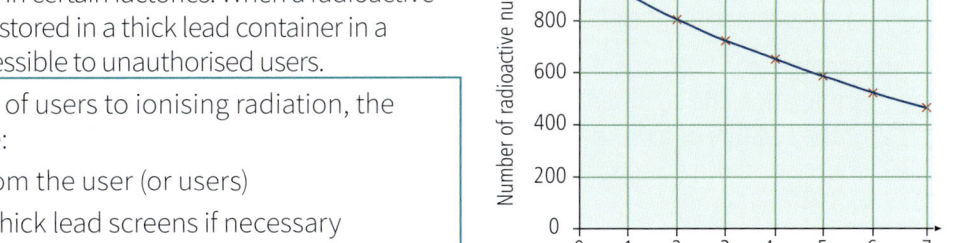

Figure 15.5.2 Half-life

A film badge

SUMMARY QUESTIONS

1 a A radioactive isotope has a half-life of 15 h. A sealed tube contains 8 mg of the isotope. What mass of this isotope is in the tube **i** 15 h later, **ii** 45 h later?

b In Figure 15.5.1 what will the count rate be 135 min after the start?

c In Figure 15.5.2, use the graph to measure the half-life of this radioactive isotope.

S 2 A Geiger counter recorded a background count of 162 counts in 300 s. When a sample of a radioactive isotope was placed at a fixed distance from the counter, the number of counts in 100 s was recorded every 1000 s for 6000 seconds as 950, 792, 662, 561, 466, 394 and 334.

a For each recording, calculate the count rate due to the sample.

b Plot a graph to find the half-life of the sample.

KEY POINTS

- The half-life of a radioactive isotope is the time it takes for the number of parent nuclei in a sample to halve.

- The number of unstable atoms and the activity decreases to half in one half-life.

15.6 Radioactivity at work
Supplement

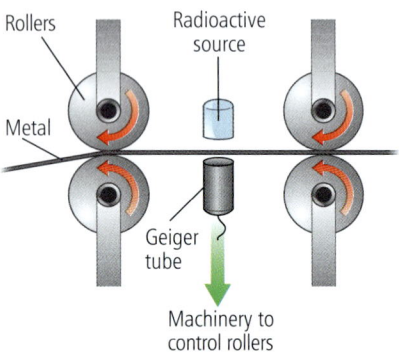

Figure 15.6.1 Thickness monitoring

Isotopes are atoms with the same number of protons but different numbers of neutrons. We use radioactive isotopes for many purposes. For each purpose, we need a radioactive isotope that emits a certain type of radiation and has a suitable half-life.

1. **Automatic thickness monitoring** is used in the production of metal foil, as shown in Figure 15.6.1. A source that emits β-radiation is used. The amount of β-radiation passing through the foil depends on the thickness of the foil. A detector on the other side of the metal foil measures the amount of radiation passing through it.

   - If the thickness of the foil increases, the detector reading drops.
   - The detector sends a signal to the rollers to increase the pressure on the metal sheet. This makes the foil thinner.

   The source needs to have a long half-life so its activity hardly changes while it is being used. This ensures variations in the count rate are due only to variations in the thickness of the foil.

   A beta-emitting source is used because changes in the foil thickness affect the amount of beta radiation passing through the foil and reaching the detector. In comparison, changes in the foil thickness would not change the detector signal because alpha radiation would all be absorbed by the foil and gamma radiation would all pass straight through it.

2. **Radioactive tracers** are used to trace the flow of a substance through a system. For example, doctors use radioactive iodine to find out if a patient's kidney is blocked.

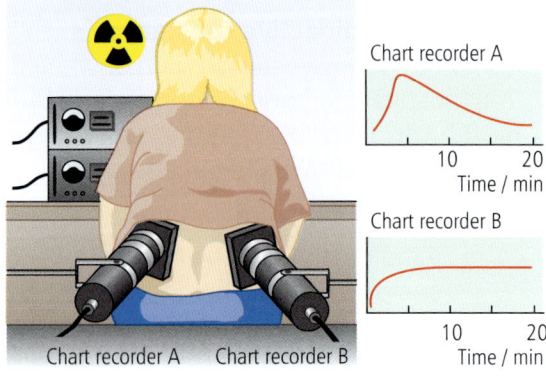

Figure 15.6.2 Using a tracer

   Before the test, the patient drinks water containing a tiny amount of the radioactive substance. A detector is then placed against each kidney. Each detector is connected to a chart recorder as shown in Figure 15.6.2. The radioactive substance flows in and out of a normal kidney. So the detector reading goes up then down as shown by chart recorder A.

   For a blocked kidney, the reading goes up and stays up because the radioactive substance that enters the kidney doesn't leave it. The radioactive isotope used for the test must

   - be a gamma emitter so it can be detected outside the body

- decay into a safe stable product
- have a half-life long enough for the test to be carried out and short enough for it to decay almost completely within a few weeks.

3 **Radioactive dating** is used to find how old ancient material is (its age). **Carbon dating** is used to find the age of ancient wood. The carbon content of living wood is made up mainly of the non-radioactive carbon isotope $^{12}_{6}C$ and a tiny proportion of the radioactive carbon isotope $^{14}_{6}C$. This has a half-life of 5600 years. When a tree dies, the amount of radioactive carbon in it decreases. To find the age of a sample, we need to measure the count rate from it. This is compared with the count rate from the same mass of living wood. For example, suppose the sample count rate is half the count rate of an equal mass of living wood. Then the age of the sample must be 5600 years.

4 **Gamma radiation** is used to kill bacteria in food and to sterilise equipment to be used, for example, in operating theatres. Gamma-emitting isotopes such as cobalt-60 ($^{60}_{27}C$) are used for such purposes because gamma radiation penetrates vacuum-packed sealed containers containing food or medical items. Cobalt-60 is prepared by exposing the stable isotope cobalt-59 to neutron radiation in a nuclear reactor. It is then used in specially designed factories to irradiate products such as medical items and food. Because cobalt-60 has a half-life of about five years, it does not need to be replaced frequently.

Gamma radiation is also used in hospitals to kill cancerous tumours. A narrow beam of gamma radiation from a cobalt-60 source is directed at the tumour. The beam is directed at it from different directions to destroy the tumour but not the surrounding tissue (see Figure 15.6.3).

SUMMARY QUESTIONS

1 Copy and complete the following sentences using words from the list below.

alpha   beta   gamma

a In the continuous production of thin metal sheets, a source of _____ radiation should be used to monitor the thickness of the sheets.

b A radioactive tracer given to a hospital patient needs to emit _____ or _____ radiation.

c The radioactive source used to trace a leak in an underground pipeline should be a source of _____ radiation.

2 a Why is alpha radiation not suitable for kidney tests?

b Why couldn't carbon dating be used to find how old a living tree is?

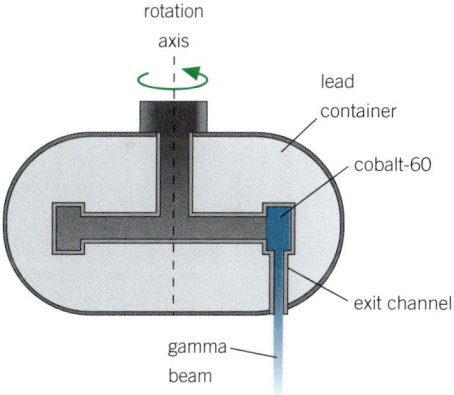

Figure 15.6.3

1 (a) Explain the meaning of the term *background radiation*.

(b) Suggest three sources of background radiation.

2 (a) State the nature of:

(i) alpha radiation

(ii) beta radiation

(iii) gamma radiation.

(b) For each of the three types of radiation describe:

(i) the charge (if any)

(ii) the ionising properties

(iii) the penetrating power

(iv) the effect of an electric or magnetic field.

3 (a) Describe, with the aid of a diagram, an experiment to find the contributions of alpha, beta and gamma radiation coming from a small sample of radioactive material.

(b) List the safety precautions that you would take when carrying out the experiment.

4 (a) Define the term *half-life*.

(b) Define the term *isotope*.

(c) Explain what is meant by the penetrating power of a radioactive emission.

S (d) Suggest two uses for radioactive isotopes and explain the importance of the half-life and penetrating power for each of the uses you choose.

5 The diagram shows the apparatus used by Ernest Rutherford when using alpha particles to study the structure of the atom.

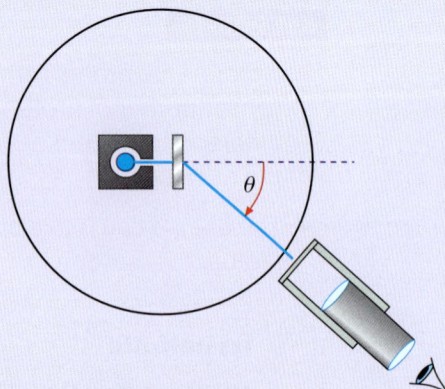

Practice Questions

1 Which of the following involve emissions of electromagnetic radiation?
1: Alpha decay, **2:** Beta decay, **3:** Gamma decay

A 1 only

B 2 only

C 3 only

D 1 and 3 only

2 Which of these is a definition of a radioactive half-life?

A The amount of material that remains after a certain period of time

B The time taken for the number of nuclei of an isotope to fall to half

C The time taken for half of the mass of a sample to decay

D The length of time an atom will last for in a decay

3 An isotope of lead has an atomic mass of 206 and an atomic number of 82. How many neutrons does an atom of this isotope have?

A 124

B 206

C 124

D 82

4 Which of the three radiations emitted during radioactive decay has the greatest ionising effect? **S**

A Alpha particles

B Beta particles

C Gamma rays

D They all have an equal ionising effect.

(a) Describe the main findings about the paths of the alpha particles.

(b) Explain the conclusions that the experiment provided.

(c) How were the alpha particles detected?

5 The half-life of a radioactive sample is 30 minutes. What fraction of the original sample will remain after 2 hours?

A $\frac{1}{2}$

B $\frac{1}{4}$

C $\frac{1}{8}$

D $\frac{1}{16}$

6 During radioactive decay, sources can emit alpha, beta and gamma radiation.

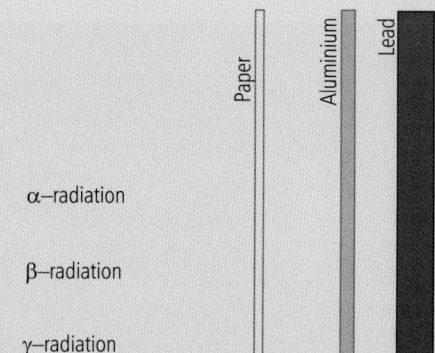

(a) Name a device which can be used to detect the radiation. [1]

(b) Copy and complete the figure to show the penetrating powers of the three radiations. [3]

All three types of radiation can cause ionisation.

(c) What happens to an atom when it becomes ionised by radiation? [1]

(d) Why is ionisation harmful to living tissue? [2]

7 A teacher is demonstrating the range of different radioactive sources before using the equipment shown in the figure below. The teacher demonstrates an alpha source, a beta source and a gamma source. Before the sources are brought into the room, the counter detects 40 counts per minute.

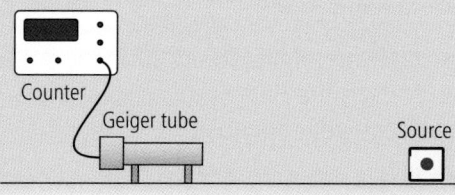

(a) Explain why the counter is detecting radiation before the sources have been used. [2]

(b) The teacher handles each of the sources using long tongs and seals them away in a lead-lined box when they are not in use. Explain why the teacher takes both of these safety precautions. [4]

(c) Explain why the detector does not measure an increase in count rate from the alpha source unless the source is placed very close to the tube. [1]

8 The isotope of plutonium $^{241}_{94}\text{Pu}$ almost always decays by emitting a beta particle but can sometimes decay by emitting an alpha particle. The half-life for $^{241}_{94}\text{Pu}$ is 14.3 years. **S**

(a) Copy and complete the decay equation for beta particle emission: [2]

$$^{241}_{94}\text{Pu} \longrightarrow\ ^{\square}_{\square}\text{Am} + \ ^{0}_{-1}\beta$$

(b) Copy and complete the decay equation for alpha particle emission: [3]

$$^{241}_{94}\text{Pu} \longrightarrow\ ^{\square}_{\square}\text{U} + \ ^{4}_{\square}\alpha$$

(c) How long will it take for the activity of a sample of $^{241}_{94}\text{Pu}$ to fall to 1/8th of its initial activity? [2]

9 (a) A Geiger counter was used to measure the background radioactivity in a laboratory. The number of counts in 600 seconds measured three times were 275, 320 and 293.

(i) Explain what is meant by background radioactivity. [2]

(ii) Calculate the average background count rate in counts per minute. [2]

(b) The following measurements were made of the count rate in counts per minute when the Geiger counter was placed at a fixed distance from a sealed radioactive source.

Time (hours)	0	0.5	1.0	1.5	2.0	2.5
Count rate (counts per minute)	539	452	373	320	266	226
Count rate due to the source only (counts per minute)						

(i) Plot a graph of the count rate (on the vertical axis) against time. [4]

(ii) Use your graph to find the half-life of the source. [2]

16.1 The Earth in space

The Earth is one of eight planets that orbit a fiery ball of gas we call the Sun. The Sun is a star that appears much brighter than any other star because we are so close to it. Light from the Sun takes about 500 seconds to reach the Earth. Light from the next nearest star takes over 4 years to reach us.

As shown in Figure 16.1.1, the Earth spins about an axis through its poles as it orbits the Sun. The Earth has a natural satellite, the Moon, that orbits the Earth as the Earth moves round the Sun.

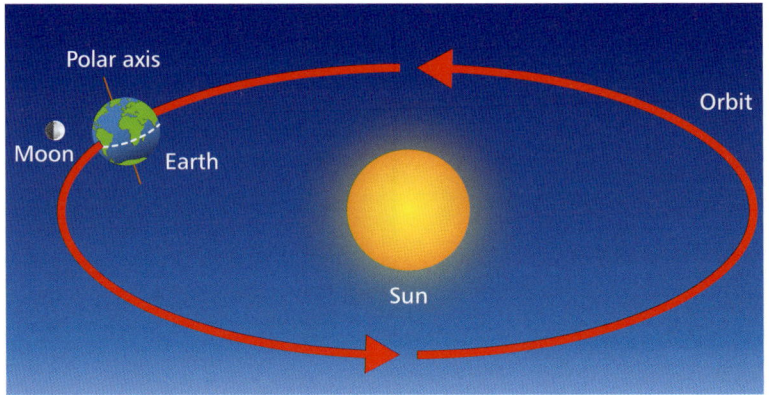

Figure 16.1.1 The Earth's orbit

Day and night

At any point on its orbit round the Sun, half the Earth's surface is always in sunlight and the other half is in darkness. As the Earth spins about its polar axis, your part of the Earth moves through daylight then through darkness every 24 hours.

One full day is the time it takes from the Earth to rotate (spin) once on its axis.

In each full day, at any point on the Earth (except at the poles), the Sun rises in the east, gradually rises to its highest point and then sets in the west. After the Sun sets, we can see the stars if the night sky is clear.

The seasons of the year

One full year is the time is takes for the Earth to travel once round the Sun. Because the Earth makes about 365 turns in this time, there are about 365 days in a full year.

The Earth's axis is tilted at a fixed angle (about 30°) to the plane of its orbit as in Figure 16.1.2. As the Earth moves round the Sun, different seasons happen each year because:

- winter is where the Earth is tilted away from Sun

- summer is where the Earth is tilted towards the Sun.

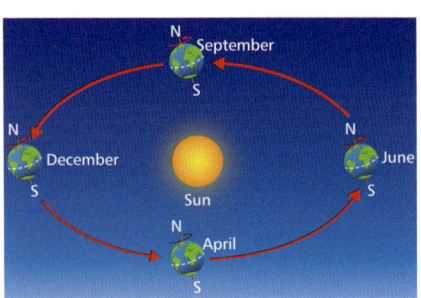

Figure 16.1.2 The seasons

The Moon and its phases

The Moon orbits the Earth approximately once every month, always keeping the same face towards the Earth. As seen by us on Earth, the Moon passes through a cycle of **phases** each time it orbits the Earth. This is because the amount of its sunlit face seen from the Earth changes gradually as it moves round the Earth. Figure 16.1.3 shows that in each cycle, the view from Earth of the Moon changes from:

new Moon → half Moon → full Moon → half Moon → new Moon.

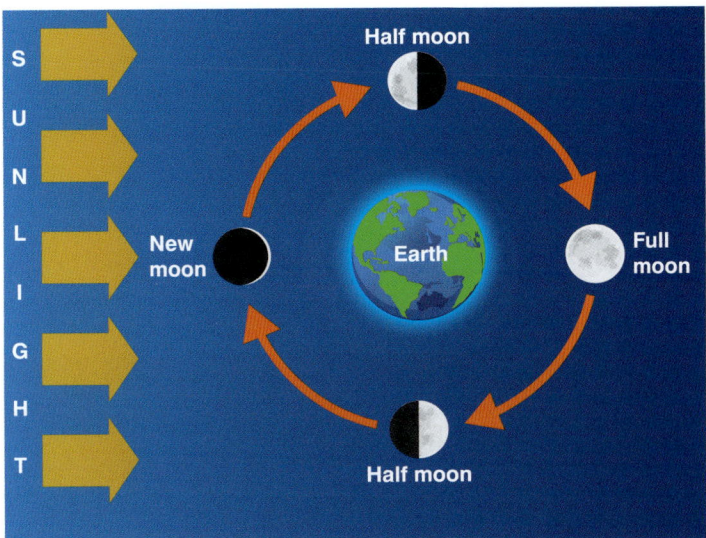

Figure 16.1.3 Phases of the Moon

Supplement

How fast do the Earth and the Moon move on their orbits?

For an orbit of average radius, R, the circumference of each orbit (distance travelled) is equal to $2\pi R$.

Using the formula $speed = \dfrac{distance}{time}$ from Topic 1.2, we can calculate the speed of a body moving round the orbit using the equation

$$v = \frac{2\pi R}{T}$$

where T is the time taken T for each orbit.

The Earth is a distance of 150 million kilometres from the Sun. It takes one year to orbit the Sun which is 31 million seconds. Therefore, for the Earth on its orbit:

its speed $v = \dfrac{2\pi R}{T} = \dfrac{2\pi \times 150 \text{ million kilometres}}{31 \text{ million seconds}} = 3000 \text{ km/s}$

The Moon is a distance of 380 000 kilometres from the Earth. It takes 27¼ days to orbit the Earth which is 2.4 million seconds. Therefore, for the Moon on its orbit,

its speed $v = \dfrac{2\pi R}{T} = \dfrac{2\pi \times 380\,000 \text{ kilometres}}{2.4 \text{ million seconds}} = 1.0 \text{ km/s}$

SUMMARY QUESTIONS

1 Copy and complete sentences **a** to **c** using words from the list below.

Earth   Moon   Sun

a The _____ rises and sets each day because the _____ spins.

b When there is a new Moon, the _____ is nearer the _____ than the _____ is.

c The southern hemisphere of the _____ at mid-summer is tilted towards the _____.

2 The average speed of the Earth on its orbit round the Sun is 3000 km/s. The planet Mars orbits the Sun once every 1.88 years. The average distance from Mars to the Sun is 1.52 times the distance from the Earth to the Sun. Use this information to calculate the average speed of Mars.

16.2 The Solar System

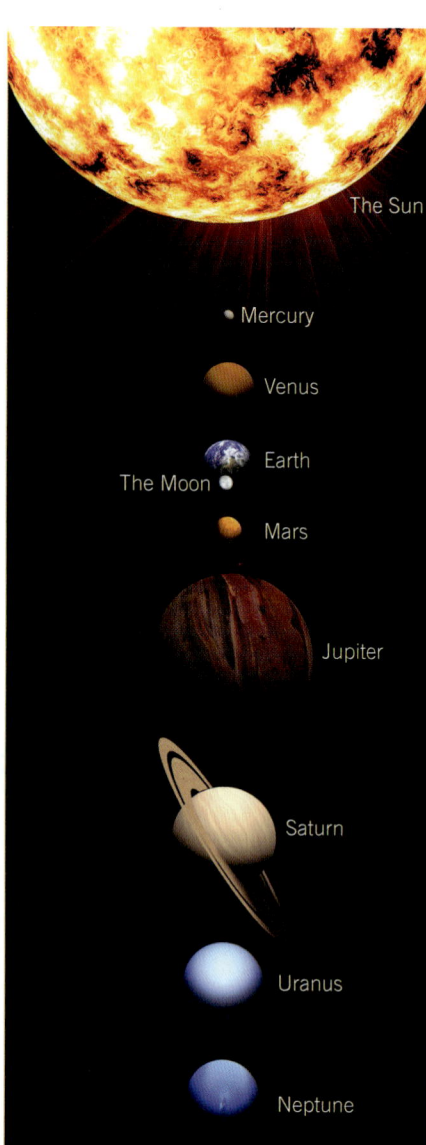

Figure 16.2.1 The solar system

How the Solar System formed

When you look at the night sky you sometimes see unexpected objects such as comets or meteors. The Solar System contains lots of objects as well as the Sun, the planets, and their moons. Comets are frozen rocks that move around the Sun in orbits that are elliptical in shape (like squashed circles). These elliptical orbits take them far away from the Sun. You only see them when they return near the Sun because then they heat up so much that they emit light. Meteors or shooting stars are small bits of rocks that burn up when they enter the Earth's atmosphere. The Solar System also includes minor (dwarf) planets and asteroids hundreds of kilometres in size orbiting the Sun, mostly between the orbits of Mars and Jupiter.

The birth of a star

The Sun formed billions of years ago from clouds of dust and gas pulled together by gravitational attraction (Figure 16.2.2).

Figure 16.2.2 The formation of the Solar System. The planets move round the Sun in circular or almost-circular orbits that are in the same plane as each other. The Earth is the third planet from the Sun. Its orbit is in the 'habitable' zone round the Sun where the temperature is between 0°C and 100°C, so liquid water can exist on a planet there.

All stars including the Sun form out of clouds of dust and gas.

- The particles in the clouds are pulled together by their own gravitational attraction so the particles speed up. The clouds merge together and become more and more concentrated to form a **protostar**, which is a star-to-be.
- As a protostar becomes denser, its particles speed up more and collide more, so its temperature increases and it gets hotter. The process transfers energy from the protostar's gravitational potential energy store to its thermal energy store.

If the protostar becomes hot enough, the nuclei of hydrogen atoms fuse together, forming helium nuclei. Energy is released in this fusion, so the protostar gets hotter and brighter and starts to shine. A star is born!

- Most of the mass of the Solar System is in the Sun. The planets formed by the **accretion** (gathering together) into a disc of spinning cloud of gas, dust and rocks. The Sun's gravity pulled rocks and dust towards the inner parts of the disc where they formed small rocky planets, leaving the giant planets of gas to form further away. The table below shows the main properties of the planets.

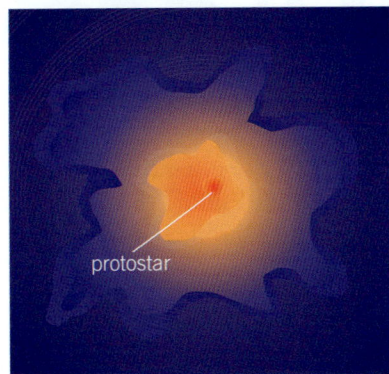

Figure 16.2.3 The birth of a star

Planet	Distance to sun relative to Earth	Diameter relative to Earth	Density kg/m³	Mean temperature on the surface / °C	Surface gravitational field strength N/kg
Mercury	0.39	0.38	5430	−170°C (max) to 390°C (min)	3.8
Venus	0.72	0.95	5240	480°C	5.1
Earth	1.00	1.00	5510	20°C	9.8
Mars	1.52	0.53	3940	−60°C	3.7
Jupiter	5.20	11.2	1.33	−110°C	24.5
Saturn	9.56	9.45	0.70	−180°C	10.8
Uranus	19.22	4.01	1.30	−218°C	8.8
Neptune	30.11	3.88	1.76	−218°C	10.8

Notes

1 The inner 4 planets are solid and relatively small. They have an atmosphere except for Mercury. The outer 4 planets are giant balls of gas. Their surface temperature is the temperature of their outer layers of gas.
2 The gravitational field strength, g, of a planet
   - at its surface depends on its mass and its diameter
   - decreases with increasing distance from the surface.

SUMMARY QUESTIONS

1 a Write which planet in the Solar System is

   i   the largest   ii nearest the Sun

   b Explain why the Earth is likely to be the only planet in the Solar System where liquid water is always present on the surface.

2 The Earth's orbit is almost circular. Give and explain two ways in which conditions on the Earth would differ if the Earth's orbit was like the orbit of a comet.

3 Describe how the Sun formed from dust and gas clouds in space.

KEY POINTS

- The Solar System formed from gas and dust clouds that gradually became more and more concentrated because of gravitational attraction.

- A protostar is a concentration of gas and dust that becomes hot enough to cause nuclear fusion.

- Energy is released inside a star because of hydrogen nuclei fusing together to form helium nuclei.

- The Sun is stable because gravitational forces acting inwards balance the forces of nuclear fusion energy in the core acting outwards.

16.3 Planetary orbits

Most of the planets orbit the Sun on orbits that are ellipses or slightly squashed circles. The Moon is a natural satellite that orbits the Earth on a circular orbit. Artificial satellites also orbit the Earth. In each case, a body orbits a much bigger body. The force on the orbiting body is the force of gravitational attraction between it and the larger body. This force decreases the further apart the two bodies are.

Circular orbits

Figure 16.3.1 shows the force of gravity acting on a planet in a circular orbit around the Sun.

Supplement

The Earth's orbit about the Sun is almost circular with the Sun at the centre of the orbit as in Figure 16.3.1. The Moon and most artificial satellites move along circular orbits with the Earth at the centre of their orbits.

For any object in a circular orbit about a central body, the force of gravitational attraction between the orbiting object and the central body keeps the orbiting object moving along its orbit.

- The force of gravity on the object is always towards the centre of the circle
- The direction of motion of the object (i.e. its velocity direction) is always at right angles to the direction of the force.

The speed of an object in a circular orbit does not change. This is because the force on it is at right angles to its direction of motion. Therefore no work is done by the force on the object. So its kinetic energy and its speed does not change.

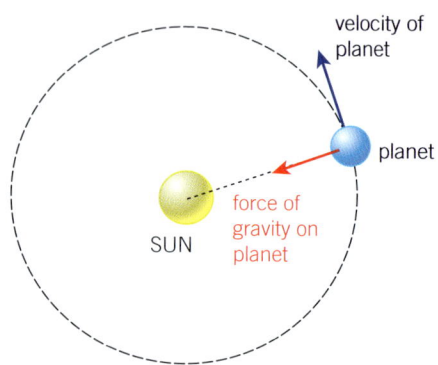

Figure 16.3.1 A circular orbit

Elliptical orbits

Most of the planets, minor planets and all the comets that orbit the Sun move in orbits that are **ellipses**. Figure 16.3.2 shows the orbit of a comet. The Sun is not at the centre of the comet's orbit. The force of gravitational attraction between the comet and the Sun is the only force acting on the comet. Because this force always acts towards the Sun, it keeps the comet on its elliptical orbit.

- **As the comet approaches the Sun**, the force of gravity on the comet increases its speed, increasing its kinetic energy and reducing its potential energy.
- **As the comet moves away from the Sun**, the force of gravity reduces its speed, reducing its kinetic energy and increasing its potential energy.

At any point of the orbit, the total energy of an orbiting body is constant.

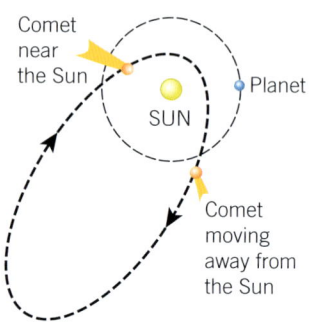

Figure 16.3.2 An elliptical orbit

Into orbit

A satellite in a circular orbit above the Earth's atmosphere moves around the Earth at a constant height above the surface. The satellite is in a stable orbit. To stay in an orbit of a particular radius, the satellite has to move at a particular speed around the Earth. The same is true for a planet moving in a circular orbit around the Sun.

Imagine a satellite is launched from the top of a very tall mountain. What would happen if the satellite was launched too fast or too slow? If the launch speed is too slow, the satellite falls to the surface. If the launch speed is too high, the satellite flies off into space. At the correct speed, the satellite moves around the Earth in a circular orbit at a constant height and a constant speed.

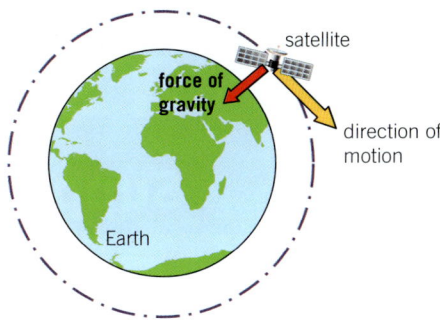

Figure 16.3.3 Launching a satellite

Using satellites

The further a satellite is from the Earth or a planet is from the Sun:

- the less the particular speed needed for it to stay in a circular orbit. The force of gravity on a satellite is weaker when the satellite is further from the Earth, and it does not need to travel as fast to stay in orbit. The same is true for a planet moving around the Sun.

- the longer the orbiting body takes to move around the orbit once. This is because the circumference of the orbit is bigger and the orbiting body moves slower in order to remain in orbit. So the time for each complete orbit (= circumference ÷ speed) is longer.

Communications satellites, including direct broadcasting satellites, are usually in an orbit at about 36 000 kilometres above the equator with a period of 24 hours. They orbit the Earth in the same direction as the Earth's spin. So they stay above the same place on the Earth's surface as they go around the Earth. These kinds of orbits are described as **geostationary**.

Monitoring satellites are fitted with TV cameras pointing to the Earth. Their uses include weather forecasting, and monitoring the environment. Monitoring satellites are in much lower orbits than geostationary satellites and they orbit the Earth once every two or three hours. Satellite phones use low-orbit satellites or geostationary satellites.

SPACE STATION

The International Space Station (ISS), launched in 1998, orbits the Earth once every 90 minutes. On a clear night, it can be seen from the ground if you know when and where to look. Use an internet search engine to look up the details on the ISS website.

KEY POINTS

- The force of gravity between:
  - a planet and the Sun keeps the planet moving along its orbit
  - a satellite and the Earth keeps the satellite moving along its orbit.

- The Earth orbits the Sun in an almost circular orbit.

- The central body in a circular orbit is at the centre of the circle. For an elliptical orbit, the central body is not at the centre of the ellipse.

- To stay in a circular orbit at a particular distance, a small body must move at a particular speed around a larger body.

SUMMARY QUESTIONS

1 A satellite is in a circular orbit around the Earth.

  a Write the direction of its acceleration

  b Explain why its velocity continually changes even though its speed is constant.

2 Light from the Sun takes 3 minutes to reach Mercury, about 8 minutes to reach Earth, and about 40 minutes to reach Jupiter. Jupiter takes 11 years to orbit the Sun, and Mercury takes about 3 months. Use this information to determine which of the three planets travels a slowest b fastest in its orbit. Explain your reasoning.

16.4 The life history of a star

Supplement

LEARNING OUTCOMES

- Explain why stars eventually become unstable
- Recall the stages in the life of a star
- Know what will eventually happen to the Sun
- Describe what a supernova is

EXAM TIP

Make sure you can recall the life cycle of stars of different sizes.

THE FUTURE OF THE SUN

The Sun is about 5000 million years old and will probably continue to shine for another 5000 million years.

The Sun will turn into a red giant bigger than the orbit of Mercury. By then, the human race will probably have long passed into history.

Figure 16.4.2 The Crab Nebula is the remnants of a supernova explosion that was observed in the 11th century

Stars such as the Sun radiate energy because of hydrogen fusion in the core. They are called **main sequence** stars because this is the main stage in their 'lifetime'. Such stars maintain their energy output for billions of years until there are no more hydrogen nuclei left to fuse together. A main sequence star is stable because the inward force of gravity on the outer layers of each star is balanced by the outward force of the radiation from the star's core.

When a star runs out of hydrogen nuclei to fuse together in its' core, it reaches the end of its main sequence stage. Its core collapses, and its outer layers swell out.

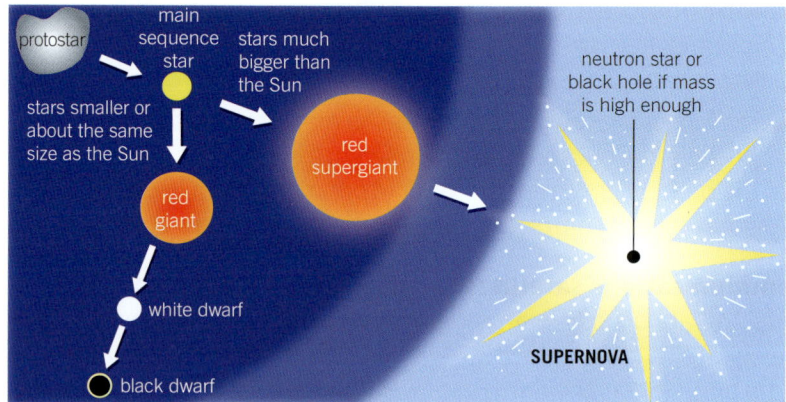

Figure 16.4.1 The life cycle of a star

Stars about the same size as the Sun (or smaller) swell out, cool down, and turn red.

- These stars are now **red giants**. At this stage, helium and other light elements in the core fuse to form heavier elements.
- When there are no more light elements in the core, fusion stops, and no more radiation is released. Because of its own gravity, the star collapses in on itself. As it collapses, it heats up and turns from red to yellow to white. It becomes a **white dwarf** star surrounded by a cloud of dust, gas and rocks called a **nebula** where planets may eventually form. This is a hot, dense, white star much smaller in diameter than it was before. Stars such as the Sun then fade out, go cold, and become **black dwarfs**.

Stars much bigger than the Sun end their lives after the main-sequence stage much more dramatically. These stars swell out to become **red supergiants**. They then collapse.

- In the collapse, the matter surrounding the star's core compresses the core more and more. Then the compression suddenly reverses in a cataclysmic explosion called a **supernova**. This event can outshine an entire galaxy for several weeks.

- The explosion creates a massive nebula containing hydrogen and new heavier elements, leaving behind a **neutron star** or a **black hole** at its centre. The nebula may form new stars with orbiting planets.

The birthplace of the elements

Light elements are formed from fusion in stars. Stars such as the Sun fuse hydrogen nuclei (i.e., protons) into helium and similar small nuclei, including carbon nuclei. When the star becomes a red giant, it fuses helium and the other small nuclei into larger nuclei.

Nuclei larger than iron nuclei cannot be formed by this process because too much energy is needed.

Heavy elements are formed when a massive star collapses then explodes as a supernova. The enormous force of the collapse fuses small nuclei into nuclei bigger than iron nuclei. The explosion scatters the elements throughout the universe.

The debris from a supernova contains all the known elements, from the lightest to the heaviest. Eventually, new stars form as gravity pulls the debris together. Planets form from debris surrounding a new star. Because of this, these planets will be made up of all the known elements too.

What's left after a supernova?

The explosion compresses the core of the star into a **neutron star**. This is an extremely dense object made up only of neutrons. If the star is massive enough, it becomes a **black hole** instead of a neutron star. The gravitational field of a black hole is so strong that nothing can escape from it. Not even light, or any other form of electromagnetic radiation, can escape a black hole.

Figure 16.4.3 M87 is a galaxy that spins so fast at its centre that it is thought to contain a black hole with a billion times more mass than the Sun

EXAM TIP

Make sure you know how the heavier elements were formed and that they were **not** formed during the Big Bang.

KEY POINTS

- Stars become unstable when they have no more hydrogen nuclei that they can fuse together.

- Stars with about the same mass as the Sun: protostar → main-sequence star → red giant → white dwarf → black dwarf.

- Stars much more massive than the Sun: protostar → main-sequence star → red supergiant → supernova → neutron star (or black hole if enough mass).

- The Sun will eventually become a black dwarf.

- A supernova is the explosion of a red supergiant after it collapses.

SUMMARY QUESTIONS

1 a The list below shows some of the stages in the life of a star such as the Sun. Put the stages in the correct sequence.
   A white dwarf   C red giant
   B protostar   D main sequence.
  b i   Name the stage in the above list that the Sun is at now.
     ii Describe what will happen to the Sun after it has gone through the above stages.

2 a i   Write the force that makes a red supergiant collapse.
     ii Write the force that prevents a main sequence star from collapsing.
  b Explain why a white dwarf eventually becomes a black dwarf.
  c Describe two differences between a red giant star and a neutron star.

3 a Explain why all the uranium in the Earth has not decayed by now.
  b Plutonium-239 has a half-life of 24 000 years. It is formed in a nuclear reactor from uranium-238. Explain why plutonium-239 is not found naturally.

16.5 The expanding universe

LEARNING OUTCOMES

- Describe what is meant by the red-shift of a light source
- Recall how red-shift depends on speed
- Know why the distant galaxies are moving away from Earth
- Recognise why we think the universe is expanding

ABOUT THE LIGHT YEAR

- One light year is the distance travelled by light in empty space in 1 year.

S - 1 light year = 9.5×10^{15} m. Check this for yourself given that the speed of light is 300 000 km/s and there are 31 million seconds in 1 year.

Red-shift

The Earth is the third planet from the Sun. The Sun is a star on the outskirts of the Milky Way galaxy. A galaxy is an enormous collection of stars that stay together because of the force of gravity between them. The Milky Way galaxy contains about 100 billion stars. Its size is about 100 000 light years across. This means that light takes 100 000 years to travel across it. But it's just one of billions of galaxies in the universe. The furthest known galaxies are about 13 billion light years away.

Figure 16.5.1 In places where there is little light pollution you can see the Milky Way galaxy

People can find out lots of things about stars and galaxies by studying the light from them. You can use a prism to split the light into a spectrum. The wavelength of light increases across the spectrum from blue to red. You can tell from its spectrum if a star or galaxy is moving towards Earth or away from Earth. This is because:

- the light waves are stretched out if the star or galaxy is moving away from you. The wavelength of the waves is increased. This is called a **red-shift** because the spectrum of light is shifted towards the red part of the spectrum.
- the light waves are squashed together if the star or galaxy is moving towards you. The wavelength of the waves is reduced. This is called a blue-shift because the spectrum of light is shifted towards the blue part of the spectrum.

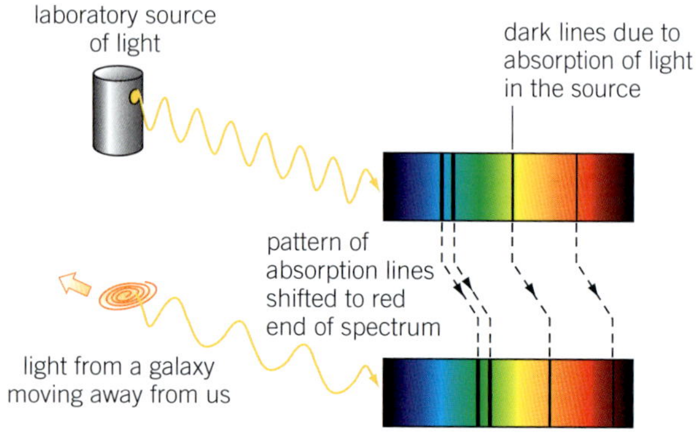

The dark spectral lines shown in Figure 16.5.2 are caused by absorption of light by specific atoms such as hydrogen that make up a star or galaxy. The position of these lines tells you if there is a shift, and if there is a shift, whether it is a red-shift or a blue-shift.

The bigger the shift, the more the waves are squashed together or stretched out. So the faster the star or galaxy must be moving towards or away from you. In other words:

The faster a star or galaxy is moving (relative to you), the bigger the shift is.

laboratory source of light

dark lines due to absorption of light in the source

pattern of absorption lines shifted to red end of spectrum

light from a galaxy moving away from us

Figure 16.5.2 Red-shift

Expanding universe

In 1929, the astronomer Edwin Hubble discovered that:

1 the light from distant galaxies was red-shifted
2 the further a galaxy is from Earth, the bigger its red-shift is.

He concluded that:

- the distant galaxies are moving away from Earth (i.e., **receding**)
- the greater the distance a galaxy is from Earth, the greater the speed at which it is moving away from Earth (its speed of recession).

Why should the distant galaxies be moving away from Earth? Humans have no special place in the universe, so all the distant galaxies must be moving away from each other. In other words, *the whole universe is expanding*.

Supplement

Hubble measured the 'red shift' in the light from distant galaxies at known distances from Earth. He used this measurement to calculate the speed of recession of each of these distant galaxies.

His results showed that for these distant galaxies,

$\dfrac{\text{its speed of recession, } v}{\text{its distance from Earth, } d}$ has the same value.

The value of v/d is called the Hubble constant, H_o. Its value is $2.2 \times 10^{-18}\ \text{s}^{-1}$.

Hubble's discovery can be summed up as

$H_o = \dfrac{v}{d}$ or $v = H_o d$ (Hubble's law)

Notes

1 Many further measurements of H_o have been made since Hubble's discovery and the value above is the currently accepted value.

2 The measurement of the distance to the distant galaxies is determined using the brightness of supernova events in these galaxies.

KEY POINTS

- The red-shift of a distant galaxy is the shift to longer wavelengths (and lower frequencies) of the light from the galaxy because it is moving away from you.
- The faster a distant galaxy is moving away from you, the greater its red-shift is.
- All the distant galaxies show a red-shift. The further away a distant galaxy is from you, the greater its red-shift is.
- The distant galaxies are all moving away from you because the universe is expanding.

SUMMARY QUESTIONS

1 Identify whether each of the following is approaching the Earth or receding from the Earth:
   a a distant galaxy
   b a galaxy that shows a blue-shift in its light.

2 Galaxy X has a larger red-shift than galaxy Y.
   a Explain what is meant by a red-shift.
   b Write which galaxy, X or Y, is:
      i nearer to Earth
      ii moving away faster.

16.6 The beginning and future of the Universe

LEARNING OUTCOMES

- Recall what the Big Bang theory of the universe is
- Explain why the universe is expanding
- **s** Describe what cosmic microwave background radiation is
- Recognise what evidence there is that the universe was created in a Big Bang

The universe is expanding, but what is making it expand? The **Big Bang theory** was put forward as a model to explain the expansion. This says that:

- the universe is expanding after exploding suddenly (the Big Bang) from a very small and extremely hot and dense region
- space, time, and matter were created in the Big Bang.

Many scientists disagreed with the Big Bang theory. They put forward an alternative theory called the Steady State theory. These scientists said that the galaxies are being pushed apart. They thought that this is caused by matter entering the universe through 'white holes' (the opposite of black holes).

Which theory is weirder – everything starting from a Big Bang, or matter leaking into the universe from outside? Until 1965, most people supported the Steady State theory.

Supplement

More about CMBR

Cosmic microwave background radiation was created as high-energy gamma radiation just after the Big Bang. It has been travelling through space since then. As the universe has expanded, it has stretched out to longer and longer wavelengths and is now microwave radiation. It has been mapped out using microwave detectors on the Earth and on satellites.

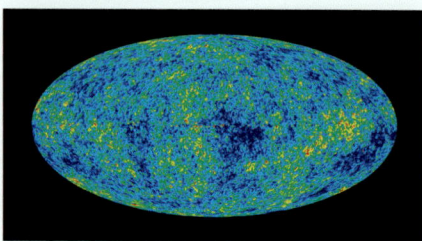

Figure 16.6.1 A microwave image of the universe from the Cosmic Background Explorer satellite

Evidence for the Big Bang

Scientists had two conflicting theories about the evolution of the universe: it was in a Steady State or it began at some point in the past with a Big Bang. Both theories could explain why *distant* galaxies are moving apart, so scientists needed to find some way of deciding which theory was correct. They worked out that if the universe began in a Big Bang, then high-energy electromagnetic radiation should have been produced very soon after the universe began. This radiation would have stretched as the universe expanded and become lower-energy radiation. Scientists thought up experiments to look for this trace energy as extra evidence for the Big Bang model.

It was in 1965 that scientists first detected microwaves coming from every direction in space. The existence of this **cosmic microwave background radiation (CMBR)** can be explained only by the Big Bang theory.

The age of the Universe

No objects can travel as fast as the speed of light in free space, c. By substituting the value of c for the speed into the Hubble equation $v = H_o d$, we obtain $c = H_o d$.

Since no galaxy (or any other object) can travel faster than c, rearranging $c = H_o d$ gives an estimate for the age of the Universe, T, as follows

$T = d/c = 1/H_o = 1/2.2 \times 10^{-18}$ s $= 4.5 \times 10^{17}$ s $= 14\,000$ million years.

The future of the universe

Will the universe expand forever? Or will the force of gravity between the distant galaxies stop them moving away from each other? The answer to this question depends on the total mass of the galaxies, how much matter is between them, and how much space they take up – in other words, the density of the universe.

Astronomers think that the stars in a galaxy account for only a small percentage of the total mass of a galaxy. They know that galaxies would spin much faster if their stars were the only matter in galaxies. The missing mass is called **dark matter** because it can't be seen. Its presence means that the average density of the universe is much bigger than if dark matter didn't exist.

- If the density of the universe is less than a particular amount, it will expand forever. The stars will die out, and so will everything else as the universe heads for a Big Yawn!
- If the density of the universe is more than a particular amount, it will stop expanding and go into reverse. Everything will head for a Big Crunch!

Observations since 1998 of supernovae in distant galaxies suggest that the distant galaxies are accelerating away from each other. These observations have been checked and confirmed by other astronomers. So astronomers have concluded that the expansion of the universe is accelerating. It could be that the universe is in for a Big Ride followed by a Big Yawn.

The discovery that the distant galaxies are accelerating is puzzling astronomers. Scientists think some unknown source of energy, called dark energy, must be causing this accelerating motion. The only known force on the distant galaxies, the force of gravity, can't be used to explain dark energy, because it's an attractive force and so it acts against the outward motion of the distant galaxies away from each other.

SUMMARY QUESTIONS

1 Put the following events **A–D** in the correct time sequence:
   A the distant galaxies were created
   B cosmic microwave background radiation was first detected
   C the Big Bang happened
   D the expansion of the universe began.

2 a Hubble estimated that the speed of recession of a distant galaxy increases by about 22 km/s for every extra distance of one million light years. A distant galaxy is receding away at a speed of 150 000 km/s. Estimate how far away the galaxy is in light years.

   b The Milky Way galaxy is about 100 000 light years across. Make an order of magnitude estimate of the ratio of the distance to the galaxy in **a** to the distance across the Milky Way.

Supplement

The origin of the Universe

The fact that H_o has the same value for all distant galaxies is evidence that all the matter in the Universe originated at a single point.

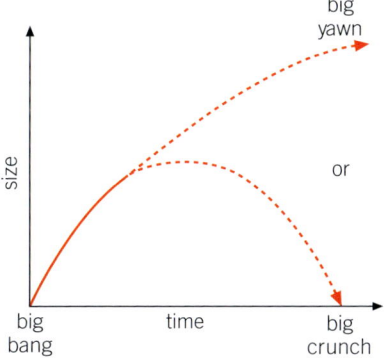

Figure 16.6.2 The future of the universe?

EXAM TIP

Make sure you know more about the Big Bang theory than just 'the universe started with a big bang'.

KEY POINTS

- The universe started with the Big Bang, which was a massive explosion from a very small point.
- The universe has been expanding ever since the Big Bang.
- Cosmic microwave background radiation (CMBR) is electromagnetic radiation that was created just after the Big Bang.
- The red shifts of the distant galaxies provide evidence that the universe is expanding. CMBR can be explained only by the Big Bang theory.

1 (a) Complete the following sentence using words from the list below:

> **day   Earth   month   Moon   second   Sun   year**

   (i) The Earth spins once about its axis in a _____ and it orbits the _____ once in a _____.

   (ii) When a full moon is seen from the Earth, the _____ is between the Sun and the _____.

(b) (i) List the planets in order of increasing distance from the Sun.

   (ii) Identify which planets are giant gas planets and which are solid planets.

   (iii) Explain why the solid planets formed nearer the Sun than the giant gas planets.

2 (a) The stages in the development of the Sun are listed below. Put the stages in the correct sequence.

A	dust and gas	**D**	red giant
B	main sequence	**E**	white dwarf
C	protostar		

(b) (i) Describe what will happen to the Sun after its present stage.

   (ii) Describe what will happen to a star that has much more mass than the Sun.

3 (a) (i) Define a supernova.

   (ii) Explain how you could tell the difference between a supernova and a distant star like the Sun at present.

(b) (i) Define a black hole.

   (ii) Describe what would happen to stars and planets near a black hole.

   (iii) Define a neutron star, and describe how it is formed.

4 (a) (i) Which element as well as hydrogen, was formed in the early universe?

   (ii) Which of the two elements in part **i** is formed from the other one in a star?

5 (a) Put these events in the correct sequence with the earliest event first:

   1 cosmic microwave background radiation was released

   2 hydrogen nuclei were first fused to form helium nuclei

   3 the Big Bang took place

   4 neutrons and protons formed.

(b) The stars were formed from clouds of dust and gas.

   (i) What force can cause dust and gas particles to attract each other?

   (ii) Where did the energy that heated the stars come from?

6 Light from a distant galaxy has a change of wavelength because of the motion of the galaxy.

(a) (i) Is this change of wavelength an increase or a decrease?

   (ii) Explain why this effect is called a red-shift.

   (iii) Explain what the change of wavelength tells you about the motion of the galaxy.

(b) Light from a particular nearby galaxy is found to have undergone a blue-shift because of the motion of the galaxy. Explain what this tells you about the motion of this galaxy.

(c) (i) Edwin Hubble discovered that the further a distant galaxy is from Earth, the greater the red-shift of the light from it. Explain what this tells you about the universe.

   (ii) Give the crucial observational evidence that led scientists to accept the Big Bang theory of the universe.

7 (a) Galaxy **A** is further from us than galaxy **B**.

   (i) Which galaxy, **A** or **B**, produces light with a greater red-shift?

   (ii) Galaxy **C** gives a bigger red-shift than galaxy **A**. Describe the distance to galaxy **C** compared with galaxy **A**.

(b) All the distant galaxies are moving away from each other.

   (i) Explain what this tells you about the universe.

   (ii) Explain what it tells you about your place in the universe?

Practice Questions

1 One theory for the origin of the universe is that it began from a single point.

   a Which of these is the correct name of the theory?

   Red Bang Big Bang Loud Bang [1]

   b Observation of distant galaxies shows that the universe is expanding. Evidence for this is the light from the galaxies appears stretched. Give the name of this evidence. [1]

   c Complete the sentence using the correct words from the list.

   the Sun   all directions   the centre of the Earth

   Cosmic background radiation has been found in the universe. It appears to come from _____ . [1]

2 In 1929 Edwin Hubble investigated the light from distant galaxies. He stated that distant galaxies were moving away from us. Figure 1 shows a graph of the Hubble Data.

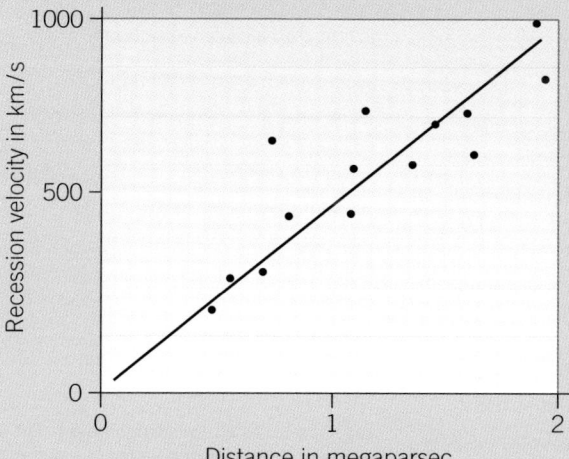

Figure 1

   a State what is meant by a distance of 1 light year. [1]

   b Describe the relationship between the recession velocity of a galaxy and the distance from the Earth. [2]

c In 1929 Hubble used a large reflecting telescope to determine the distance of galaxies from the Earth. He estimated the value of H_o to be about seven times the current accepted value.

In 2013 NASA researchers surveyed more than 125 000 galaxies and claim to have measured the Hubble Constant with an uncertainty of less than 5%. Suggest two reasons why modern researchers can claim to be more accurate in measuring the Hubble Constant. [2]

3 **a** Describe how the Sun was formed and became a star. You should include details of how the Sun releases energy. [3]

   b Give two reasons why the Sun is a main sequence star and is stable. [2]

   c A star many times bigger than the Sun will eventually complete its life cycle and become a black hole. Explain what a black hole in space is. [2]

   d Complete the following sentence.

   Elements heavier than iron are formed in a _____ . The explosion of this massive star distributes _____ throughout the _____ . [3]

4 Table 1 shows some information about the Earth, the Moon, and satellites.

Table 1

Name	Mass (kg)	Altitude (km)	Orbit time (hours)	Speed (km/h)
Earth	5.98×10^{24}	0	0	0
Moon	7.45×10^{22}	384 000	672	3 600
GPS satellite	2.2×10^2	20 000	12	13 900
Weather satellite	2.4×10^2	36 000	24	11 100
Spy satellite	2.8×10^2	705	1.45	27 500

Using the information given in Table 1, evaluate the differences and similarities between planets, moons, and artificial satellites. [6]

Glossary

A

absolute scale of temperature
temperature scale in kelvin (K) with a fixed point at absolute zero (0 K) such that $0\,°C = 273\,K$ and $100\,°C = 373\,K$

absolute zero the lowest possible temperature, 0 kelvin (K) or $-273\,°C$

ac generator a device consisting of a coil that spins in a magnetic field and generates an alternating emf as a result

acceleration change of velocity per second, measured in metres per second^2 (m/s^2)

accuracy of a measurement the difference between the measured reading and the true reading

alpha radiation radiation consisting of particles emitted by certain radioactive substances; an α-particle consists of 2 protons and 2 neutrons; α-particles are stopped by paper, deflected by an electric or magnetic field and are very ionising.

ammeter electrical instrument used to measure electric current in amperes

amplitude the maximum distance of any part of a wave from its undisturbed position; for a transverse wave, this is the height of a wave crest or the depth of a wave trough from the middle

analogue circuit circuits in which the voltage at any point can be at any value between the maximum and minimum values of the power supply (e.g. any value between +5 V or zero in a circuit with a 5 V power supply)

analogue signal signal that varies and can have any value between a maximum and a minimum

angle of incidence angle between the incident ray and the normal

angle of reflection angle between the reflected ray and the normal

average speed total distance travelled ÷ total time taken, measured in metres per second (m/s)

B

background radiation alpha (α) or beta (β) or gamma (γ) radiation emitted by radioactive substances that occur naturally all around us

beta radiation radiation consisting of electrons emitted by certain radioactive substances; a β-particle is emitted by an unstable nucleus when a neutron in the nucleus changes into a proton to enable the nucleus to become more stable; β-particles are stopped by several mm of aluminium paper, easily deflected by an electric or magnetic field and are ionising

Big Bang theory theory that the universe was created in a massive explosion (the Big Bang) and that the universe has been expanding ever since

black hole An object in space that has so much mass that nothing, not even light, can escape from its gravitational field

boiling change of state from liquid to vapour at the boiling point

Boyle's Law for a fixed mass of gas at constant temperature, its pressure × its volume = constant

Brownian motion the random motion of particles in a fluid (such as smoke particles in air) due to collisions with individual fluid molecules

C

carrier waves waves used to carry a signal

Celsius scale temperature scale based on ice point (the temperature of pure melting ice) defined as $0\,°C$ and steam point (the temperature of steam at standard atmospheric pressure) defined as $100\,°C$

centre of gravity every object behaves as if all its weight were concentrated at one point called its centre of gravity

centripetal force the force needed to keep an object moving around in a circle

charging by direct contact process in which an insulated conductor is charged by direct contact with a charged body

charging by induction process in which an insulated conductor is charged by earthing it briefly in the presence of a charged body; the insulated conductor gains the opposite type of charge to that of the charged body

chemical energy energy released when a chemical reaction occurs

circuit breaker an electromagnetically-operated switch that opens if too much current passes through it and which needs to be reset manually once it has opened.

compression in a wave part of a sound wave where air molecules are pushed together

condensation change of state from vapour to liquid at the boiling point or from vapour to solid directly

conduction electrons *see* free electrons

conduction of heat transfer of energy due to a temperature difference within a substance as a result of vibrations of atoms in any substance and the movement of free electrons in a metal

conservation of energy the total amount of energy before and after a change is the same

convection transfer of energy in a liquid or a gas due to a temperature difference causing circulation within the liquid or gas

converging lens a lens that focuses parallel rays of light to a point

cosmic microwave background radiation (CMBR) electromagnetic radiation that has been travelling through space ever since it was created shortly after the Big Bang

critical angle the angle of incidence of a light ray in a transparent substance when it is refracted along the boundary

D

dark energy unknown force releasing hidden energy thought to be causing the expansion of the Universe to accelerate

decelerates slows down

density mass per unit volume of an object, measured in kilograms per cubic metre (kg/m^3)

diffraction spreading of waves when they pass through a gap or move around an obstacle

digital circuit circuits in which the voltage at any point is either high, '1', (e.g. +5 V) or low, '0', (e.g. zero)

digital signal a sequence of pulses (i.e. 0s and 1s) with no in-between levels

diode a component that allows current to pass through it in one direction only (its 'forward' direction)

dispersion splitting of white light (or any non-monochromatic light) into separate colours using a glass prism; the effect occurs because the refractive index of glass varies with the colour of light

diverging lens lens with a shape that causes a parallel beam of light to diverge; used to correct short-sight

drag friction-like force acting on an object when it moves through a fluid

E

earth wire a word used to describe a wire used to connect a metal object to the ground so the object cannot retain charge

echo reflection of sound from a smooth surface

efficiency useful energy transferred by a device ÷ energy supplied to the device

elastic limit the limit to which a material can be stretched without being permanently extended

elastic object an object that regains its shape after being distorted is said to be elastic

elastic strain energy energy stored in an object due to a change of its shape

electric charge there are two types of electric charge: positive and negative; charged objects attract or repel one another according to whether or not they have the same type of charge, in which case they repel, or they have different (i.e. unlike or opposite) types of charge, in which case they attract. Charge is measured in coulombs (C)

electric current a flow of charge; measured in amperes (A) where 1 ampere is a rate of flow of charge of 1 coulomb per second

electric field the space surrounding a charged object in which any other charged object experiences a force due to its charge

electric line of force a line in an electric field in which a small positively charged object would move if free to do so

electrical conductor a substance through which an electric current can flow

electrical energy energy transferred by an electric current

electrical insulator a substance through which an electric current cannot flow

electrical power electrical energy transferred per second, measured in watts (W). Electrical power = current × pd

electricity grid network of cables and transformers used to distribute mains electricity produced by power stations to buildings

electromagnet a device consisting of insulated wire wrapped around an iron core; when there is a current in the wire, the iron core is magnetic and will attract unmagnetised ferrous objects

electromagnetic induction the generation of an emf in a wire when the wire cuts across the lines of a magnetic field; the emf increases the faster the wire moves across the field lines and reverses if the field or the direction of motion of the wire is reversed

electromagnetic spectrum spectrum of electromagnetic waves; in order of increasing frequency (and decreasing wavelength): radio waves, microwaves, infrared radiation, light, ultraviolet radiation, X radiation, gamma radiation

electromagnetic waves electric and magnetic disturbances that transfer energy from one place to another

electromotive force (emf) the work done or energy transferred per unit charge by a cell or battery (or other electrical power supply) in moving charge around a complete circuit. It is measured in volts

electron a negatively charged particle which has a much smaller mass than the proton and orbits the nucleus in every atom

electroscope an instrument used to detect small amounts of charge

electrostatic precipitator an electro-static device used to remove particles of ash and dust from the flue gases produced in a coal-fired power station

endoscope a medical instrument used to view objects and surfaces in cavities inside the body

energy, dissipation of energy tends to spread out, for example where energy is transferred to the surroundings by heating or by sound waves

energy transfers ways of storing and transferring energy

equilibrium an object at rest is said to be in equilibrium

evaporation the process in which a liquid turns to vapour below its boiling point

extension increase of the length of an object

F

ferrous material that contains iron

fission reactor nuclear reactor in which energy is released due to nuclear fission of uranium

fixed points standard 'degrees of hotness', usually melting or boiling points of pure substances, used to define a temperature scale

fluorescent substance substance that glows when high-energy radiation or particles such as electrons moving at high speed are directed at it

focal length the distance from the centre of a lens to its principal focus

force a force changes the motion of the object if it is the only force acting on the object. Force is measured in newtons (N). The resultant force on an object = its mass × its acceleration

free electrons electrons in a metal that move about freely inside the metal; free electrons have broken free from the atoms in the metal; also referred to as conduction electrons

free-fall falling at constant acceleration

frequency number of complete waves passing a point in one second, measured in hertz (Hz)

fuse a component that contains a thin wire that melts if too much current passes through it

fusion, thermal process of fusing different solids together when they melt

fusion reactor nuclear reactor (not yet developed) in which energy is released due to nuclear fusion

G

galaxy an enormous collection of stars that stay together because of the force of gravity between them. The Sun is one of billions of stars in the Milky Way galaxy

gamma radiation high energy electromagnetic radiation electrons emitted by certain radioactive substances, usually after an unstable nucleus emits an α- or a β- particle;

Glossary

γ-radiation needs several cm of lead to stop it; it is not deflected by an electric or magnetic field and is weakly ionising

global warming increase of average global temperature

gravitational field strength force of gravity per unit mass on an object at a point in a gravitational field

gravitational potential energy energy due to position; for an object of mass m which is moved up or down through height h, its change of gravitational potential energy $= mgh$

H

half-life of an isotope the time taken for the number of nuclei of an isotope in a sample to decay by half

hard magnetic material magnetic material that is hard to magnetise and demagnetise (e.g. steel)

Hooke's law the extension of a spring is directly proportional to the weight it supports

Hubble's law the speed of recession of a distant galaxy = the Hubble constant × its distance from Earth. The accepted value of the Hubble constant is 2.2×10^{-18} s^{-1}.

hydraulic pressure pressure in a hydraulic system that enables a force to be exerted

I

infrared radiation electromagnetic radiation just beyond the red part of the visible spectrum; it can be detected by its heating effect

input sensor a sensor device (or a sensor circuit) that produces an electrical signal in response to a change of a physical property

input transducer a sensor circuit that provides an input signal to another circuit

internal energy energy of an object due to the motion and positions of its molecules

ionisation the process of creating charged atoms (i.e. ions); ions are created when X-rays or radiation from a radioactive source pass through a substance

isotope atoms that have the same number of protons and different numbers of neutrons in the nucleus

K

kelvin unit of temperature; see absolute scale of temperature

kinetic energy energy due to motion; for a mass m moving at speed v, its kinetic energy $= \frac{1}{2}mv^2$

L

law of force between electric charges like charges repel; unlike charges attract

law of force between magnets like poles repel; unlike poles attract

law of reflection for a light ray reflected by a mirror, the angle of incidence = the angle of reflection

law of refraction for a light ray that is refracted when it travels from air into a transparent substance of refractive index n, $\sin i \div \sin r = n$

light dependent resistor (LDR) a resistor which has a resistance that depends on the incident light intensity; most LDRs have a resistance that decreases as the light intensity increases

light energy energy transfer by light

limit of proportionality the limit to which a spring can be stretched and still obey Hooke's Law

live wire the wire in a mains circuit that is at a high voltage and is lethal to touch

long sight eye defect in which the eye is unable to see near objects clearly; corrected by means of a converging lens

longitudinal waves waves in which the direction of vibration is parallel to the direction in which the waves are travelling, for example sound waves

loudness the louder a sound, the greater the amplitude of the sound waves

loudspeaker a device that converts an electrical signal into sound waves

M

magnetic field the space surrounding a magnet in which a plotting compass would be affected

magnetic induction magnetism induced in an unmagnetised ferrous bar by holding a permanent magnet near it

magnetic lines of force a line in a magnetic field along which a plotting compass points

magnetic poles the ends of a magnet where the lines of force are concentrated

magnification an optical image of an object is said to be magnified if it is larger than the object itself

magnifying glass a converging lens used to view a magnified image of an object by holding the lens near the object and viewing the object through the lens

main sequence main stage in the life of a star during which it radiates energy because of fusion of hydrogen nuclei in its core

mains electricity alternating current at high voltage supplied to buildings from a distribution network

mass measure of the amount of matter in an object, measured in kilograms (kg)

mass number A, the number of protons and neutrons in the nucleus of an atom; the mass number of a nucleus is approximately equal to the mass of the nucleus relative to the mass of the proton

measuring cylinder instrument used for measuring volume

melting change of state from solid to liquid at the melting point

micrometer instrument used to measure lengths up to 30 mm to within 0.01 mm

microwaves electromagnetic waves between radio waves and infrared radiation in the electromagnetic spectrum used in communications and in microwave ovens

molecule the smallest particle of a substance that can be identified with the substance

moment of force force × perpendicular distance from the pivot measured in newton metres (N m)

momentum this equals mass × velocity; the unit of momentum is the kilogram metre per second (kg ms^{-1})

monochromatic light light of a single colour

motor effect the force exerted on a current-carrying wire when it is in a magnetic field aligned at right angles (or at any non-zero angle) to the magnetic field lines

N

neutral wire the wire in a mains circuit that is earthed at the local sub-station neutron an uncharged particle of about the same mass as the proton that is in the nucleus of an atom

neutron star highly compressed core of a massive star that remains after a supernova explosion

newtonmeter a meter used to measure force in newtons

noise unwanted variations in the intensity of a signal

normal line at right angles to a boundary

nuclear energy energy released when the nucleus of an atom splits or disintegrates

nuclear fission process in which a uranium nucleus splits in two to form two smaller nuclei

nuclear fusion process in which two small nuclei fuse together to form a larger nucleus

nucleus the positively charged object at the centre of every atom which contains most of the mass of the atom and around which electrons in the atom move in orbits; the nucleus of an atom is composed of protons and neutrons

nuclide the name for different types of nuclei, each with a certain number of protons and a certain number of neutrons in its nucleus

O

Ohm's Law the current in a resistor at constant temperature is proportional to the potential difference across the resistor; the resistance does not depend on the current or the pd

ohmic conductor a conductor that obeys Ohm's Law (i.e. its resistance is constant)

optical fibre a thin transparent fibre which a light ray can travel along from one end to the other, undergoing total internal reflection if it reaches the surface

oscilloscope electrical instrument used to display electrical signals on a screen to show the variation of the signal amplitude with time; also used to measure pds and the frequency of electrical waveforms

P

parallel components components connected between the same two points in an electrical circuit; the pd across components in parallel is the same. The total current through two or more

components in parallel is the sum of the current through each component

parallelogram of forces geometrical method for finding the resultant force of two or more forces

partial reflection a light ray that is partly reflected and partly refracted at a boundary

peak value the maximum (positive or negative) value of an alternating pd

pitch the higher the pitch of a sound, the greater the frequency of the sound waves

plane waves waves with straight wave fronts

potential capacity to deliver energy

potential difference (pd) work done or energy transferred per unit charge by charge passing through a component. It is measured in volts (V)

potential divider two or more resistors in series connected to a source of fixed pd; each resistor has a share of the fixed pd in proportion to its resistance

potentiometer a potential divider with a variable output pd

power rate of transfer of energy, measured in watts (W); power = energy transferred ÷ time taken

pressure force per unit area, measured in pascals (Pa); pressure = force ÷ area

primary (P) waves seismic waves that travel faster than secondary (S) waves; longitudinal in nature

principal focus, also focal point the point to which parallel rays of light directed straight at a converging lens are focused by the lens

principal of conservation of momentum in a closed system the total momentum before an event is equal to the total momentum after the event; momentum is conserved in any explosion or collision provided no external forces act on the objects that collide or explode.

principle of moments for any object in equilibrium, the sum of the clockwise moments about any point is equal to the sum of the anticlockwise moments about the same point

projector an optical instrument used to focus an image of an object onto a screen

proton a positively charged particle that is in the nucleus of every atom

proton number Z, the number of protons in the nucleus of an atom; all atoms of the same element have the same number of protons in the nucleus

protostar concentration of dust clouds and gas in space that forms a star

R

radial electrical field the pattern of the lines of force surrounding a point charge (i.e. a very small charged sphere); the lines of force are straight lines that 'radiate' from the point charge

radio waves electromagnetic waves at the long wavelength end of the electromagnetic spectrum, used in communications

radioactive dating method of finding the age of a sample by measuring the proportion of unstable nuclei of a radioactive isotope that have decayed

radioactive decay the change that happens to an unstable nucleus when it emits alpha (α) or beta (β) or gamma (γ) radiation

radioactive tracer radioactive isotope used to trace the flow of a substance through a system

radioactivity property of substances such as uranium that give out radiation all the time; the radiation is emitted because the nuclei of the atoms are unstable and become stable by emitted alpha (α) or beta (β) or gamma (γ) radiation

radiograph a photographic image taken using X-rays instead of light

random changes events that cannot be predicted

range of a thermometer the range of temperatures that can be measured from the lowest to the highest

rarefaction in a wave part of a sound wave where air molecules are spaced further apart than when they are undisturbed

real image image of an object formed by focusing light onto a screen

red giant star that has expanded and cooled, resulting in it becoming red and much larger and cooler than it was before it expanded

Glossary

red supergiant star much more massive than the Sun and will swell out after the main sequence stage to become a red supergiant before it collapses

red shift increase in the wavelength of electromagnetic waves emitted by a star or galaxy due to its motion away from us. The faster the speed of the star or galaxy, the greater the red-shift is

reflection return of waves after they reach a smooth surface

refraction change of direction of waves when they travel across a boundary where their speed changes

refractive index speed of light in air ÷ speed of light in a substance; symbol n

relay an electrically operated switch

resistance potential difference across a component ÷ the current through it, measured in ohms (Ω)

resistor an electrical component designed to have a certain resistance

resistors in parallel for two or more resistors in parallel, the combined resistance is less than the smallest individual resistance; the total resistance R of two or more resistors in parallel is given by an equation of the form:

$$\frac{1}{R} = \frac{1}{R_1} + \frac{1}{R_2}$$

resistors in series the total resistance of two or more resistors in series is equal to the sum of their separate resistances

resultant force a single force on an object that has the same effect as the forces acting on it

S

Sankey diagram diagram showing the energy transfers in and out of a device or a system

saturated vapour vapour in air that can hold no more vapour

scalar a physical quantity that has magnitude only

seismic waves shock waves that travel through the Earth from an earthquake

secondary (S) waves seismic waves that are slower than primary (P) waves; transverse in nature

seismometer instrument used to detect and record seismic waves

sensitivity of a thermometer change of the thermometric property for a 1°C rise of temperature

sensor circuit a circuit with an output pd that changes in response to a change of a physical property (e.g. temperature or light intensity)

series components components connected in the same circuit such that the same current flows through them; the total pd across two or more components in series is the sum of the pds across each component

short-circuit a fault in a circuit that provides a low-resistance path allowing a dangerously high current to flow

short sight eye defect in which the eye is unable to see distant objects clearly; corrected by means of a diverging lens

soft iron iron that loses its magnetism when the magnetising current is switched off

solar cell a cell (usually in a panel with other solar cells) that generates electricity when sunlight falls on it

solar heating panel a panel in which water flowing through it is heated by solar energy

solenoid a long coil of wire wound around a tube that may contain an iron core or be air-filled

solidifying change of state from liquid to solid at the melting point, also referred to as freezing

sonar sound waves used for echo sounding

sound waves vibrations of layers of air that travel through the air

specific heat capacity energy needed by 1 kg of a substance to raise its temperature by 1°C

speed distance travelled per second, measured in metres per second (m/s)

speed of a wave, wavespeed distance travelled by a wave crest or a wave trough per second, measured in metres/second (m/s); speed of a wave = wavelength × frequency

split-ring commutator electric motor coil connections in the form of a split ring used to connect the spinning coil to the motor power supply; the split- ring arrangement ensures the current in the coil reverses

every time the coil turns through a half turn so the coil continues to spin in one direction only

spontaneous change change that happens without any external cause

spring constant force/extension for a spring that obeys Hooke's Law

standard atmospheric pressure the mean pressure of the Earth's atmosphere at sea level

state of matter physical state of a substance, either solid or liquid or gas (also referred to as vapour)

static electricity electric charge on an object

sublimation change of state directly from solid to vapour explosion of a massive star after fusion in its core ceases and the matter surrounding its core collapses on to the core and rebounds.

supernova explosion of a massive star after fusion in its core ceases and the matter surrounding its core collapses on to the core and rebounds

T

terminal speed speed reached by a falling object

thermal energy energy transfer from a hot object to a cold object

thermal expansion increase of length of a solid or increase of volume of a liquid or gas due to an increase of temperature

thermal radiation electromagnetic radiation emitted by an object due to its temperature

thermionic emission process in which free electrons in a heated metal filament in a vacuum are emitted from the filament

thermistor a resistor which has a resistance that depends on temperature; most thermistors have a resistance that decreases as its temperature increases

thermocouple thermometer electrical thermometer that makes use of the voltage between two different metals in contact with each other

thermometer instrument used to measure temperature

thermometric property physical property of a thermometer that varies with temperature

total internal reflection a light ray in a transparent substance is totally internally reflected when it reaches the surface if its angle of incidence is greater than the critical angle of the substance

transformer a device consisting of a primary coil and a secondary coil wound on the same iron core; the ratio of the secondary pd to the primary pd is equal to the ratio of the number of turns of the secondary coil to the number of turns of the primary coil; a step-up transformer has more secondary turns than primary turns so the secondary pd is greater than the primary pd; a step-down transformer has fewer secondary turns than primary turns so the secondary pd is less than the primary pd used to change the peak value of an alternating pd

transformer efficiency ratio of power delivered by a transformer to the power supplied to it. For a transformer that is 100% efficient, the primary current × primary voltage = secondary current × secondary voltage

transverse waves waves in which the direction of vibration is perpendicular to the direction in which the wave travels; example include electromagnetic waves and waves on a vibrating string or rope

U

ultrasonic waves or **ultrasound** sound waves at frequencies above 20 000 Hz, the upper frequency limit of the normal human ear

ultraviolet radiation electromagnetic radiation just beyond the violet part of the visible spectrum; it has no heating effect and is harmful to the eye

uniform electric field the pattern of the lines of force between two oppositely charged parallel plates; between the plates, the lines are straight lines directed from the positive to the negative plate

uniform wire a wire with the same diameter all along its length

unsaturated vapour vapour in air that can hold more vapour

useful energy energy transferred for a purpose

V

Van de Graaff generator a machine that generates electrostatic charge on its metal dome

vaporisation change of state from liquid to vapour below or at the boiling point

vector any physical quantity that has magnitude and direction, such as force, velocity and acceleration

velocity speed in a given direction, measured in metres per second (m/s)

virtual image image of an object viewed through a lens or a mirror formed where light appears to come from

voltmeter electrical instrument used to measure emf or potential difference (i.e. voltage) in volts

W

wavefront points along a crest or trough of a wave as it progresses

wavelength the distance from one wave crest to the next; symbol λ (lambda); measured in metres (m)

waves disturbances that transfer energy from one place to another

weight the force of gravity on an object, measured in newtons (N)

white dwarf star that has collapsed from the red giant stage to become much hotter and denser than it was

white light light which can be split into all the colours of the spectrum, namely red, orange, yellow, green, blue, indigo and violet

work done energy transferred by a force when it moves an object or changes its shape; work done = force × distance moved in the direction of the force; work is measured in joules

X

X-rays electromagnetic waves that ionise substances and are produced by an X-ray tube

Index

References to tables are given in **bold** type. References to pictures are given in *italic* type.

A

absolute scale of temperature 90–1
absolute zero 90–1
acceleration 8, 10–13
    force and 24–5
    in free fall 15
age of the Universe 242
air
    pressure and volume 84
    sound speed **149**
air cooling systems 83, 105
air resistance 13, 15
alcohol 149
alpha radiation 219, 220
    emission 224–9
    scattering 222
    smoke alarms 229
alternating current 198–3
    generators 210–15
ammeter 93, 168, 184
ampere 168
amplitude (of a wave) 108, 109
analogue meters 174
analogue signal 139
atmospheric pressure 84–90
atomic number 224
atoms
    magnetic 158
    nucleus 222–7
    structure *56*, 162

B

background radiation 219
batteries 172–3, 190
Becquerel, Henri 218
beta radiation 219–24, 225, 228
Big Bang theory 242
bimetallic strips 89
black hole 239
body temperature 91
boiling 76–7, 95
Boyle's law 85
Brownian motion 81

C

cables, electrical 196–7
calibration, thermometers 90
carbon dating 229
carbon dioxide, sound speed **149**
carrier waves 139
cells (power)
    in series 190
Celsius scale 90
centre of gravity 42–3
charge 162, 166–72
chemical energy 50
circuit breakers 198–9
circuit diagrams, symbols 184

circular motion 27
climate change 63
comets 234, **236**
compasses (magnetic) 154
compression waves 143
condensing 76
conduction (electrical) 166
conduction (thermal) 96–7
conservation of energy 52–3
control circuits 194–5
convection 98–9
convector heater 99
converging lens 128–33
cooling 82, 104
copper, sound speed **149**
cosmic microwave background
    radiation 242
coulomb (unit) 162, 169
count rate 226
critical angle 126–7
Curie, Marie 218
cycle dynamo 55

D

dark matter 243
deceleration 8, 13
density 20–1, 75
diffraction 114–15
digital meters 174
digital signal 139
diodes 184, 185, 194
direct current 194
dispersion 134
displacement can 21
distance–time graphs 6–8
diverging lens 128, 133
drag force 15

E

Earth 232–234
    seasons 232
    spin 232
earthing 166–5, 199
echo sounding 147–8
echoes 147
efficiency 55
elastic strain energy 50
elasticity 22–3
electric charge 166–8
electric current
    charge and 168–9
    energy and 181
    magnetic fields and 204–5
    'normal' for a wire 185
    parallel circuits 188
    resistors 191
    series circuits 186
electric fields 164–5
electric motors 200–5, 202–07
electrical circuits
    components 184
    earthing 199

    parallel 189–191
    sensor 192–3,
    series 186–7, 190
    switching circuits 194–5
electrical energy 50
electricity
    batteries 172–3
    generation 54–5
    generators 55
    geothermal 61
    nuclear 56–7
    potential difference 174–5
    resistance 176–7
    solar 60–1
    static 162–3
    transmission 214–19
    wind and wave 58–9
electromagnetic radiation
    applications 136–7
    infrared 100–3
    light 110
    microwaves 115
    ultraviolet *135*, **136**, 137
    wavelength 135
    white light 134
electromagnets 182
electromotive force 172–3, 208
    alternating current generators 210–15
electrons
    atoms 162, 219
    beams 205
    beta radiation 220, 221
    electrical conduction 181
    static electricity 162
    thermal conduction 97
electroscope 163
energy
    changes of state and 95
    conservation 52–3
    dissipated 52
    transfers 50–1
equilibrium 42–3
evaporation 82–3, 95
expansion gaps 89
extension 22

F

fire alarms 89
fission (nuclear) 56
Fleming's left hand rule 204
Fleming's right hand rule 209
focal point 128
force 18–19
    elasticity and 22–3
    moment 36–7, 38–9
    motion and 24–5
    resultant 24
    vectors 46–7
forces
    moment 36–7, 40–1
    representation 26–7
free fall 14–15

freezing 76–7
frequency (of a wave) 99
  electromagnetic radiation 135
friction 24, 162–3
fuel 54–5
fuses 198
fusion (nuclear) 57
fusion (thermal) 94–95

G
g (gravitational field strength) 15
galaxy 240
gamma radiation *135*, **136**, 219, 219–5, 225
gases
  arrangement of molecules 79
  convection 98–9
  pressure 78, 80–1
  properties 76
  temperature 78, 80–1, 91
Geiger counter 218, 226
generators 208–13
  alternating current 210–15
geostationary satellite 237
geothermal energy 60
global warming 102–3
gravitational field strength 15
gravitational potential energy 50, 64–3
gravity 15, 19

H
half-life 226–7
heating 78–79
hertz (unit) 109
Hooke's law 23
Hubble constant 241
Hubble's law 241
hydraulic pressure 72
hydroelectricity 58–9
hydrogen 57
  sound speed **149**

I
image formation 119–121, 128–133
induction 167, 208–3
infrared radiation 100–3, *135*, **136**
insulation (electrical) 196
insulation (thermal) 96–7, 104
internal energy **50**
ionisation 162, 165, 221
  by X-rays 137
ions 163
iron 155, 158–9
isotopes 224

J
joulemeters 92
joules 55

K
kelvins 90–1
kilogram 2, 18
kinetic energy **50**, 65

L
lamps 187–8
  parallel circuits 188
  series circuits 187
LDR (light-dependent resistor) 193
lead, sound speed **149**
lenses 128–31
light 110, **136**
  ray diagrams 128–131
  reflection 118–120
  refraction 122–7
  speed 122–124, 134
light-emitting diodes 127, 178, 196
light energy **50**
light-dependent resistors 193
light-gate timer 4
lighting circuits 189
lightning conductor 165
limit of proportionality 23
lines of force 164
line of sight error 3
liquid-in-glass thermometer 90–91
liquids
  arrangement of molecules 79
  boiling 94, 95
  convection 98–99
  density 21, 75
  evaporation 82–3
  pressure 74–5
  properties 76
  thermal expansion 88
longitudinal waves 110–11, 142
long-sight 132–3
loudspeakers 142, 159

M
magnetic fields 156–7, 202–7
  motor effect 204–7
  solenoids 202–3
magnetic materials 155, 158–9
magnetic poles 154–8
magnetic resonance scanners 203
magnetisation 156–7
magnets 154, 158
  electron beams 204
  electromagnets 159, 184
main sequence star 238
manometer 74
mass 18–20
matt surfaces 101–2
measurements
  at work 2
  energy 55
  free fall 14
  length 2–3
  pressure 74–5
  specific heat capacity 92–93
  speed 9
  temperature 90–91
  time 4–5
  volume 4

measuring cylinder 4
melting 94
mercury barometer 74–5
metals
  crack detection 150–1
  thermal conduction 97
metre rule 2
micrometer 3
microwaves *135*, **136**, 137
  diffraction 115
mirrors 118–121
molecules 77
moment (of a force) 36–9
momentum 28
monochromatic light 134
Moon's phases 233
motion, force and 24–5
motor effect 204–7
motors, electric 206–7

N
nebula 238
neutral wire 189
neutrons 224
neutron star 239
newton (unit) 18
newtonmeters 18, 36
noise (electrical) 139
nuclear energy **50**, 56–7
nuclear equations 224–9
nuclear fission 56
nuclear fusion 57
nuclear radiation 218–21
nucleus 56–7, 222–7
nuclides 224

O
ohm (unit) 176
orbits 236–7
optical fibres 127
oscilloscopes 150, 194, 211
output power 66

P
parallel circuits 188–190
parallelogram of forces 46–7
pascal (unit) 70
pendulums 5
planets 234–5
  formation 234
  orbits 235–6
plasma 57
potential difference 174, 192–3
  energy and 181
  parallel circuits 188–9
  series circuits 186–7
potential divider 192–3
potentiometer 192–3
power 66–7
  electrical 180–1
power stations 54–5, 214–5

Index

pressure 70–9
  examples 72–1
  gases 78
  temperature and 80–1
  liquids 74–5
principle of conservation of
  momentum 28
principle of moments 40
proton number 224
protons 224
protostar 234

R
radiation
  infrared 100–3
  nuclear 218–21
radio waves *135*, **136**, 137
radioactive tracers 228–9
radioactivity 218
  at work 228–9
  decay 219, 226–9
  half-life 226–7
radium 219
rarefaction waves 143
rectifier circuits 194
red giant 238
red shift 240–1
reflection 118–121
  image formation 119
  law 119
  light 118–121
  total internal 126–7
  sound 147
refraction 112–15
  light 122–6
  refractive index 124–5
relays 184, 194
resistance (electrical) 176–7, 178–9
  parallel circuits 188–9
  series circuits 186–7
resistors 178, 181, 184, 193
  in parallel 190–1
  in series 186–7
resultant force 24
Rutherford, Ernest 219, 220, 222–3

S
safety
  electrical 196–3
  radioactive materials 227
  X-rays 137
salt pans 83
Sankey diagram 55
satellites 237
scalars 47
seismic waves 111
  -primary (P) waves 111
  -secondary (S) waves 111
seismometer 111
sensor circuits 192–3
series circuits 186–7
short-circuits 196

short-sight 132–3
smoke 81
smoke alarms 229
soft iron 158
solar energy 60, 61
solar system 234–5
  formation 234
  planetary orbits 236–7
solenoids 202–3
solids
  arrangement of molecules 78
  melting 94
  properties 76
  sound speed 148–9
  thermal expansion 88–9
sonar 147–8
sound
  speed 146
  waveforms 151
sound energy **50**
sound waves 111, 144–5
specific heat capacity 92–3
speed 7
  distance–time graphs and 6
  light 122–124
  sound 146
speed–time graphs 10, 13, 24
spontaneous change 219
spring constant 23
springs 22–23
stability 44–5
states of matter 76–7
  arrangement of molecules 78–9
  changes 94–5
static electricity 162–3
steel 158–9
stretch tests 22–3
supernova 238
surfaces, infrared radiation 102–5
switching circuits 194–5

T
temperature
  gases 78
  pressure and 80–1
  scales 90
  unit of 2
temperature sensors 192
terminal speed 15, 26
thermal energy 50
  transfer 104–5
  conduction 96–7
  convection 98–9
  radiation 100–3
thermal conduction 96–7
thermal expansion 88–9
thermistors 192–5
thermometers 90–1
thermometric property 90
ticker timer 14
tidal energy 59
tidal power **58**

time intervals, measurement 4
total internal reflection 126–7
transformers 184, 212–15
  efficiency 214

U
U-tube manometer 74
ultrasonic waves 144, 150–1
ultraviolet radiation *135*, **136**, 137

V
vacuum, sound and 144
vacuum flasks 104
Van de Graaff generator 164
vaporisation 76, 77
vapour 82
variable resistor 184
vectors 26–7, 46–7
velocity 9
  acceleration and 11
virtual images 121, 131
voltage 172
voltmeter 93, 184
volts 172–4
volume, measurement 4

W
water
  apparent depth 123
  sound speed 147, **149**
  thermal conductivity 99
watts 66
wave power 58
wavefronts 108
  diffraction 115
  reflection and refraction 112–115
wavelength 109
  electromagnetic radiation 135
waves 108–9
  diffraction 114–15
  longitudinal 111
  production 142–3
  reflection 112–13
  refraction 112–13
  sound 111, 143
  speed 109
  transverse 110–11
weight 18–19
white dwarf 238
wind energy 58
wind turbines **58–9**
wire-wound resistor 178–9
wires
  normal current 185
  safety 196
  uniform 178, 179
work 64–5

X
X-rays 135, **136**, 136–7